Catalysis in Asymmetric Synthesis

Postgraduate Chemistry Series

A series designed to provide a broad understanding of selected growth areas of chemistry at postgraduate student and research level. Volumes concentrate on material in advance of a normal undergraduate text, although the relevant background to a subject is included. Key discoveries and trends in current research are highlighted, and volumes are extensively referenced and cross-referenced. Detailed and effective indexes are an important feature of the series. In some universities, the series will also serve as a valuable reference for final year honour students.

Titles in the Series:

Catalysis in Asymmetric Synthesis
Jonathan M.J. Williams

Protecting Groups in Organic Synthesis
James R. Hanson

Organic Synthesis with Carbohydrates
Geert-Jan Boons and Karl J. Hale

Catalysis in Asymmetric Synthesis

JONATHAN M. J. WILLIAMS
Professor of Organic Chemistry
University of Bath

**Blackwell
Science**

First published 1999
Copyright © 1999 Sheffield Academic Press

Published by
Sheffield Academic Press Ltd
Mansion House, 19 Kingfield Road
Sheffield S11 9AS, England

ISBN 1-85075-984-7

Published in the U.S.A. and Canada (only) by
Blackwell Science, Inc.
Commerce Place
350 Main Street
Malden, MA 02148-5018, U.S.A.
Orders from the U.S.A. and Canada (only) to Blackwell Science, Inc.

U.S.A. and Canada only:
ISBN 0-6320-4504-3

NOTICE: The authors of this volume have taken care that the information contained herein is accurate and compatible with the standards generally accepted at the time of publication. Nevertheless, it is difficult to ensure that all the information given is entirely acccurate for all circumstances. The publisher and authors do not guarantee the contents of this book and disclaim liability, loss, or damage incurred as a consequence, directly or indirectly, of the use and application of any of the contents of this volume.

Trademark Notice: Product or corporate names may be trademarks or registered trademarks, and are used only for identification and explanation, without intent to infringe.

British Library Cataloguing-in-Publication Data:
A catalogue record for this book is available from the British Library

Library of Congress Cataloging-in-Publication Data:
A catalog record for this book is available from the Library of Congress

Preface

Asymmetric synthesis has become a major aspect of modern organic chemistry. The importance of stereochemical purity in pharmaceutical products has been one driving force in the quest for improved control over the stereochemical outcome of organic reactions. The fact is, like it or not, stereochemistry is hard to avoid.

Asymmetric catalysis is a very important aspect of asymmetric synthesis, and one that has seen tremendous activity during the 1990's. This book aims to capture the latest results in asymmetric catalysis, and covers the literature up until June/July 1998. A few references as late as October 1998 have also been included. The emphasis in this book has been on non-enzymatic methods for asymmetric catalysis, although key references to enzyme-catalysed reactions have been incorporated where appropriate.

It cannot be possible to be comprehensive in a book of this size, although hopefully there are no major omissions. I apologise if I have omitted any 'favourite' reactions, or if a topic has not been treated with the depth that it may deserve. The emphasis, in general, has been on asymmetric catalytic reactions which are as current as possible.

I am grateful to my former mentors, Professor S.G. Davies (Oxford) and Professor D.A. Evans (Harvard), who introduced me to asymmetric synthesis and asymmetric catalysis. I am also indebted to my current co-workers and students who help to keep organic chemistry alive for me with their enthusiasm for the subject.

My thanks also go to Mrs. J.W. Curtis, who has typed the bulk of the manuscript, and Miss H.L. Haughton, who helped with many of the chemical structures.

Finally, I am especially grateful to my wife, Cathy, who has not complained about the extra time I have needed to work during the preparation of this book.

J. M. J. Williams
University of Bath

**Dedicated to my children
Charlotte, Sam, Harry and Alice**

Contents

1   Introduction

There are several ways of producing compounds as single enantiomers. The resolution of a mixture of enantiomers can often be the cheapest, most practical way of obtaining enantiomerically pure material. The conversion of a cheap enantiomerically pure starting material into another derivative is another useful technique in some cases.

However, asymmetric synthesis can provide a more general approach to the preparation of enantiomerically-enriched compounds. Asymmetric synthesis is limited by the range and scope of methodology available. Fortunately, with so much research in the area, there are now many suitable methods, with further asymmetric reactions being developed each year. There may still be cases where asymmetric synthesis does not provide the best method for the preparation of a particular enantiomerically pure compound; however, it certainly allows for the preparation of a more diverse range of structures.

Asymmetric catalysis is an especially appealing aspect of asymmetric synthesis. Small amounts of a catalyst (frequently less than 1 mol%) can be used to control the stereochemistry of the bulk reaction. The use of a catalyst often makes the isolation of a product easier, since there is less unwanted material to remove at the end of a reaction.

1.1   Reactions amenable to asymmetric catalysis

From a synthetic viewpoint, the titles of individual chapters in the present book give a fair impression of the scope of processes involved. The types of reaction amenable to asymmetric catalysis can also be considered from a more stereochemical perspective.

Most reactions involving asymmetric catalysis are based around the conversion of a planar sp^2 carbon atom into a tetrahedral sp^3 carbon atom. This category of reactions includes asymmetric hydrogenation of alkenes and ketones, as well as the addition of other reagents to these groups, as presented in Figure 1.1.

Substrates containing enantiotopic groups can be converted into enantiomerically-enriched compounds using asymmetric catalysis. Representative examples are presented in Figure 1.2. These reactions break the symmetry (e.g. *meso* or achiral) of the starting material.

Asymmetric catalysis may also be achieved by the kinetic resolution of racemic substrates, where one enantiomer of starting material is selectively converted into product, leaving the other enantiomer unreacted. In some

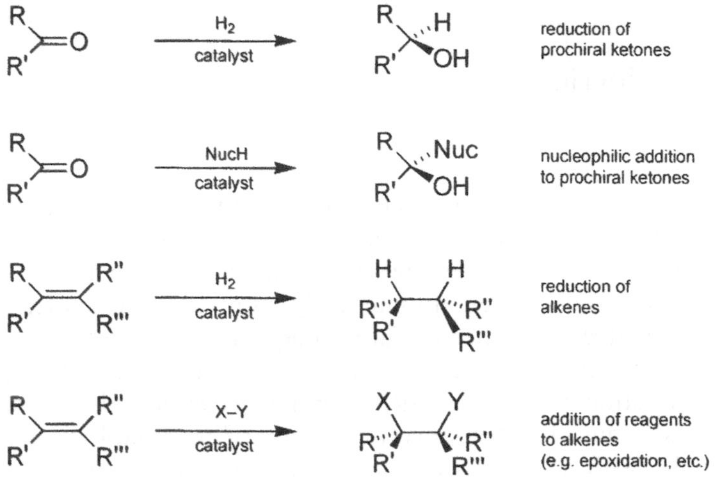

Figure 1.1 Asymmetric catalytic reactions involving conversion of a planar sp^2 carbon atom into a tetrahedral sp^3 carbon atom.

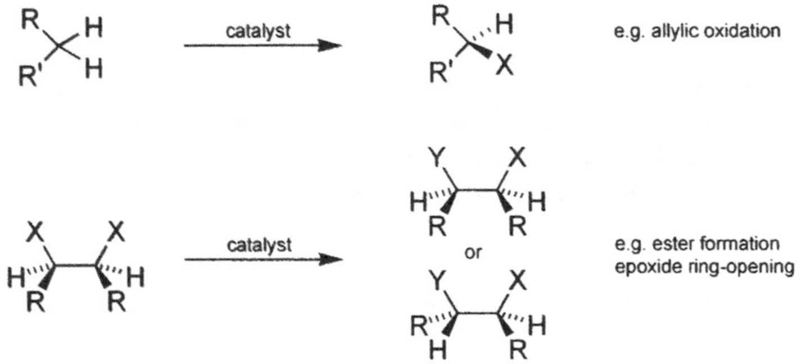

Figure 1.2 Asymmetric catalytic reactions, which involve breaking the symmetry of the starting material.

instances, both enantiomers of a starting material are converted into the same enantiomer of product, i.e. dynamic resolution. Most reactions fit into one of the categories identified in Figures 1.1–1.3, even if the exact structure is not represented.

1.2 Assignment of (*R*) and (*S*) stereochemical descriptors

The Cahn-Ingold-Prelog (CIP) system for describing the stereochemistry of chiral molecules is universally accepted.[1,2] Simple molecules contain-

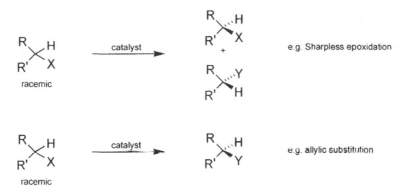

Figure 1.3 Kinetic and dynamic resolution of racemic substrates.

ing one chiral centre are described as (*R*) (from the Latin, *rectus*) or (*S*) (from the Latin, *sinister*). A short account of how to distinguish between (*R*) and (*S*) follows, although more detailed information has been published previously.[3]

For a tetrahedral carbon-based chiral centre, the priority of the four attached groups must be determined according to the sequence rules. The two enantiomers in Figure 1.4 have groups attached, where a > b > c > d. The structure must be viewed from the side opposite the lowest priority group. In these cases, the group d must, therefore, point away from the viewer. If the priority of the remaining groups, a–b–c, is in an

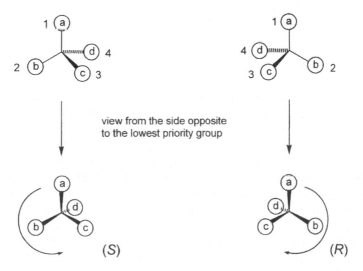

Figure 1.4 Assignment of (*R*) and (*S*) stereochemical descriptors to tetrahedral chiral centres.

anticlockwise sense, the stereochemical descriptor is (S). When the a–b–c sequence is clockwise, the stereochemical descriptor is (R).

Of course, for real molecules, it is necessary to decide which groups have priority over other groups. In general, the most important sequence rule is that groups of higher atomic number precede groups of lower atomic number. Hence, molecule (1.01) has the (R)-configuration.

- lowest priority group pointing away
- Br > Cl > F > H
- clockwise priority, hence (R)

(1.01)

However, not all molecules are so simple! When the first atoms attached to the chiral centre have the same priority, the groups may need to be considered in more detail. In the case of molecule (1.02), the lowest priority group, the H atom, can be rapidly identified. However, the three remaining groups each contain a carbon attached to the chiral centre; therefore, the next 'sphere' of atoms must be considered. In the second sphere, the Cl atom takes priority over the O, which takes priority over the H. The compound therefore has the (S)-configuration.

(S)

(1.02)

For functional groups which contain multiple bonds, 'phantom' atoms are incorporated, making it easier to assign priorities. The 'phantom' atoms are not attached to any further groups. Some examples of expanded functional groups are presented in Figure 1.5.

The expansion of functional groups makes it possible to apply the sequence rules to more complex structures, (1.03–1.08). A few comments may be helpful concerning some of these assignments. In structure (1.04), the -CH$_2$Cl group takes priority over the CF$_3$ group; and the chloride takes precedence over the fluoride, even though there is only one chloride, because the fluorides are not added together. In the phosphine (1.07), the lone pair of electrons is considered to have the lowest priority (lower than any atom). In structure (1.08), the -CH$_2$Cl group takes priority over the

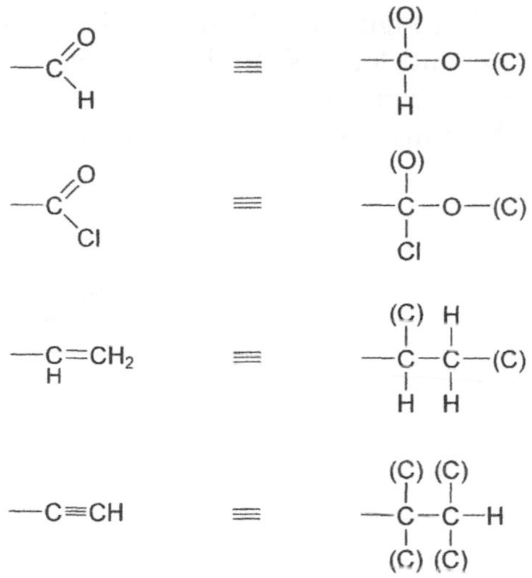

Figure 1.5 Expanded functional groups with phantom atoms.

-CH_2CH_2I group, because the Cl atom comes in the second sphere, whereas the I atom does not appear until the third sphere.

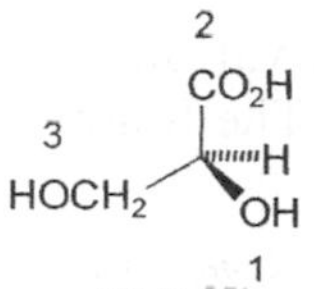

(S)-(1.03)

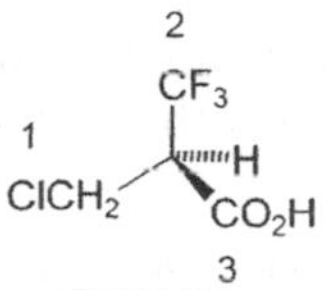

(R)-(1.04)

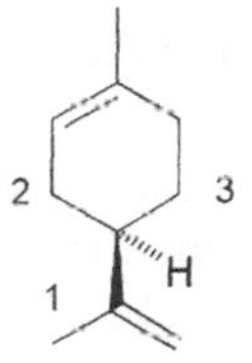

(R)-(1.05)

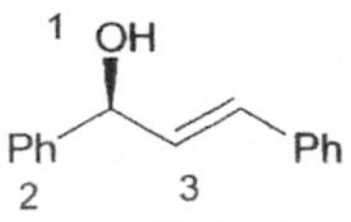

(S)-(1.06)

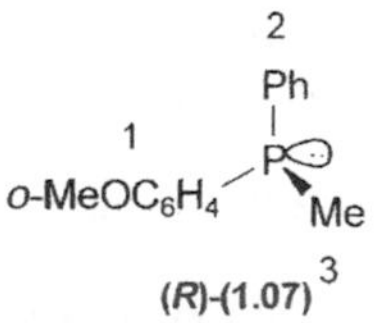

(R)-(1.07)

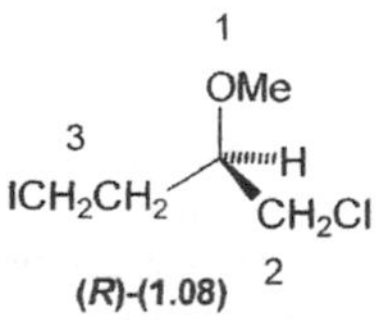

(R)-(1.08)

Another aspect of the (*R*,*S*)-nomenclature involves assignment of the stereochemical descriptors to structures that possess axial chirality or planar chirality. Because of the number of ligands that possess axial chirality (BINOL, BINAP and related ligands), the assignment of (*R*) and (*S*) to these structures will be considered briefly. An additional sequence rule is helpful here: the nearer end of an axis (or plane) precedes the further end. For the axially chiral molecule, BINOL (**1.09**), a simple way to assign stereochemistry is by looking down the chiral axis. The nearer groups take priority over the further groups and, therefore, sequence is assigned as presented in Figure 1.6. Again, clockwise corresponds to the (*R*)-configuration and anticlockwise to the (*S*)-configuration. This is like a helix spiralling away from the point of view. The assignment will be made by looking along the chiral axis from the other end.

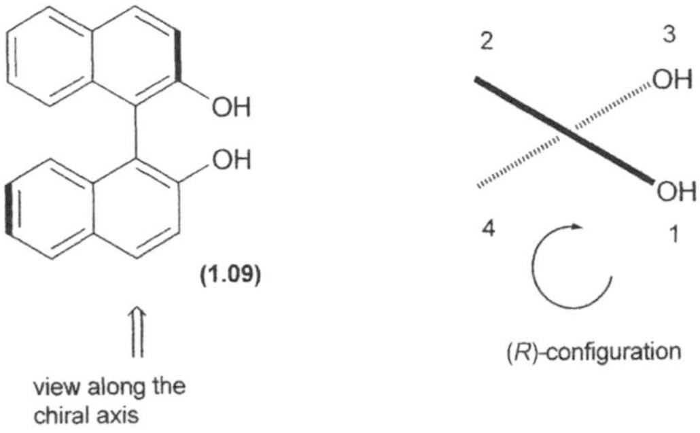

Figure 1.6 Assignment of axial chirality.

Axial chirality can also be assigned using (*M*,*P*)-nomenclature[2]. For axial chirality, (*R*) is equivalent to (*M*) and (*S*) is equivalent to (*P*).

Further reading

Several other textbooks are available, dealing with various aspects of stereochemistry, asymmetric synthesis and catalysis. The following titles may be useful.

E.L. Eliel, S.H. Wilen and L.N. Mander, *Stereochemistry of Organic Compounds*, John Wiley and Sons, New York, **1994**.

R.E. Gawley and J. Aubé, *Principles of Asymmetric Synthesis*, Tetrahedron Organic Chemistry Series, Volume 14, Pergamon, Oxford, **1996**.

R.A. Sheldon, *Chirotechnology: Industrial Synthesis of Optically Active Compounds*, Marcel Dekker, New York, **1993**.

Stereoselective Synthesis, (G. Helmchen, R.W. Hoffman, J. Mulzer and E. Schaumann, eds.) Georg Thieme, Stuttgart, **1995**.

Catalytic Asymmetric Synthesis, (I. Ojima, ed.) VCH, New York, **1993**.

R. Noyori, *Asymmetric Catalysis in Organic Synthesis*, John Wiley, New York, **1994**.

M. Nógrádi, *Stereoselective Synthesis*, 2nd edition, VCH, New York, **1995**.

G. Procter, *Asymmetric Synthesis*, Oxford University Press, Oxford, **1996**.

References

1. R.S. Cahn, C. Ingold and V. Prelog, *Angew. Chem., Int. Ed. Engl.*, **1966**, *5*, 385.
2. V. Prelog and G. Helmchen, *Angew. Chem., Int. Ed. Engl.*, **1982**, *21*, 567.
3. E.L. Eliel, S.H. Wilen and L.N. Mander, *Stereochemistry of Organic Compounds*, John Wiley and Sons, New York, **1994**, Chapter 5.

2   Reduction of alkenes

The reduction of alkenes has been the earliest type of catalytic reaction to be exposed to a substantial research effort directed at achieving an asymmetric variant. Direct hydrogenation reactions and related processes, such as hydrosilylation and hydroboration, are considered in the present chapter. Hydroformylation is also considered, as well as hydroacylation and hydrocyanation reactions of alkenes.

2.1   Asymmetric hydrogenation with rhodium complexes

The reduction of an alkene to an alkane is a particularly important synthetic transformation, since construction of a $C{=}C$ bond is often straightforward and this can subsequently be converted into the corresponding $C{-}C$ bond. The ability to achieve this transformation with asymmetric induction has been the most widely studied of all transition metal catalysed asymmetric reactions, and the area has been the focus of many reviews.[1]

Achiral Wilkinson's catalyst, $RhCl(PPh_3)_3$, is an effective homogeneous catalyst for the hydrogenation of alkenes. The first examples of homogenous asymmetric hydrogenation were reported, independently, by Horner and co-workers[2] and Knowles and Sabacky.[3] These were variants of Wilkinson's catalyst using enantiomerically pure monodentate phosphine ligands, and provided definite, but low, asymmetric induction. In the early 1970s, Kagan and Dang prepared the bidentate ligand DIOP (2.01), which possesses C_2-symmetry, and was found to give good enantioselectivity in the rhodium-catalysed hydrogenation of α-(aryl-amino)acrylic acids, such as substrate (2.02), which affords the α-amido acid product (2.03).[4]

Numerous enantiomerically pure phosphines have been prepared and examined in the hydrogenation of alkenes, and the α-(acylamino)acrylic acids have remained popular substrates. New ligands, usually bidentate and often possessing C_2-symmetry, continue to be tested. Any new ligand must be demonstrated to be effective for several types of alkene in order for it to join the 'asymmetric hydrogenation establishment'. Many ligands have provided very high enantioselectivity in the reduction of enamides. A representative set of structures (**2.04–2.19**) is shown, all of which have given over 90% ee (and often higher).[5] Libraries of enantiomerically pure phosphine ligands, prepared using combinatorial methods, have been generated and examined for their ability to provide asymmetric induction in rhodium-catalysed alkene hydrogenation reactions.[6] This approach has considerable scope as a method for producing highly enantioselective ligands but further research is required in order to achieve this potential.

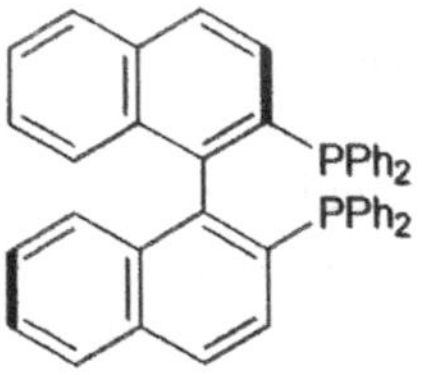

(2.04) BINAP

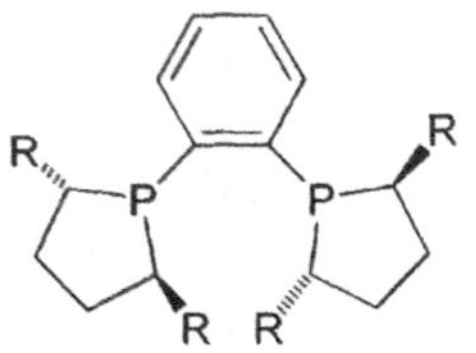

(2.05) DuPHOS

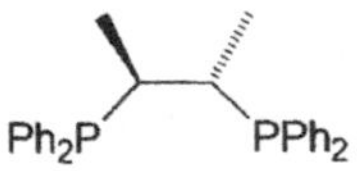

(2.06) CHIRAPHOS

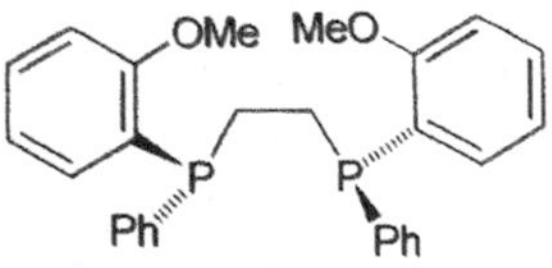

(2.07) DIPAMP

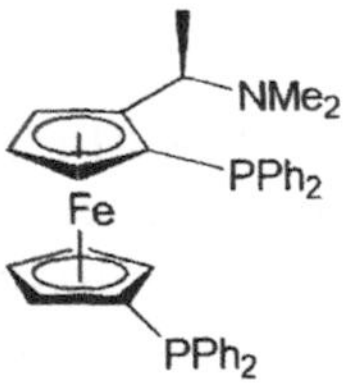

(2.08) BPPFA

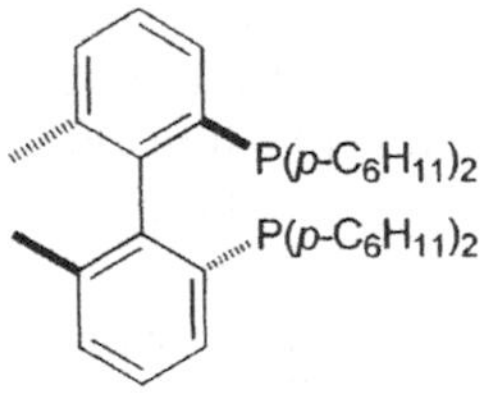

(2.09) (R)-BICHEP

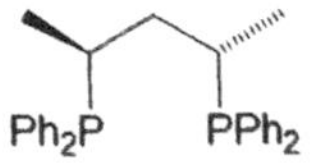

(2.10) SKEWPHOS

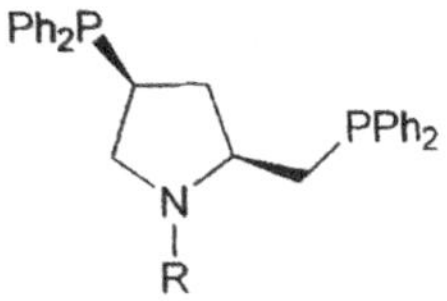

(2.11)

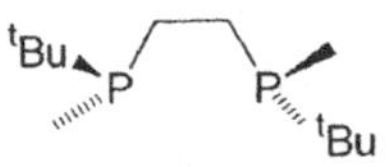

(2.12)

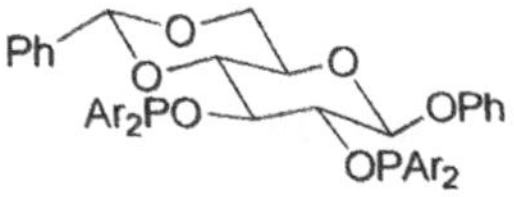

(2.13)
Ar = 3,5-(CH$_3$)$_2$C$_6$H$_3$-

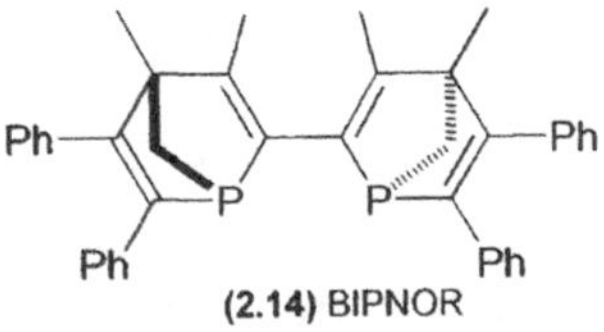

(2.14) BIPNOR

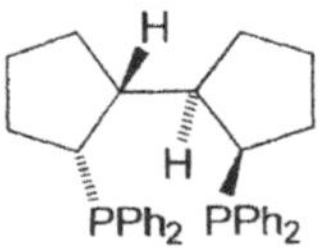

(2.15) BICP

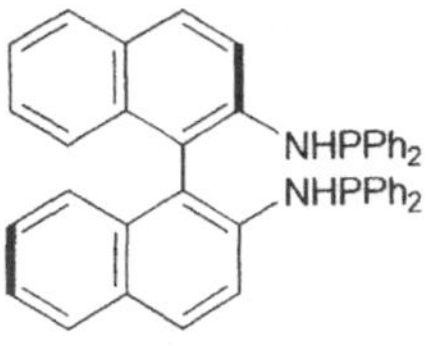

(2.16)

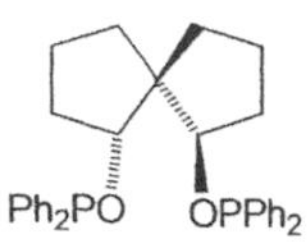

(2.17) (R)-SIROP

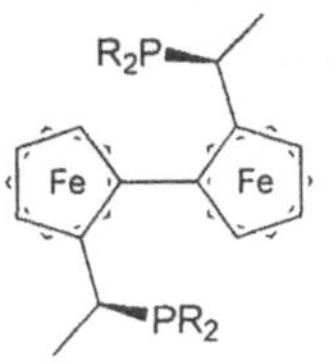

(2.18) (R)-TRAP

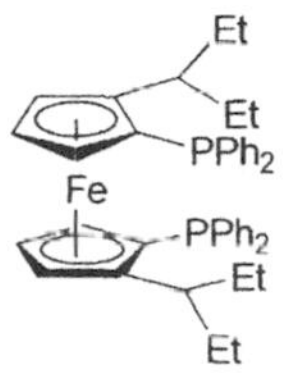

(2.19) (S,S)-FerroPHOS

The reduction of α-(acylamino)acrylic acids is certainly a useful process, since it leads to α-amino acid derivatives. Monsanto synthesised the drug, L-dopa (2.20), which is used in the treatment of Parkinson's disease, by means of rhodium-catalysed asymmetric hydrogenation.[7] The α-(acyl-amino)acrylic acid (2.21) was reduced by an Rh/DiPAMP combination to give the L-dopa precursor (2.22), with high enantioselectivity. This was the first commercial application of a transition metal catalysed asymmetric reaction.

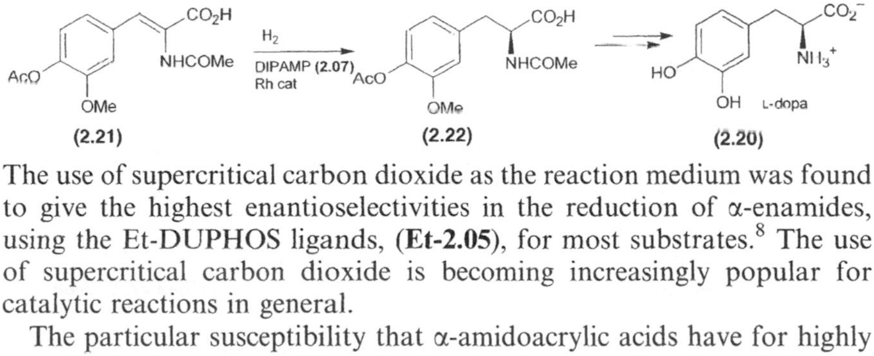

(2.21)

(2.22)

(2.20)

The use of supercritical carbon dioxide as the reaction medium was found to give the highest enantioselectivities in the reduction of α-enamides, using the Et-DUPHOS ligands, (Et-2.05), for most substrates.[8] The use of supercritical carbon dioxide is becoming increasingly popular for catalytic reactions in general.

The particular susceptibility that α-amidoacrylic acids have for highly enantioselective asymmetric reduction has been attributed to their two-point binding. There are two diastereomeric intermediates (2.23) and (2.24) involved when an enantiomerically pure phosphine is employed. Mechanistic studies have shown that oxidative addition of H_2 occurs more quickly to the minor diastereomer.[9, 10] The reaction, therefore, proceeds via the minor diastereomer to give the enantioselectivity observed. The same principles hold true for DuPHOS ligands (2.05), where the higher reactivity of these ligands is attributed to a low binding constant between the rhodium and enamide, facilitating rapid ligand exchange.[11]

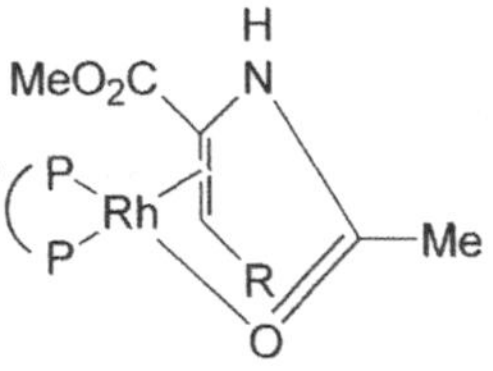

minor diastereomer
(2.23)

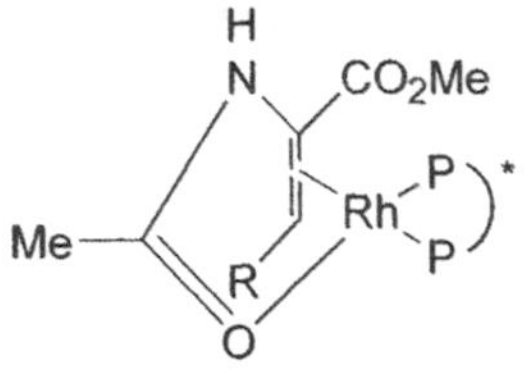

major diastereomer
(2.24)

[Rh(BINAP)]ClO$_4$ was amongst the first catalysts used for the asymmetric reduction of enamides and is still one of the best available. The geometry of the enamide has been shown to be important.[12] The (Z)-alkene substrate (2.25) is converted into the α-amid acid (2.27), with the (R)-enantiomer predominating. However, the alternative (E)-alkene substrate geometry (2.26) provides the opposite (S) configuration in the product (2.27).

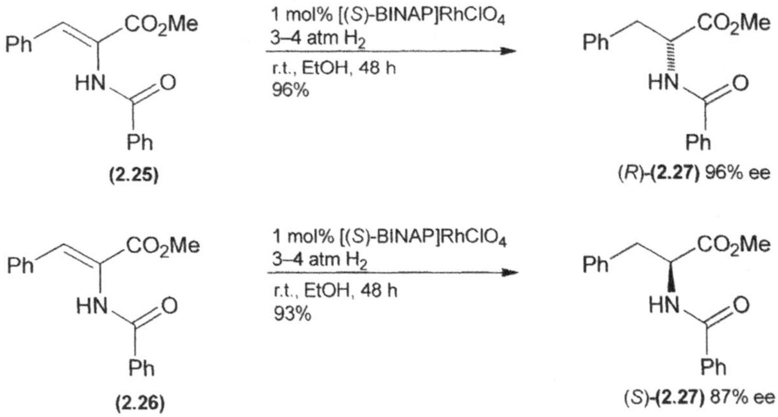

(2.25)

1 mol% [(S)-BINAP]RhClO$_4$
3–4 atm H$_2$

r.t., EtOH, 48 h
96%

(R)-(2.27) 96% ee

(2.26)

1 mol% [(S)-BINAP]RhClO$_4$
3–4 atm H$_2$

r.t., EtOH, 48 h
93%

(S)-(2.27) 87% ee

The diene (2.28) undergoes selective hydrogenation of the enamide double bond to give an amino acid derivative (2.29) containing an alkene functionality.[13] The most enantioselective catalyst for this reaction was found to be the cationic rhodium complex of DuPHOS (2.05), which yielded less than 1% of the fully-hydrogenated product as a by-product.

β,β-Disubstituted enamides are also substrates for enantioselective hydrogenation reactions. In such cases, the Me-BPE catalyst (2.30) generally gave better results than the normal DuPHOS ligands.[14] The enamide (**2.31**) is converted into the β-branched α-amido ester (**2.32**) with high enantioselectivity. By using the correct geometry of enamide substrate, the stereochemistry can also be controlled in the β-position.

Thus, the (*Z*)-enamide (**2.33**) yields one diastereomer of the amido ester (**2.35**), whereas the (*E*)-enamide (**2.34**) yields the alternative diastereomer (**2.36**). TRAP ligands (**2.18**) have also been used to control the stereochemistry in the β-position, including the reduction of oxygenated enamides.[15] Reduction of the (*Z*)-alkene (**2.37**) yields the *anti* diastereomer of amido ether product (**2.39**), whereas the (*E*)-alkene yields the *syn* diastereomer (**2.40**).

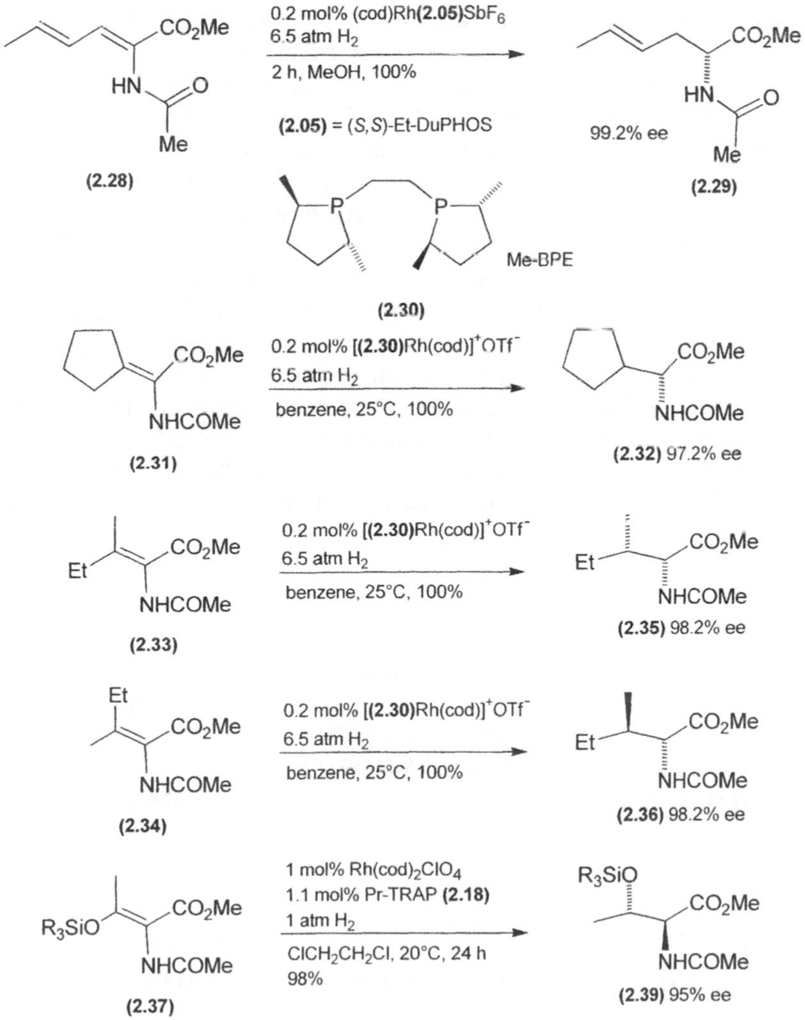

^{t}BuCO$_2$ · · · CO$_2$Me, NHCOMe

1 mol% Rh(cod)$_2$ClO$_4$
1.1 mol% Pr-TRAP **(2.18)**
1 atm H$_2$

ClCH$_2$CH$_2$Cl, 20°C, 24 h
99%

^{t}BuCO$_2$ CO$_2$Me NHCOMe

(2.38)

(2.40) 97% ee

Rhodium-catalysed asymmetric hydrogenation is not restricted to the use of enamides as substrates, although a coordinating group other than the alkene is needed in the substrate for high enantioselectivities.

The reduction of α-(aryloxy)acrylates with the Rh/DuPHOS **(2.05)** catalyst has recently been reported.[16] Interestingly, the substrate **(2.41)** was used as a 3:1 (*E*)/(*Z*) mixture but the enantioselectivity was still very high in the product **(2.42)**. When this reaction was attempted using benzene as solvent, no product was formed, even though benzene is a suitable solvent for enamide reduction. This was attributed to coordination of the benzene to the cationic rhodium complex. The enol ester is not able to displace the benzene, whilst an enamide can do so.

Other non-enamide substrates, which have been reduced enantioselectively with rhodium catalysts, include dimethyl itaconate **(2.43)**[17] and trisubstituted acrylic acids, such as alkene **(2.44)**, where the carboxylic acid was believed to bind to the amino groups of the ligand **(2.47)**, giving a more rigid transition state.[18] Some enol acetates are also good substrates. Boaz has shown that, whilst aliphatic enol esters react with only moderate selectivity, alkynyl enol esters are good substrates.[19] Thus, the enol ester **(2.48)** gives rise to the acetate product **(2.50)** with only 64% ee, whilst the corresponding alkynyl substrate **(2.49)** provides a product **(2.51)** with 98.5% ee. Evidently, two-point binding is again important. The alkyne is reduced to a (*Z*)-alkene under the reaction conditions.

CO$_2$Et, O, Ph, O

0.4 mol% [(cod)Rh**(2.05)**]$^+$OTf$^-$
4 atm H$_2$

12 h, r.t., CH$_2$Cl$_2$, 100%

(2.05) = (*S,S*)-Et-DuPHOS

CO$_2$Et, O, Ph, O

(2.41) (E)/(Z) 3:1

(2.42) 99.8% ee

MeO$_2$C CO$_2$Me

(2.43)

1.1 mol% **(2.09)** Rh$^+$
5 atm H$_2$

EtOH

(2.09) = (*R*)-BICHEP

MeO$_2$C · · · CO$_2$Me

(2.45) 99% ee

2.2 Asymmetric hydrogenation with ruthenium catalysts

The ruthenium catalysts most frequently used for alkene hydrogenation are BINAP derivatives, such as complex (2.52).[20] Alternative BINAP complexes, including $RuBr_2(BINAP)$ and cationic BINAP complexes, e.g. [RuCl(arene)(BINAP)]Cl, can also be used. Ligands other than BINAP have been used, including TolBINAP (2.53) and H_8-BINAP (2.54)[21] which sometimes provide a small increase in enantioselectivity. Ruthenium-catalysed hydrogenation reactions have also been reported using a polymer-supported BINAP ligand, with only slightly lower enantioselectivities than in the free homogeneous system.[22]

Genêt and co-workers screened a series of ligands in ruthenium-catalysed alkene and ketone reductions.[23] The ligands which provided the highest selectivities were those which possessed axial chirality, such as BINAP, but also other axially chiral ligands (2.55). Other chelating diphosphines generally provided lower enantioselectivity, although a Ru/Me-DuPHOS catalyst yielded up to 80% ee in the hydrogenation of tiglic acid.

Ruthenium/BINAP complexes can be used in the asymmetric reduction of enamides, although enantioselectivity is often worse than for the corresponding rhodium/BINAP complexes. Interestingly, using the same enantiomer of BINAP provides the opposite enantiomer of product, depending on whether rhodium or ruthenium is employed as the catalyst.[24] The ruthenium/BINAP catalysts were found to be highly effective in the reduction of the enamide (2.56).[25] The products, tetrahydroisoquinolines (2.57), are useful in the synthesis of isoquinoline alkaloids.

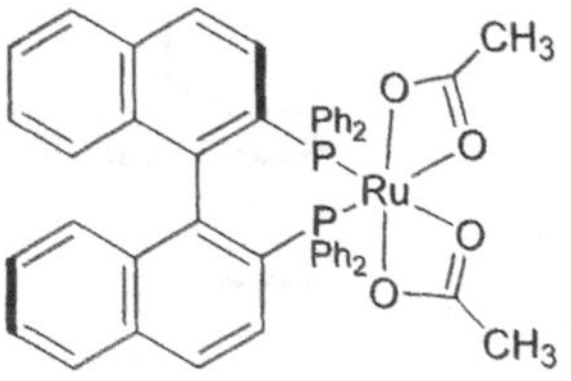

Ru[(*R*)-BINAP](OCOCH$_3$)$_2$ **(2.52)**

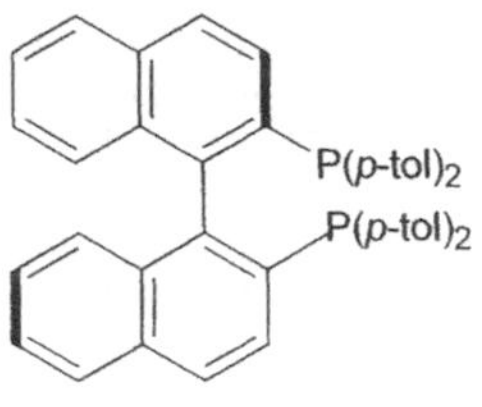

(*R*)-TolBINAP **(2.53)**

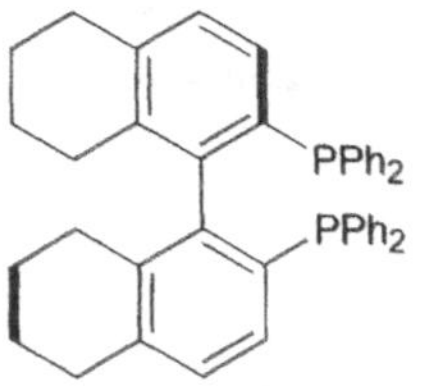

H$_8$-BINAP **(2.54)**

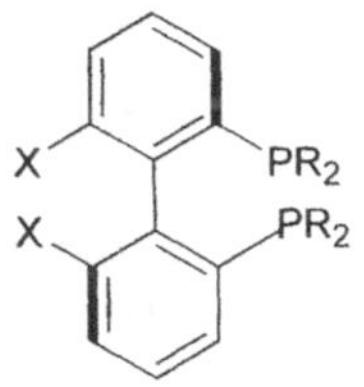

(2.55)

Ruthenium/BINAP complexes have been used successfully in the asymmetric reduction of acrylic acids.[26] This methodology has been used to prepare the anti-inflammatory drug, (*S*)-naproxen (**2.59**), by reduction of the acrylic acid (**2.58**). The reaction has also been used in the preparation of α-fluorocarboxylic acids, with good enantioselectivity, including the conversion of the alkenyl fluoride (**2.60**) into α-fluorohexanoic acid (**2.61**).[27]

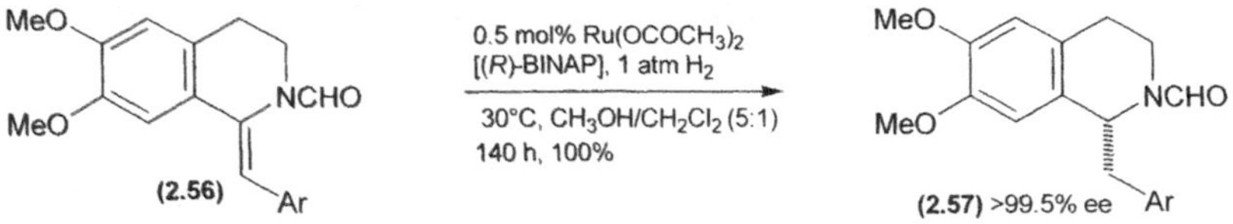
MeO
MeO
NCHO
(2.56)
Ar
0.5 mol% Ru(OCOCH3)2
[(R)-BINAP], 1 atm H2
30°C, CH3OH/CH2Cl2 (5:1)
140 h, 100%
MeO
MeO
NCHO
(2.57) >99.5% ee
Ar

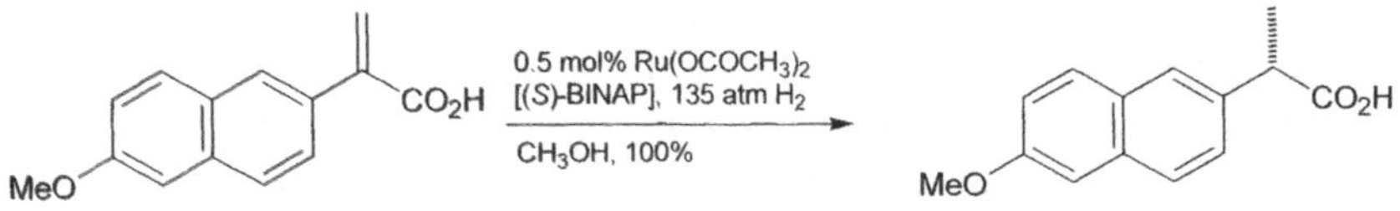
CO2H
MeO
0.5 mol% Ru(OCOCH3)2
[(S)-BINAP], 135 atm H2
CH3OH, 100%
CO2H
MeO

(2.58) **(2.59)** 97% ee

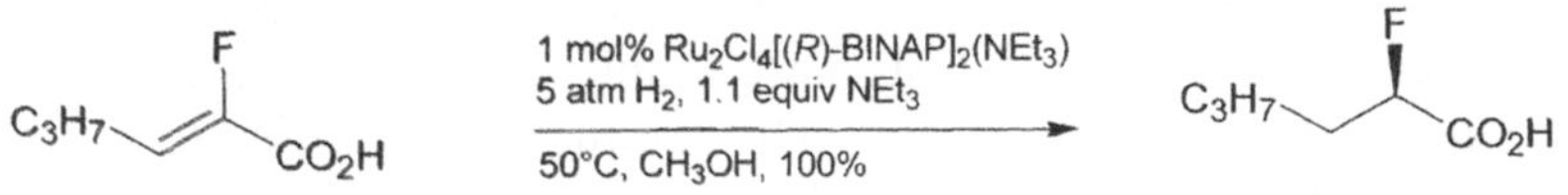
F
C3H7
CO2H
1 mol% Ru2Cl4[(R)-BINAP]2(NEt3)
5 atm H2, 1.1 equiv NEt3
50°C, CH3OH, 100%
F
C3H7
CO2H

(2.60) **(2.61)** 90% ee

Allylic alcohols make good substrates for ruthenium/BINAP catalysed hydrogenation.[28] Geraniol **(2.62)** and nerol **(2.63)** are geometrical isomers, (Z) and (E), and these substrates give rise to opposite enantiomers of the product, citronellol **(2.64)**. The allyl alcohol **(2.65)** could be hydrogenated using ruthenium/TolBINAP complexes, with exceptionally high diastereoselectivity.[29] The substrate diastereoselectivity and catalyst selectivity represent a matched pair; when the enantiomeric (S)-TolBINAP ligand was used, the opposite diastereomer was formed with only 56% de.

Kinetic resolution of racemic allylic alcohols has been achieved with a high selectivity factor.[30] Racemic allylic alcohol **(2.67)** undergoes kinetic resolution when hydrogenated with a ruthenium/BINAP complex. The product **(2.68)** and recovered substrate **(2.67)** can both be obtained with high enantioselectivity. Faller and Tokunaga demonstrated chiral poisoning in the kinetic resolution of allylic alcohol **(2.69)**.[31] Using a ruthenium catalyst made from racemic BINAP, and adding an enantiomerically pure poison, $(1R,2S)$-ephedrine **(2.70)** still gives good selectivity. It is assumed that the ephedrine preferentially deactivates one enantiomer of the ruthenium/BINAP complex, leaving the other enantiomer available to perform the kinetic resolution.

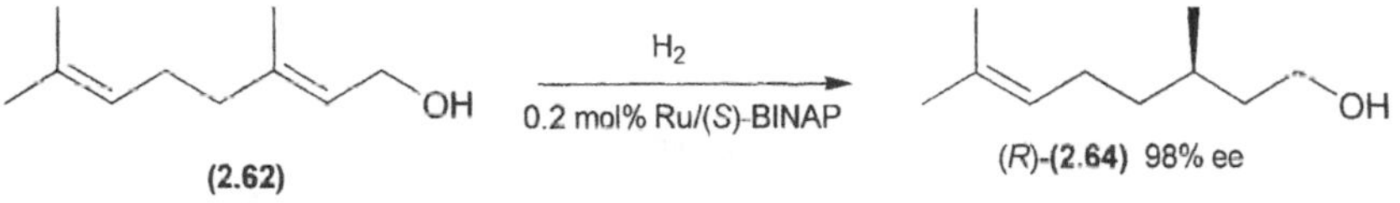

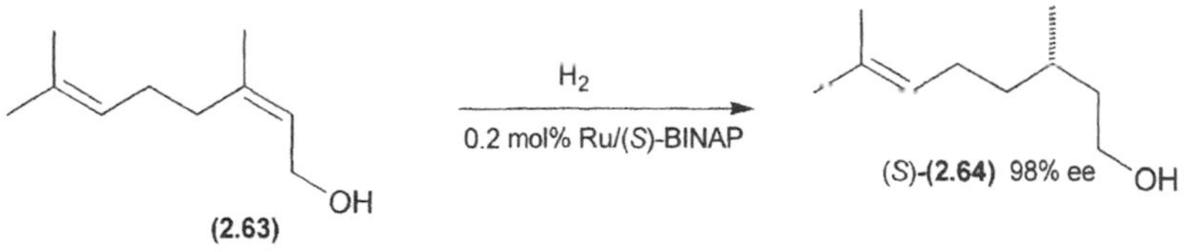

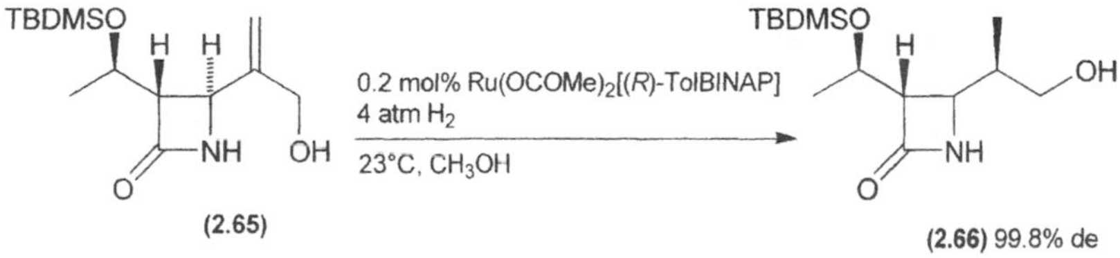

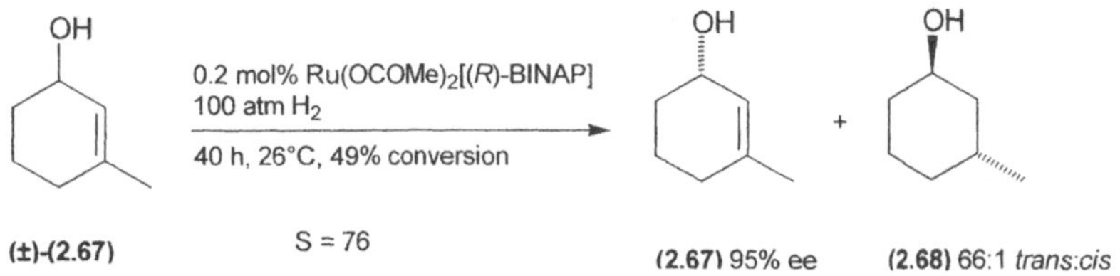

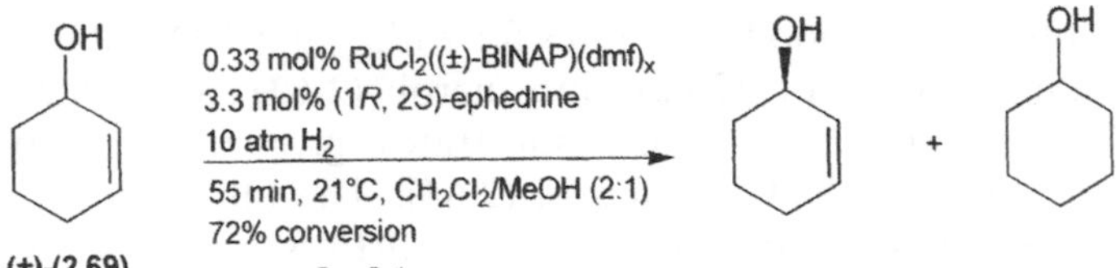

(±)-(2.69)

0.33 mol% RuCl$_2$((±)-BINAP)(dmf)$_x$
3.3 mol% (1R, 2S)-ephedrine
10 atm H$_2$
———————————————
55 min, 21°C, CH$_2$Cl$_2$/MeOH (2:1)
72% conversion

S = 6.4

(2.69) 93% ee   +   achiral

Ph — NHMe
|
OH

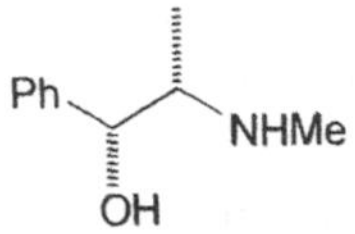

(1R,2S)-ephedrine
(2.70)

Dixneuf and co-workers reported an interesting reduction of the unsaturated acyl oxazolidinone **(2.71)**.[32] The reduction works with high yield and asymmetric induction, and the product **(2.74)** is effectively propionic acid with a chiral auxiliary attached. In a subsequent step, the chiral auxiliary was used to induce asymmetry.

Many other substrates have been subjected to ruthenium-catalysed asymmetric hydrogenation, including diketene **(2.72)**[33] and unsaturated lactones **(2.73)**.[34]

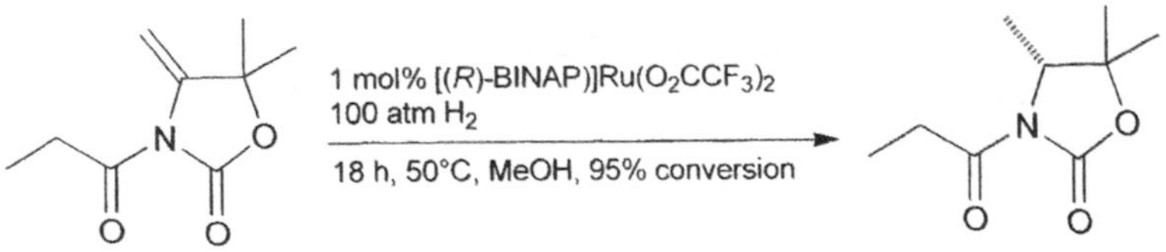

(2.71)

1 mol% [(R)-BINAP)]Ru(O$_2$CCF$_3$)$_2$
100 atm H$_2$
———————————————
18 h, 50°C, MeOH, 95% conversion

(2.74) 98% ee

(2.72)

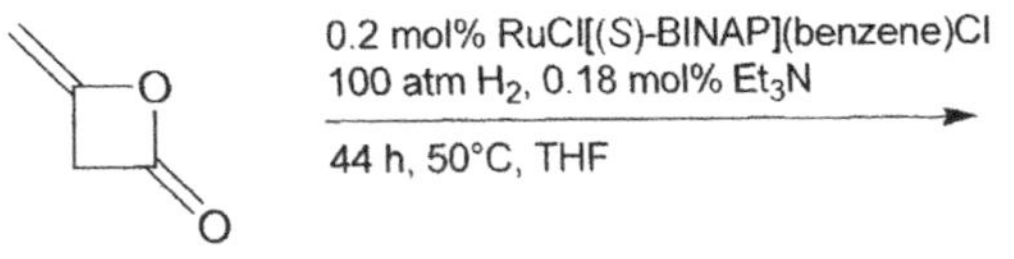

0.2 mol% RuCl[(S)-BINAP](benzene)Cl
100 atm H$_2$, 0.18 mol% Et$_3$N
———————————————
44 h, 50°C, THF

(2.75) 92% ee
+ 3% butyric acid

(2.73)

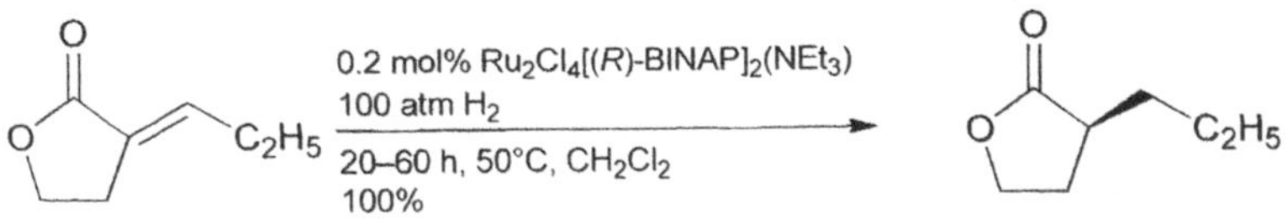

0.2 mol% Ru$_2$Cl$_4$[(R)-BINAP]$_2$(NEt$_3$)
100 atm H$_2$
———————————————
20–60 h, 50°C, CH$_2$Cl$_2$
100%

(2.76) 95% ee

2.3 Alkene hydrogenation with titanium catalysts

The use of enantiomerically pure titanocene-derived catalysts for the asymmetric hydrogenation of alkenes was first reported by Kagan and co-workers in 1979.[35] Alternative titanocene derivatives, which afforded up to 96% ee, were reported by Vollhardt and co-workers.[36]

Buchwald and Broene used the titanium version of the Brintzinger catalyst (2.77)[37] in the asymmetric reduction of trisubstituted alkenes.[38] The catalyst is reduced *in situ* to a titanium (III) hydride species (see Section 3.2). The reduction is achieved with n-butyllithium and hydrogen, whilst the silane serves to stabilise the catalyst. The catalyst works well for trisubstituted alkenes, including substrates (2.78) and (2.79). A transition state model assumes that the alkene approaches the titanium-hydride 'front on' and explains the selectivities seen. Waymouth and Pino reported earlier work with a related zirconocene complex and offered a similar rationalisation for the stereochemical outcome.[39] The enantioselective reduction of alkenes with samarium cyclopentadienyl complexes has also been reported.[40]

The same titanium catalyst (2.77) has been used in the reduction of enamines, with very good enantioselectivities (89–98% ee), including the reduction of enamine (2.80) to the amine (2.83).[41]

X,X = 1,1'-binaphth-2,2'-diolate

(2.77)

1. 2 equiv n-BuLi, 0°C, 1 atm H$_2$
2. 2.5 equiv PhSiH$_3$

activated
Ti(III) hydride

(2.77)*

(2.78)

5 mol% (2.77)*
140 atm H$_2$

9 h, 65°C, THF
94%

(2.81) >99% ee

(2.79)

5 mol% (2.77)*
140 atm H$_2$

43 h, 65°C, THF
65%

(2.82) 95% ee

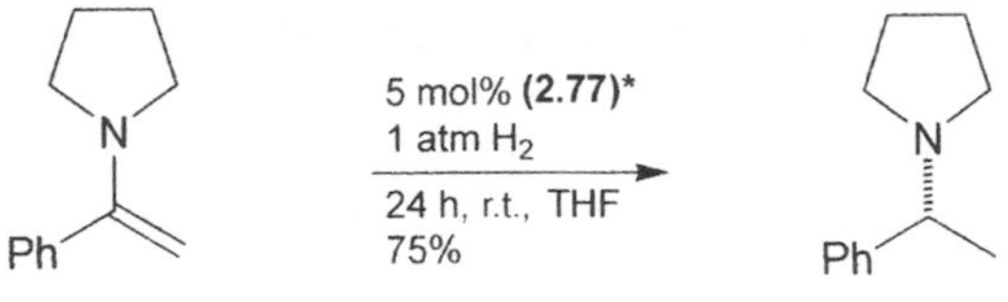

(2.80) **(2.83)** 92% ee

2.4 Alkene hydrosilylation

Amongst the various examples of catalytic enantioselective hydrosilylation reactions of alkenes that have been reported,[42,43] the MOP ligand/palladium catalyst combination of Hayashi and co-workers offers the highest level of enantioselectivity. The MOP ligands are monodentate phosphine ligands, including MeO-MOP (**2.84**),[44,45] H-MOP (**2.85**)[46] and MOP-phen (**2.86**).[47] Representative alkene hydrosilylation reactions carried out with the MOP ligands are presented in the following schemes (the low catalyst loading in many cases is noteworthy): norbornene (**2.87**), dihydrofuran (**2.88**), styrene (**2.89**) and cyclopentadiene (**2.90**) are all converted into the hydrosilylated derivatives, with good to excellent enantioselectivity. In most cases, the trichlorosilanes were derivatised into the corresponding alcohols, although the trichloroallylsilane (**2.94**) was reacted with benzaldehyde to give the addition product (**2.98**).

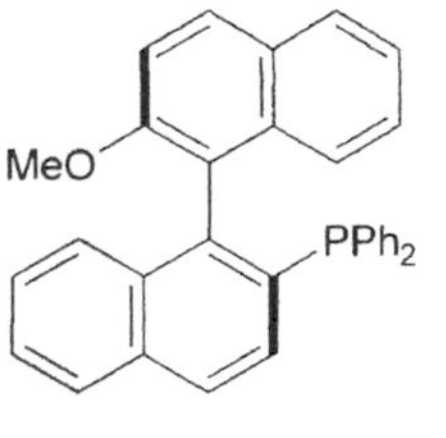

(*R*)-MeO-MOP **(2.84)**

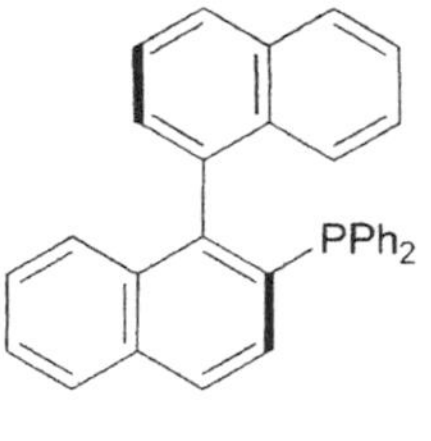

(*S*)-H-MOP **(2.85)**

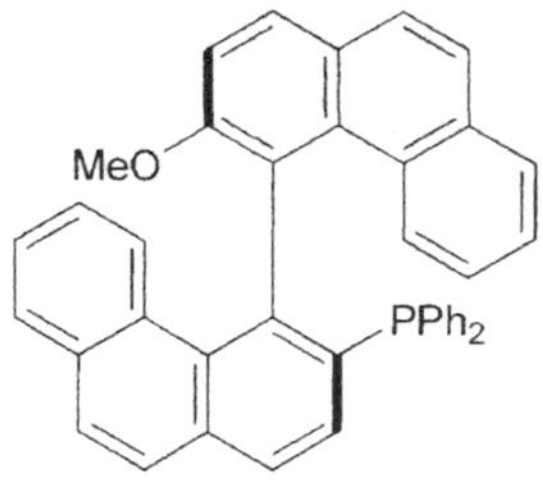

(*R*)-MOP-Phen **(2.86)**

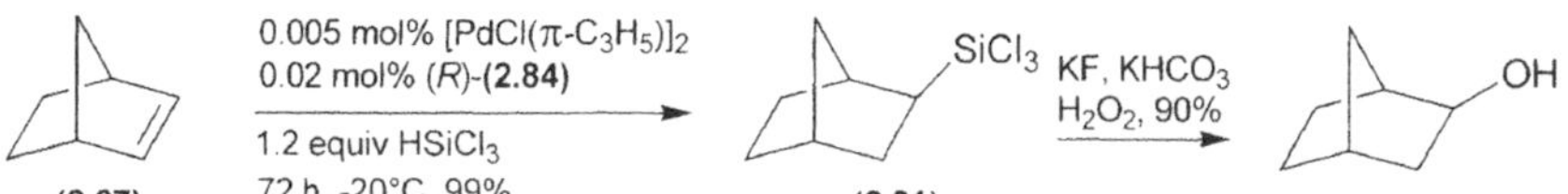

(2.87) **(2.91)** **(2.95)** 96% ee

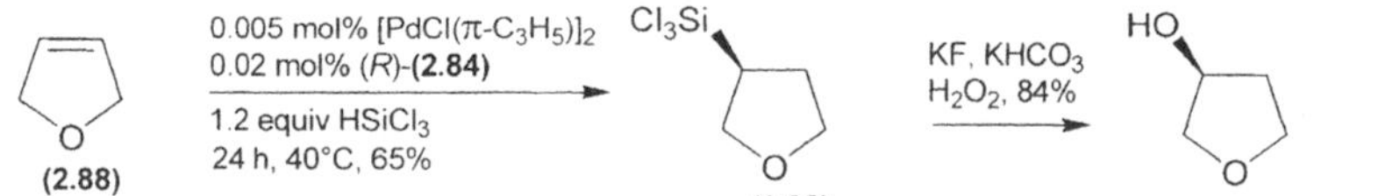

(2.88) **(2.92)** **(2.96)** 95% ee

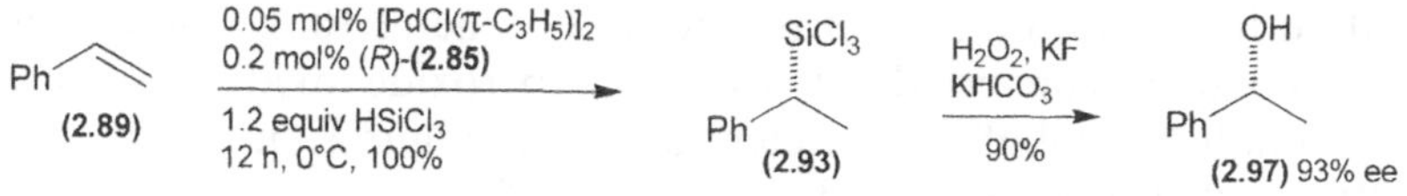

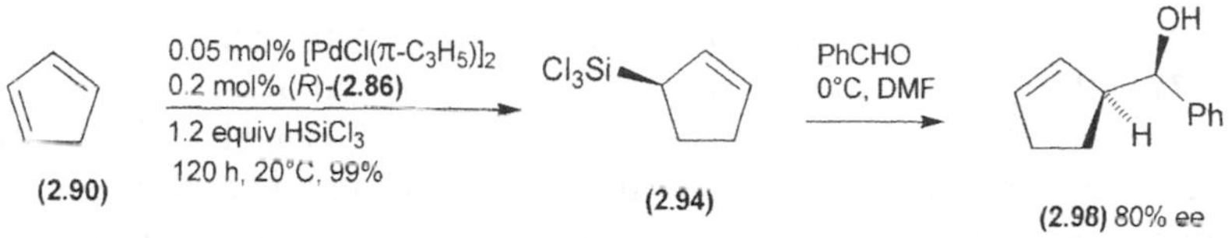

The 1,4-disilylation of α,β-unsaturated ketones using a disilane with a palladium BINAP catalyst has also been achieved, with good enantiomeric excess.[48] For example, with α,β-unsaturated ketone (2.99), the product formed initially (2.100) can be converted into the β-silyl ketone (2.101) by addition of methyllithium followed by hydrolysis. However, quenching the intermediate lithium enolate (formed on addition of MeLi) with an alkylating agent leads to an α-substituted product (2.102), with high anti-selectivity. The β-silyl ketones can be further converted into β-hydroxyketones by oxidation of the Si–C bond.

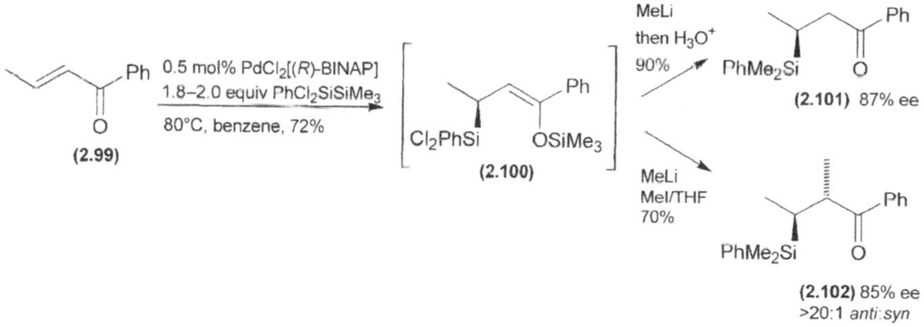

2.5   Alkene hydroboration

Whilst hydroboration of alkenes will occur without the addition of a catalyst, the reaction has been shown to be greatly accelerated by rhodium complexes.[49, 50] The use of an enantiomerically pure rhodium complex provides the opportunity for asymmetric induction in the hydroboration reactions. Some of the ligands examined for this reaction include: BINAP,[51] QUINAP (2.103)[52, 53] and the pyrazole-containing ligand of Togni and co-workers (2.104).[54, 55] A standard procedure uses 1 mol% catalyst in THF at room temperature; however, to achieve high selectivity with BINAP, a temperature of −78°C was required. The standard hydroborating agent is catecholborane (**2.106**). The intermediate boronate is converted, by hydrogen peroxide oxidation, into the corresponding alcohol, with retention of configuration. Thus, styrene (**2.89**) is converted into phenethyl alcohol (**2.97**) by this two-stage process.

The pyrazole ligand (**2.104**) can be easily prepared with other substitutents on the pyrazole ring and on the phosphine, yielding product (**2.97**) with up to 98% ee; however, the regioselectivity of the reaction is never good (21–59% of the primary alcohol, $PhCH_2CH_2OH$, is formed).

An interesting development reported by Brown and co-workers was the conversion of the intermediate boronate into a primary amine.[56] This provided a route for the one-pot conversion of alkenes into enantio-merically-enriched amines. Whilst the intermediate boronates cannot be directly converted into amines, treatment of the boronates with a Grignard reagent forms a borane, which is then converted into an amine using standard procedures. For example, alkene (**2.107**) is converted into amine (**2.108**) in a one-pot process.

(S)-QUINAP (**2.103**)

(**2.104**)

	L	
	BINAP	96% ee
	QUINAP	88% ee
	ligand (**2.104**)	95% ee

(**2.89**)   1 mol% [Rh(cod)L]BF$_4$ / (**2.106**) / 20°C, THF / then, H$_2$O$_2$/NaOH   (**2.97**)

(**2.107**)   1 mol% [(**2.103**)Rh(cod)]OTf / 1 h, THF (**2.106**) / then 2 equiv MeMgCl, 30 min / then 3 equiv H$_2$NOSO$_3$H 15 h / then, HCl/H$_2$O   (**2.108**) 98% ee / > 98% regioselectivity

2.6 Hydroformylation

Hydroformylation reactions involve the addition of H_2 and CO to an alkene, catalysed by a suitable metal complex, and are industrially important catalytic reactions. The hydroformylation of unsymmetrical alkenes can lead to regioisomeric products; for example, a monosubstituted alkene (2.109) yields either the linear (2.110) or branched (2.111) aldehyde as the product. For an asymmetric reaction, it is the branched product which must be formed preferentially as well as enantioselectively. Many ligands have been examined for their ability to provide an asymmetric environment suitable for enantioselective hydroformylation. Gladiali and co-workers reviewed progress in the area in 1995.[57] Some representative ligands are presented, including structures (2.112),[58] (2.113)[59] and (2.114).[60] However, the phosphine phosphite ligands (2.115) have been shown by Nozaki and co-workers to provide good enantioselectivity and regioselectivity across a range of alkenes.[61] This ligand forms a well-organised, active catalytic species, with a phosphine in an equatorial position and a phosphite in an apical position around a rhodium atom. In addition to rhodium-catalysed hydroformylation, the use of platinum-catalysed hydroformylation has been reported to give good enantioselectivity, although regioselectivity is typically poor.[62]

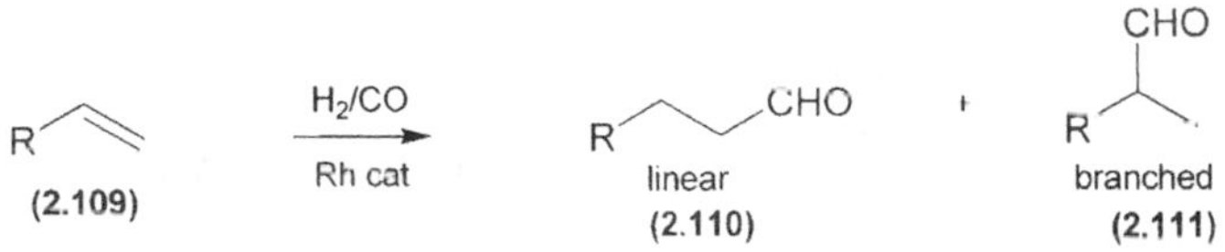

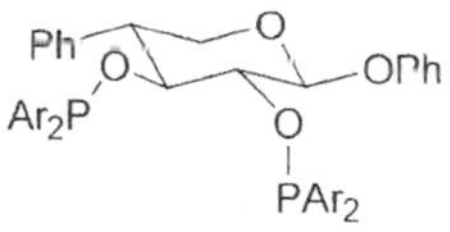

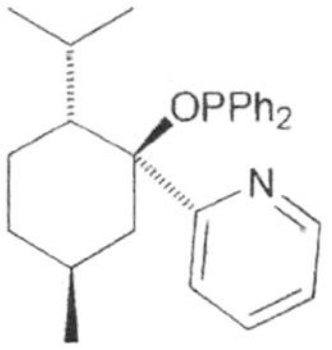

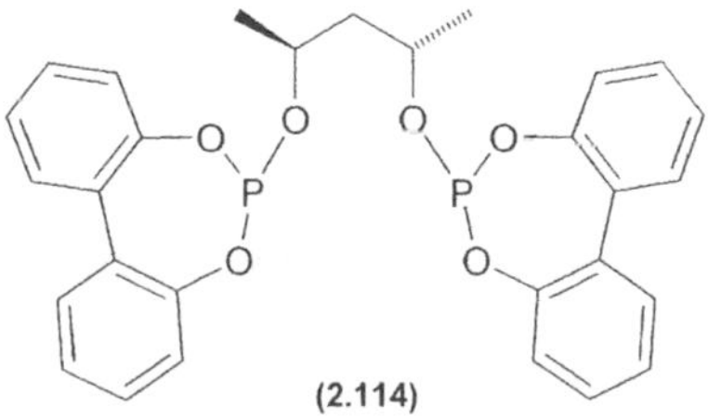

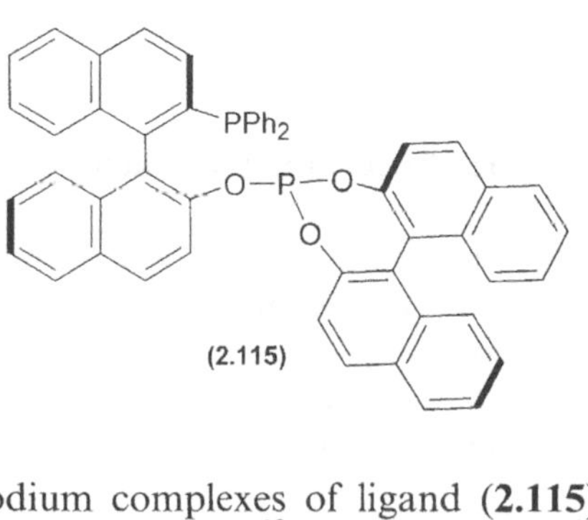

Representative examples using rhodium complexes of ligand (2.115) include the hydroformylation of styrene (2.89),[63] diene (2.116),[64]

cinnamyl alcohol **(2.117)** (which cyclises to give a hemi-acetal product **(2.123)**,[65] and indene **(2.118)**).[66] It is interesting to note that the enantiocontrol of these reactions often exceeds the regiocontrol. Recently, BINAPHOS **(2.115)** has been incorporated into a cross-linked polymer matrix. Subsequent hydroformylation reactions were found to be only slightly less selective and active than their homogeneous counterparts.[67]

Ph⎯⎯═
(2.89)

0.1–0.2 mol% (S,R)-**(2.115)**
0.05 mol% Rh(acac)(CO)₂
H₂:CO (100 atm)
⎯⎯⎯⎯⎯⎯⎯→
43 h, 60°C, benzene,
>99% conversion
(2.119/2.120) 88:12

CHO
Ph⎯⎯│   +   Ph⎯⎯⎯⎯CHO
(2.119) 94% ee   **(2.120)**

(2.116)

2 mol% (R,S)-**(2.115)**
0.5 mol% Rh(acac)(CO)₂
H₂:CO 1:1 (100 atm)
⎯⎯⎯⎯⎯⎯⎯→
18 h, 60°C, benzene,
85% conversion
(2.121/2.122) 86:14

CHO
(2.121) 94% ee   +   **(2.122)** ⎯CHO

Ph⎯⎯⎯⎯OH
(2.117)

2 mol% (R,S)-**(2.115)**
0.05 mol% Rh(acac)(CO)₂
H₂:CO 1:1 (30 atm)
⎯⎯⎯⎯⎯⎯⎯→
30 h, 60°C, benzene,
>99% conversion

Ph⎯⎯⎯O
HO⎯⎯
(2.123) 88% ee

(2.118)

4 mol% (R,S)-**(2.115)**
1 mol% Rh(acac)(CO)₂
H₂:CO 1:1 (100 atm)
⎯⎯⎯⎯⎯⎯⎯→
6 h, 60°C, benzene
(2.124/2.125) 95:5

CHO
(2.124) 97% ee   +   **(2.125)** ⎯CHO

2.7 Hydroacylation of alkenes

The addition of an aldehyde group across an alkene is a hydroacylation reaction. Whilst no hydrogen gas is needed for these reactions, the process has some similarity to hydroformylation from a synthetic viewpoint, and is therefore included in the present chapter. The intramolecular hydroacylation of alkenes has been achieved in an asymmetric sense using rhodium complexes.[68, 69] Using rhodium catalysts, alkenes (2.126) and (2.127) undergo cyclisation to provide the cyclopentanones (2.128) and (2.129).

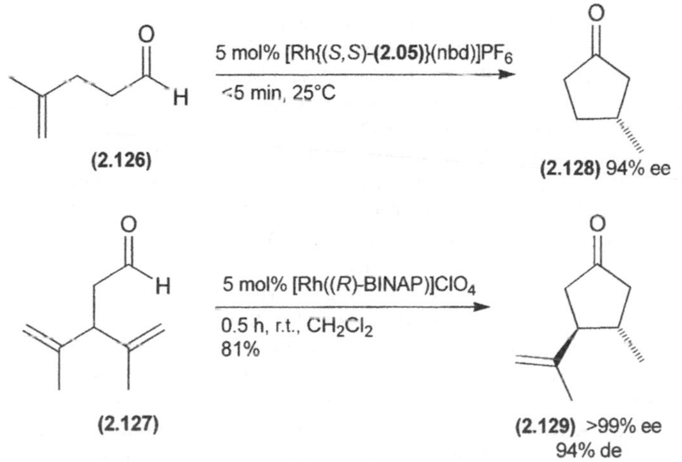

(2.126)

5 mol% [Rh{(S,S)-(2.05)}(nbd)]PF$_6$

<5 min, 25°C

(2.128) 94% ee

(2.127)

5 mol% [Rh((R)-BINAP)]ClO$_4$

0.5 h, r.t., CH$_2$Cl$_2$
81%

(2.129) >99% ee
94% de

2.8 Hydrocyanation of vinylarenes

The hydrocyanation of vinylarenes has been studied by a DuPont team using nickel catalysis.[70] The hydrocyanation of 6-methoxy-2-vinyl-naphthalene (2.130) gives rise to product (2.131), where the enantiomeric excess is strongly dependent upon the electronic nature of the bis-phosphinite ligand (2.13). Hydrolysis of the nitrile (2.131) gives rise to the nonsteroidal anti-inflammatory drug, naproxen.

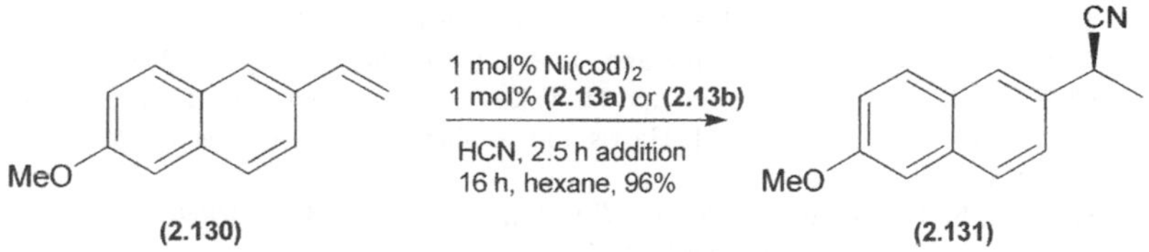

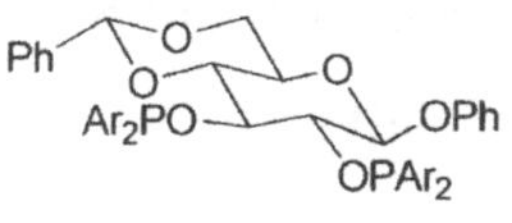

MeO

(2.130)

1 mol% Ni(cod)$_2$
1 mol% (2.13a) or (2.13b)

HCN, 2.5 h addition
16 h, hexane, 96%

MeO

(2.131)

Ph

Ar$_2$PO

OPh

OPAr$_2$

(2.13a) Ar = 3,5(CF$_3$)$_2$C$_6$H$_3$-   provides 85–91% ee
(2.13b) Ar = 3,5(CH$_3$)$_2$C$_6$H$_3$-   provides 16% ee

References

1. For some recent reviews on asymmetric hydrogenation see: (a) R. Noyori, *Asymmetric Catalysis in Organic Synthesis*, John Wiley and Sons, New York, **1994**, 16. (b) H. Takaya, T. Ohta and R. Noyori, in *Catalytic Asymmetric Synthesis*, (I. Ojima, ed.) VCH, New York, **1993**, 1. (c) D. Arntz and A. Schafer, in *Metal Promoted Selectivity in Organic Synthesis*, (A. F. Noels, M. Graziani and A.J. Hubert, eds.) Kluwer Academic, Dordrecht, **1991**, 161.

2. L. Horner, H. Siegel and H. Büthe, *Angew. Chem., Int. Ed. Engl.*, **1968**, *7*, 942.

3. W.S. Knowles and M.J. Sabacky, *J. Chem. Soc., Chem. Commun.*, **1968**, 1445.

4. (a) T.P. Dang and H.B. Kagan, *J. Chem. Soc., Chem. Commun.*, **1971**, 481. (b) H.B. Kagan and T.P. Dang, *J. Am. Chem. Soc.*, **1972**, *94*, 6429.

5. (a) A. Miyashita, A. Yasuda, H. Takaya, K. Toriumi, T. Ito, T. Souchi and R. Noyori, *J. Am. Chem. Soc.*, **1980**, *102*, 7932. (b) M.J. Burk, *J. Am. Chem. Soc.*, **1991**, *113*, 8518. (c) M. D. Fryzuk and B. Bosnich, *J. Am. Chem. Soc.*, **1997**, *99*, 6262. (d) B.D. Vineyard, W.S. Knowles, M.J. Sabacky, G.L. Bachman and D.J. Weinkauff, *J. Am. Chem. Soc.*, **1977**, *99*, 5946. (e) T. Hayashi and M. Kumada, *Acc. Chem. Res.*, **1982**, *15*, 395. (f) A. Miyashita, H. Karino, J.-I. Shimamura, T. Chiba, K. Nagano, H. Nohira and H. Takaya, *Chem. Lett.*, **1989**, 1849. (g) P.A. McNeil, N.K. Roberts and B. Bosnich, *J. Am. Chem. Soc.*, **1981**, *103*, 2280. (h) U. Nagel, E. Kinzel, J. Andrade and G. Prescher, *Chem. Ber.*, **1986**, *119*, 3326. (i) T. Imamoto, J. Watanabe, Y. Wada, H. Masuda, H. Yamada, H. Tsuruta, S. Matsukawa and K. Yamaguchi, *J. Am. Chem. Soc.*, **1998**, *120*, 1635. (j) T.V. RajanBabu, T.A. Ayers, G.A. Halliday, K.K. You and J.C. Calabrese, *J. Org. Chem.*, **1997**, *62*, 6012. (k) F. Robin, F. Mercier, L. Ricard, F. Mathey and M. Spagnol, *Chem. Eur. J.*, **1997**, *3*, 1365. (l) J. Kang, J.H. Lee, S.H. Ahn and J.S. Choi, *Tetrahedron Lett.*, **1988**, *39*, 5523. (m) F.-Y. Zhang, C.-C. Pai and A.S.C. Chan, *J. Am. Chem. Soc.*, **1988**, *120*, 5808. (n) A.S.C. Chan, W. Hu, C-C. Pai and C.-P. Lau, *J. Am. Chem. Soc.*, **1997**, *119*, 9570. (o) G. Zhu, P. Cao, Q. Jiang and X. Zhang, *J. Am. Chem. Soc.*, **1997**, *119*, 1799. (p) M. Sawamura, R. Kuwano and Y. Ito, *J. Am. Chem. Soc.*, **1995**, *117*, 9602.

6. S.R. Gilbertson and X. Wang, *Tetrahedron Lett.*, **1996**, *37*, 6475.

7. W.S. Knowles, *Acc. Chem. Res.*, **1983**, *16*, 106.

8. M.J. Burk, S. Feng, M.F. Gross and W. Tumas, *J. Am. Chem. Soc.*, **1995**, *117*, 8277.

9. J. Halpern, *Pure Appl. Chem.*, **1983**, *55*, 99.

10. J.M. Brown and P.A. Chaloner, *J. Am. Chem. Soc.*, **1980**, *102*, 3040.

11. S.K. Armstrong, J.M. Brown and M.J. Burk, *Tetrahedron Lett.*, **1993**, *34*, 879.

12. A. Miyashita, H. Takaya, T. Souchi and R. Noyori, *Tetrahedron*, **1984**, *40*, 1245.

13. M.J. Burk, J.G. Allen and W.F. Kiesman, *J. Am. Chem. Soc.*, **1998**, *120*, 657.

14. M.J. Burk, M.F. Gross and J.P. Martinez, *J. Am. Chem. Soc.*, **1995**, *117*, 9375.

15. R. Kuwano, S. Okuda and Y. Ito, *J. Org. Chem.*, **1998**, *63*, 3499.

16. M.J. Burk, C.S. Kalberg and A. Pizzano, *J. Am. Chem. Soc.*, **1998**, *120*, 4345.

17. T. Chiba, A. Miyashita, H. Nohira and H. Takaya, *Tetrahedron Lett.*, **1991**, *32*, 4745. See also: M.J. Burk, F. Bienewald, M. Harris and A. Zanotti-Gerosa, *Angew. Chem., Int. Ed. Engl.*, **1988**, *37*, 1931.

18. T. Hayashi, N. Kawamura and Y. Ito, *Tetrahedron Lett.*, **1988**, *29*, 5969.

19. N.W. Boaz, *Tetrahedron Lett.*, **1988**, *39*, 5505.

20. R. Noyori, *Chem. Soc. Rev.*, **1989**, *18*, 209.

21. T. Uemura, X. Zhang, K. Matsumura, N. Sayo, H. Kumobayashi, T. Ohta, K. Nozaki and H. Takaya, *J. Org. Chem.*, **1996**, *61*, 5510.

22. D.J. Bayston, J.L. Fraser, M.R. Ashton, A.D. Baxter, M.E.C. Polywka and E. Moses, *J. Org. Chem.*, **1998**, *63*, 3137.

23. J.P. Genêt , C. Pinel, V. Ratovelomanana-Vidal, S. Mallart, X. Pfister, L. Bischoff, M.C. Caño de Andrade, S. Darses, C. Galopin and J.A. Laffitte, *Tetrahedron: Asymmetry*, **1994**, *5*, 675.

24. (a) H. Kawano, T.I. Kariya, Y. Ishii, M. Saburi, S. Yoshikawa, Y. Uchida and H. Kumobayashi, *J. Chem. Soc., Perkin Trans. 1*, **1989**, 1571. (b) R. Noyori, T. Ikeda, T. Ohkuma, M. Widhalm, M. Kitamura, H. Takaya, S. Akutagawa, N. Sayo, T. Saito, T. Taketomi and H. Kumobayashi, *J. Am. Chem. Soc.*, **1989**, *111*, 9134.
25. M. Kitamura, Y. Hsiao, M. Ohta, M. Tsukamoto, T. Ohta, H. Takaya and R. Noyori, *J. Org. Chem.*, **1994**, *59*, 297.
26. T. Ohta, H. Takaya, M. Kitamura, K. Nagai and R. Noyori, *J. Org. Chem.*, **1987**, *52*, 3174.
27. M. Saburi, L. Shao, T. Sakurai and Y. Uchida, *Tetrahedron Lett.*, **1992**, *33*, 7877.
28. H. Takaya, T. Ohta, N. Sayo, H. Kumobayashi, S. Akutagawa, S. Inoue, I. Kasahara and R. Noyori, *J. Am. Chem. Soc.*, **1987**, *109*, 1596.
29. M. Kitamura, K. Nagai, Y. Hsiao and R. Noyori, *Tetrahedron Lett.*, **1990**, *31*, 549.
30. M. Kitamura, I. Kasahara, K. Manabe, R. Noyori and H. Takaya, *J. Org. Chem*, **1988**, *53*, 708.
31. J.W. Faller and M. Tokunaga, *Tetrahedron Lett.*, **1993**, *34*, 7359.
32. P. Le Gendre, F. Jérôme, C. Bruneau and P.H. Dixneuf, *J. Chem. Soc., Chem. Commun.*, **1998**, 533.
33. T. Ohta, T. Miyake and H. Takaya, *J. Chem. Soc., Chem. Commun.*, **1992**, 1725.
34. T. Ohta, T. Miyake, N. Seido, H. Kumobayashi and H. Takaya, *J. Org. Chem.*, **1995**, *60*, 357.
35. E. Cesaroth, R. Ugo and H.B. Kagan, *Angew. Chem., Int. Ed. Eng.*, **1979**, *18*, 779.
36. R.L. Halterman, K.P.C. Vollhardt, M.E. Welker, D. Bläser and R. Boese, *J. Am. Chem. Soc.*, **1987**, *109*, 8105.
37. F.R.W.P. Wild, L. Zsolnai, G. Huttner and H.H. Brintzinger, *J. Organomet. Chem.*, **1982**, *232*, 233.
38. R.D. Broene and S.L. Buchwald, *J. Am. Chem. Soc.*, **1993**, *115*, 12569.
39. R. Waymouth and P. Pino, *J. Am. Chem. Soc.*, **1990**, *112*, 4911.
40. V.P. Conticello, L. Brard, M.A. Giardello, Y. Tsuji, M. Sabat, C.L. Stern and T.J. Marks, *J. Am. Chem. Soc.*, **1992**, *114*, 2761.
41. N.E. Lee and S.L. Buchwald, *J. Am. Chem. Soc.*, **1994**, *116*, 5985.
42. P.-F. Fu, L. Brard, Y. Li and T.J. Marks, *J. Am. Chem. Soc.*, **1995**, *117*, 7157.
43. S. Gladiali, S. Pulacchini, D. Fabbri, M. Manassero and M. Sansoni, *Tetrahedron: Asymmetry*, **1998**, *9*, 391.
44. Y. Uozumi, S.-Y. Lee and T. Hayashi, *Tetrahedron Lett.*, **1992**, *33*, 7185.
45. Y. Uozumi and T. Hayashi, *Tetrahedron Lett.*, **1993**, *34*, 2335.
46. K. Kitayama, Y. Uozumi and T. Hayashi, *J. Chem. Soc., Chem. Commun.*, **1995**, 1533.
47. K. Kitayama, H. Tsuji, Y. Uozumi and T. Hayashi, *Tetrahedron Lett.*, **1996**, *37*, 4169.
48. Y. Matsumoto, T. Hayashi and Y. Ito, *Tetrahedron*, **1994**, *50*, 335.
49. D. Männig and H. Nöth, *Angew. Chem., Int. Ed. Engl.*, **1985**, *24*, 878.
50. D.A. Evans, G.C. Fu and A.H. Hoveyda, *J. Am. Chem. Soc.*, **1992**, *114*, 6671.
51. T. Hayashi, Y. Matsumoto and Y. Ito, *J. Am. Chem. Soc.*, **1989**, *111*, 3426.
52. J.M. Brown, D.I. Hulmes and T.P. Layzell, *J. Chem. Soc., Chem. Commun.*, **1993**, 1673.
53. J.M. Valk, G.A. Whitlock, T.P. Layzell and J.M. Brown, *Tetrahedron: Asymmetry*, **1995**, *6*, 2593.
54. A. Schnyder, L. Hintermann and A. Togni, *Angew. Chem., Int. Ed. Engl.*, **1995**, *34*, 931.
55. A. Schnyder, A. Togni and U. Wiesli, *Organometallics*, **1997**, *16*, 255.
56. E. Fernandez, M.W. Hooper, F.I. Knight and J.M. Brown, *J. Chem. Soc., Chem. Commun.*, **1997**, 173.
57. S. Gladiali, J.C. Bayón and C. Claver, *Tetrahedron: Asymmetry*, **1995**, *6*, 1453.
58. T.V. RajanBabu and T.A. Ayers, *Tetrahedron Lett.*, **1994**, *35*, 4295.
59. C.G. Arena, F. Nicolò, D. Drommi, G. Bruno and F. Faraone, *J. Chem. Soc., Chem. Commun.*, **1994**, 2251.

60. G.J.H. Buisman, E.J. Vos, P.C.J. Kamer and P.W.N.M. van Leeuwen, *J. Chem. Soc., Dalton Trans.*, **1995**, 409.
61. K. Nozaki, N. Sakai, T. Nanno, T. Higashijima, S. Mano, T. Horiuchi and H. Takaya, *J. Am. Chem. Soc.*, **1997**, *119*, 4413.
62. G. Consiglio, S.C.A. Nefkens and A. Borer, *Organometallics*, **1991**, *10*, 2046.
63. N. Sakai, S. Mano, K. Nozaki and H. Takaya, *J. Am. Chem. Soc.*, **1993**, *115*, 7033.
64. T. Horiuchi, T. Ohta, K. Nozaki and H. Takaya, *J. Chem. Soc., Chem. Commun.*, **1996**, 155.
65. K. Nozaki, W.-G. Li, T. Horiuchi and H. Takaya, *Tetrahedron Lett.*, **1997**, *38*, 4611.
66. N. Sakai, K. Nozaki and H. Takaya, *J. Chem. Soc., Chem. Commun.*, **1994**, 395.
67. K. Nozaki, Y. Itoi, F. Shibahara, E. Shirakawa, T. Ohta, H. Takaya and T. Hiyama, *J. Am. Chem. Soc.*, **1998**, *120*, 4051.
68. X.-M. Wu, K. Funakoshi and K. Sakai, *Tetrahedron Lett.*, **1993**, *34*, 5927.
69. (a) R.W. Barnhart, X. Wang, P. Noheda, S.H. Bergens, J. Whelan and B. Bosnich, *J. Am. Chem. Soc.*, **1994**, *16*, 1821. (b) R.W. Barnhart, D.A. McMorran and B. Bosnich, *J. Chem. Soc., Chem. Commun.*, **1997**, 589–590.
70. A.L. Casalnuvo, T.V. RajanBabu, T.A. Ayers and T.H. Warren, *J. Am. Chem. Soc.*, **1994**, *116*, 9869.

3   Reduction of ketones and imines

The reduction of the carbonyl group (and related functionalities) has been successfully achieved using various catalytic methods: direct hydrogenation; transfer hydrogenation; and the use of alternative stoichiometric reducing agents, including boranes and silanes. These catalytic methods are often selective for the presence of aldehydes and ketones in the presence of less reactive carbonyl groups, such as esters and amides.

3.1   Hydrogenation of ketones

The reduction of ketones into enantiomerically-enriched alcohols using hydrogen as the stoichiometric reductant is an appealing transformation. The reaction is atom economical, with no by-products, and has been achieved with very low catalyst loadings.

Ruthenium and rhodium catalysts have maintained the best track record in hydrogenation of ketones.[1] For high enantioselectivities, ketones which contain an additional functional group are often needed. Typical substrates are represented by ketones (**3.01–3.04**), where amines, esters, halides[2,3] and phosphonates[4,5] provide an additional donor group on the substrate. BINAP/ruthenium combinations, such as complex (**3.09**) have given consistently high enantioselectivities with such ketones. Variation of the counter-ions and the method of catalyst construction can affect the enantioselectivity observed.

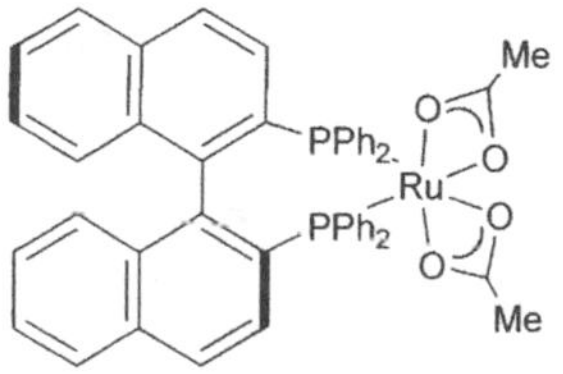

(3.09)   [(S)-BINAP]Ru(OAc)₂

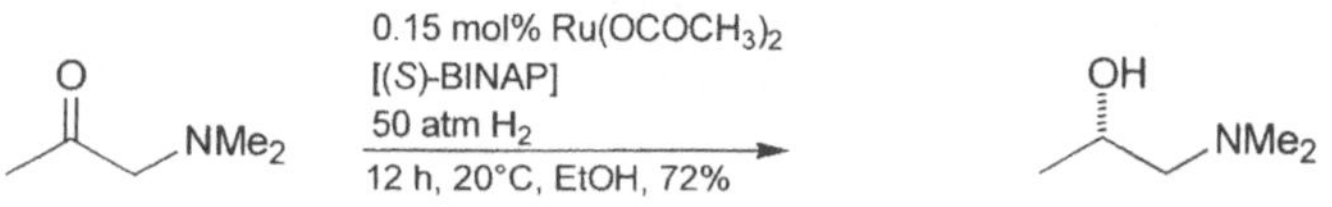

(3.01)   (3.05) 96% ee

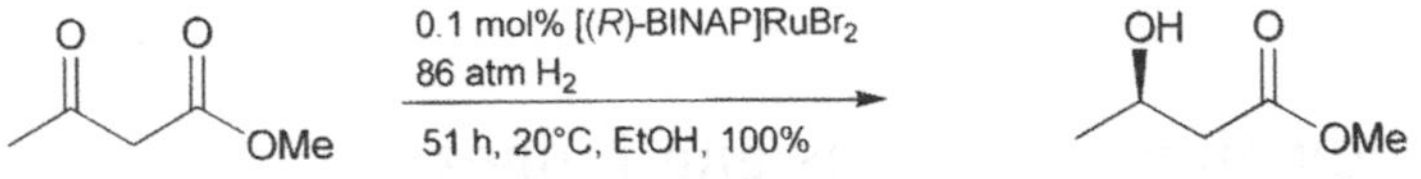

O
O
OMe
0.1 mol% [(R)-BINAP]RuBr2
86 atm H2
51 h, 20°C, EtOH, 100%
OH
O
OMe

(3.02) **(3.06)** >99% ee

O
Br
0.1 mol% Ru2Cl4[(R)-BINAP]2.NEt3
100 atm H2
62 h, 20°C, EtOH, 97%
OH
Br

(3.03) **(3.07)** >92% ee

O
O
P(OMe)2
0.2 mol% RuCl2[(R)-BINAP](dmf)n
4 atm H2
72 h, 25°C, MeOH, 99%
OH
O
P(OMe)2

(3.04) **(3.08)**

Not surprisingly, BINAP is not the only ligand to provide high selectivities in hydrogenation reactions. The alternative catalysts (**3.10**–**3.13**) also work well in related reactions. Some examples are provided in the reduction of the standard β-keto-ester (**3.02**),[6, 7, 8] as well as the α-keto ester (**3.14**) and the β-keto sulfide (**3.15**).[9]

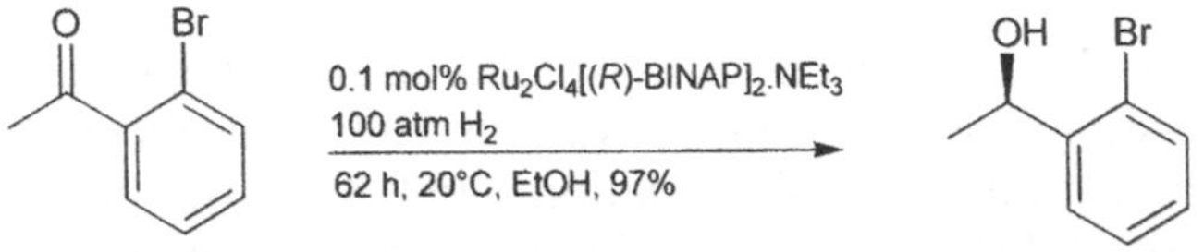

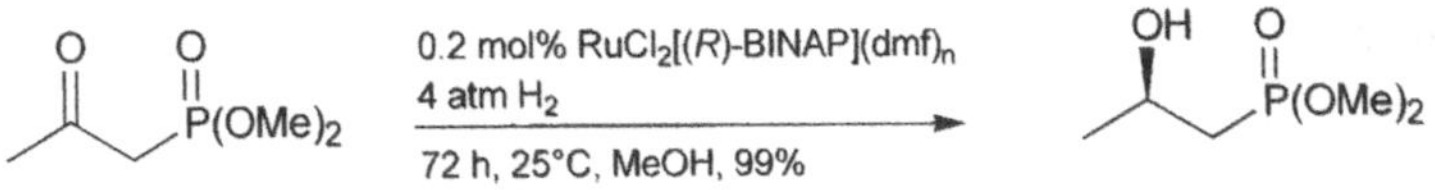

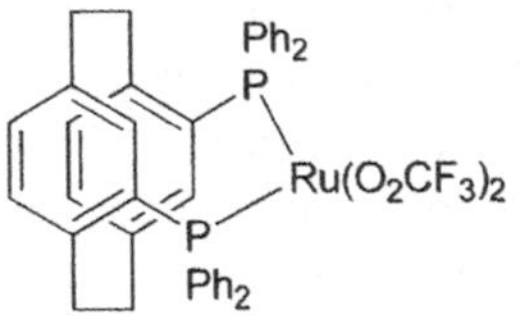

Ph2
P
Ru(O2CF3)2
P
Ph2

(3.10)

Cy2
P
Ru
I
PCy2

(3.12)

iPr
Pri
P
P
Ru
Br
Br
Pri
iPr

(3.11)

PPh2
RuBr2
PPh2

(3.13)

O
O
OMe
0.4–0.8 mol% (3.10)
3 atm H2, 5 mol% Bu4NI
18 h, -5°C, MeOH/H2O 100%
OH
O
OMe

(3.02) **(3.06)** 96% ee

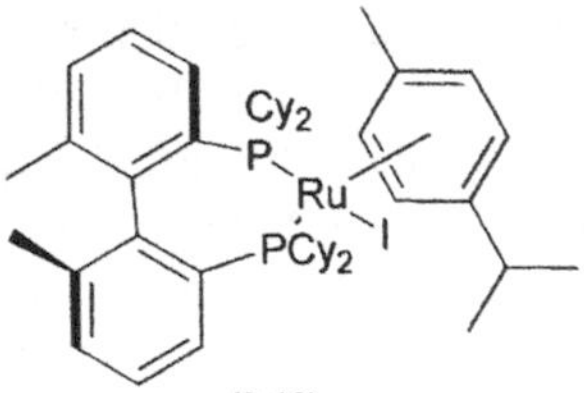

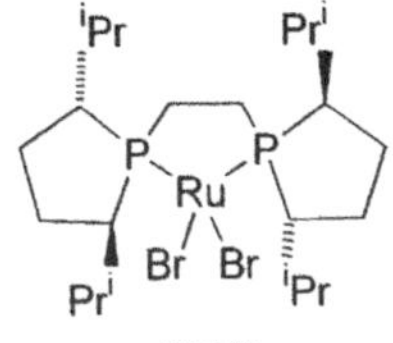

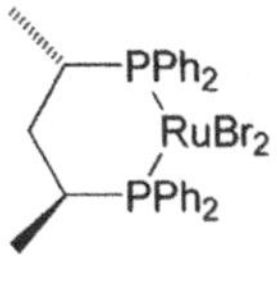

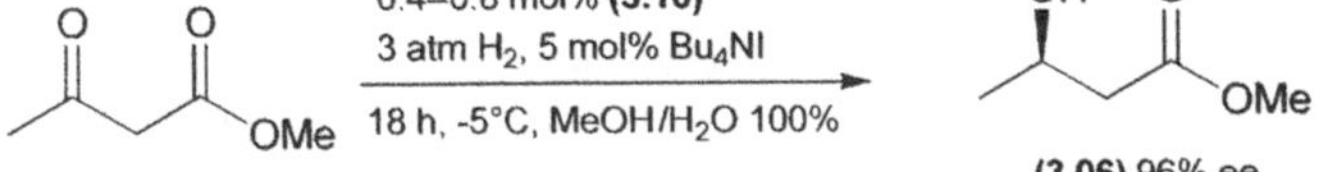

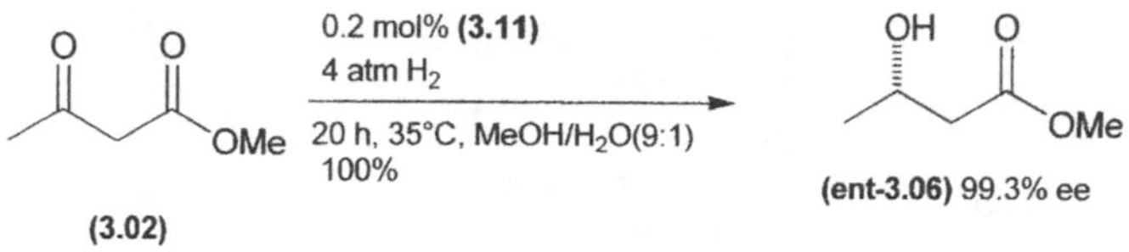

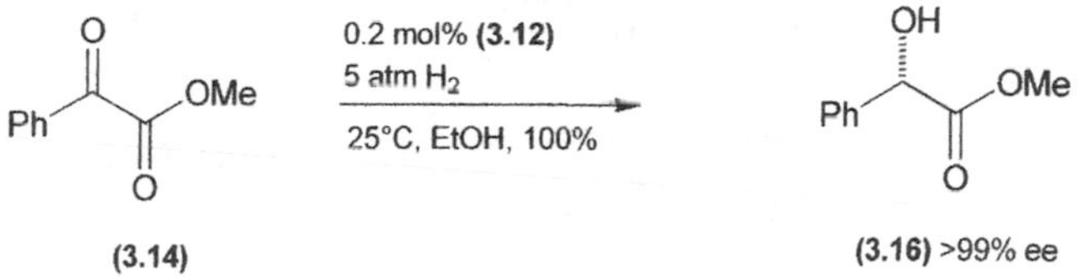

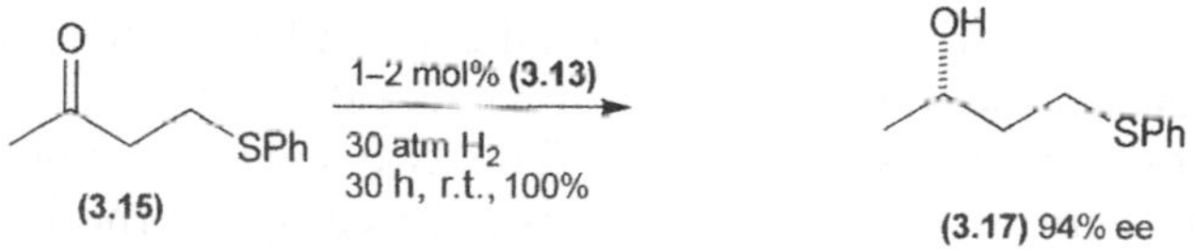

(3.02) → 0.2 mol% (3.11), 4 atm H$_2$, 20 h, 35°C, MeOH/H$_2$O(9:1), 100% → (ent-3.06) 99.3% ee

(3.14) → 0.2 mol% (3.12), 5 atm H$_2$, 25°C, EtOH, 100% → (3.16) >99% ee

(3.15) → 1–2 mol% (3.13), 30 atm H$_2$, 30 h, r.t., 100% → (3.17) 94% ee

The rhodium complexes of amidophosphine-phosphinite ligands are effective for asymmetric hydrogenation of ketones;[10] particular success being achieved with the ligand (3.18).[11] Thus, ketone (3.19) is reduced into the amino alcohol (3.20), with good enantioselectivity.

The rhodium catalyst based on the Penn-Phos ligand (3.21) is noteworthy, since it is effective for the asymmetric reduction of unfunctionalised ketones, with good to excellent enantioselectivities (72–96% ee).[12] For example, acetophenone (3.22) and the aliphatic ketone (3.23) are selectively reduced to the corresponding alcohols.

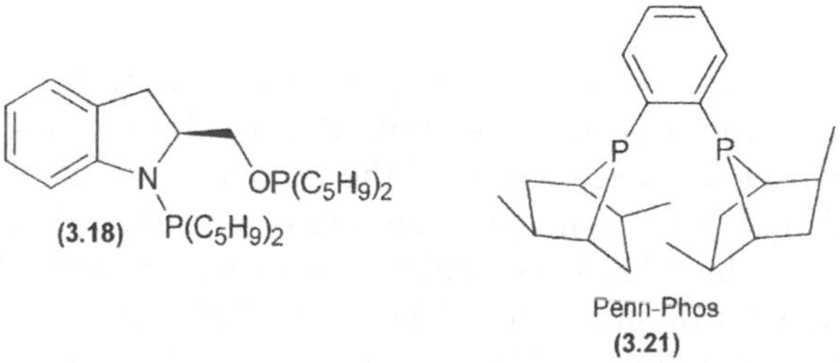

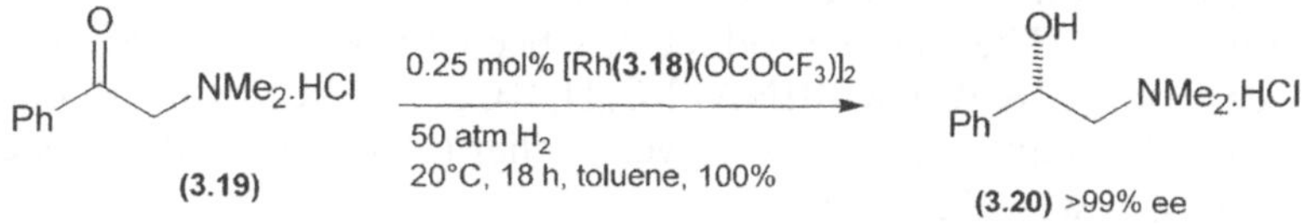

(3.18) P(C$_5$H$_9$)$_2$ OP(C$_5$H$_9$)$_2$

Penn-Phos (3.21)

(3.19) → 0.25 mol% [Rh(3.18)(OCOCF$_3$)]$_2$, 50 atm H$_2$, 20°C, 18 h, toluene, 100% → (3.20) >99% ee

(3.22) → (3.24) 95% ee

0.5 mol% [Rh(cod)Cl₂]₂
1 mol% (3.21)
30 atm H₂
0.4 mol% 2,6-lutidine
24 h, MeOH, 97%

(3.23) → (3.25) 85% ee

0.5 mol% [Rh(cod)Cl₂]₂
1 mol% (3.21)
30 atm H₂
0.4 mol% 2,6-lutidine
1.0 mol% KBr
24 h, MeOH, 97%

A remarkable hydrogenation of unsaturated keto-esters has been reported by Takahashi and co-workers.[13] The reaction mixture contains a rhodium catalyst, which reduces the alkene, and a ruthenium catalyst, which reduces the ketone. Thus, the enone (3.26) is reduced to the alcohol (3.27), with excellent stereocontrol. The two catalysts co-exist in the reaction mixture, although the alkene is reduced by the rhodium catalyst first, with an increase in pressure before the ketone is reduced by the ruthenium catalyst. The diastereoselectivity of the process has been reversed by the use of rhodium and ruthenium complexes, which possess different ligands from each other.

(3.26) → (3.27) >95% ee, >99% de

1 mol% [Rh(cod)((S)-BINAP)]⁺ClO₄⁻
1 mol% RuBr₂((S)-BINAP)
10 atm for 24 h
then, 90 atm for 24 h
40°C, EtOH, 99%

α-Substituted β-keto esters are of particular interest because the stereochemistry at the α-position is labile due to the low pKa of such substrates. However, after reduction of the ketone to an alcohol group, the stereochemistry is fixed. This provides the opportunity for a dynamic kinetic resolution, which has been achieved with remarkable efficiency.[14] The two enantiomers of substrate, (3.28) and (3.29), are in equilibrium under the reaction conditions. The ruthenium catalyst reacts selectively with the (R)-enantiomer (3.28), and provides the β-keto ester (3.30), with high selectivity. The major product has been formed with 98% de and 92% ee using {RuCl[(R)-BINAP](C₆H₆)}Cl in dichloromethane. In methanol, the enantioselectivity is even higher (99% ee) but the diastereoselectivity is lower (92% de).[15]

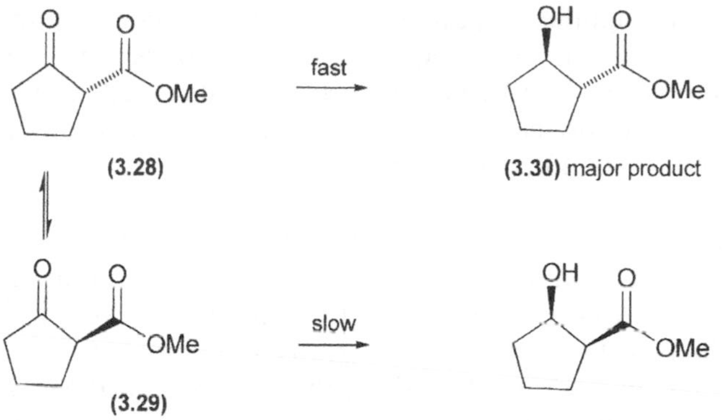

(3.28)   fast   (3.30) major product

(3.29)   slow   (3.31) minor product

The dynamic resolution strategy has been successfully applied to related systems; for example, the amide (3.32),[16] α-chloro-substituted compound (3.33),[17] and the α-amido β-ketophosphonate (3.34).[18] In each case, a similar principle operates, where the starting material is rapidly racemising and the catalyst selects only one enantiomer in the reduction process.

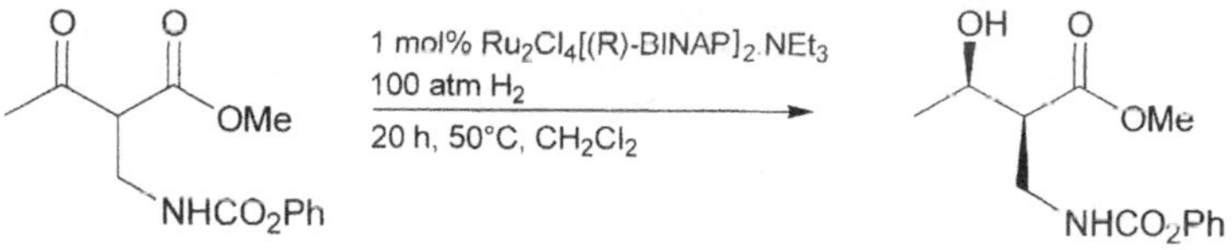

1 mol% Ru$_2$Cl$_4$[(R)-BINAP]$_2$.NEt$_3$
100 atm H$_2$
20 h, 50°C, CH$_2$Cl$_2$

(3.32)   (3.35) 88% de, 98% ee

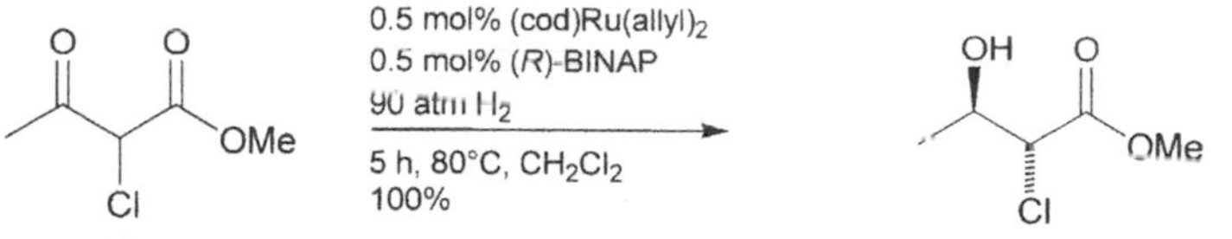

0.5 mol% (cod)Ru(allyl)$_2$
0.5 mol% (R)-BINAP
90 atm H$_2$
5 h, 80°C, CH$_2$Cl$_2$
100%

(3.33)   (3.36) 98% de, 99% ee

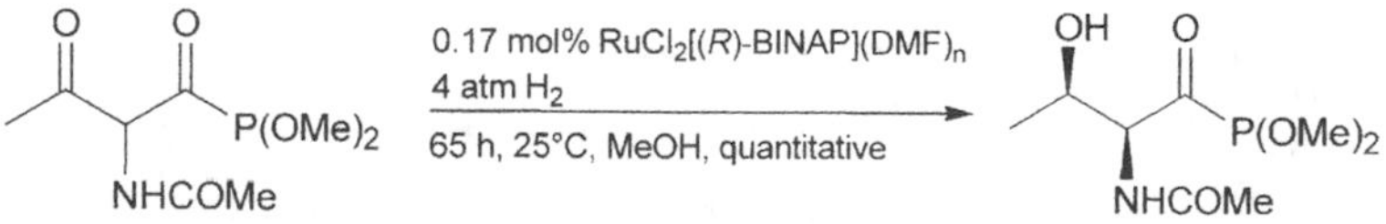

0.17 mol% RuCl$_2$[(R)-BINAP](DMF)$_n$
4 atm H$_2$
65 h, 25°C, MeOH, quantitative

(3.34)   (3.37) 94% de, >98% ee

3.2 Hydrogenation of imines and related compounds

The hydrogenation of imines and their derivatives provides a convenient route to enantiomerically-enriched amines.

In 1992, two significant papers were published concerning the asymmetric hydrogenation of the C=N group.[19] Burk and Feaster used rhodium complexes of DuPHOS (3.38) as catalysts to reduce hydrazones.[20] The levels of enantioselectivity were high for the reduction of hydrazones, such as (3.39) and (3.40), derived from arylalkyl ketones and α-keto esters.

Willoughby and Buchwald employed the titanium-based Brintzinger catalyst (3.41) for the asymmetric reduction of imines.[21] The catalyst is activated by reduction to what is assumed to be the titanium (III) hydride species (3.42). The best substrates for this catalyst are cyclic imines, which yield products with 95–98% ee. Various functional groups, including alkenes, vinyl silanes, acetals and alcohols, were not affected under the reaction conditions.[22] For example, imine (3.43) was reduced, with excellent enantioselectivity, without reduction of the alkene moiety.

Using the racemic substrate (3.45), a highly selective kinetic resolution was achieved, affording the reduced product (3.46) and recovered substrate, both with very high enantioselectivity.[23] Mechanistic studies

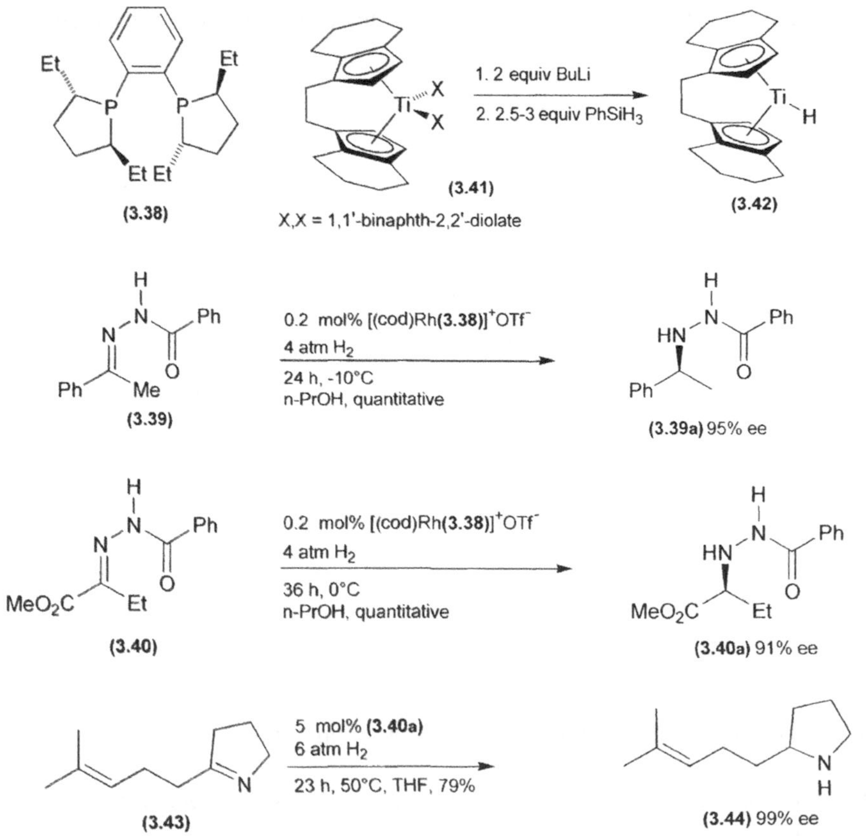

(3.38)

(3.41)

X,X = 1,1'-binaphth-2,2'-diolate

(3.42)

1. 2 equiv BuLi
2. 2.5-3 equiv PhSiH₃

(3.39)

0.2 mol% [(cod)Rh(3.38)]⁺OTf⁻
4 atm H₂
24 h, -10°C
n-PrOH, quantitative

(3.39a) 95% ee

(3.40)

0.2 mol% [(cod)Rh(3.38)]⁺OTf⁻
4 atm H₂
36 h, 0°C
n-PrOH, quantitative

(3.40a) 91% ee

(3.43)

5 mol% (3.40a)
6 atm H₂
23 h, 50°C, THF, 79%

(3.44) 99% ee

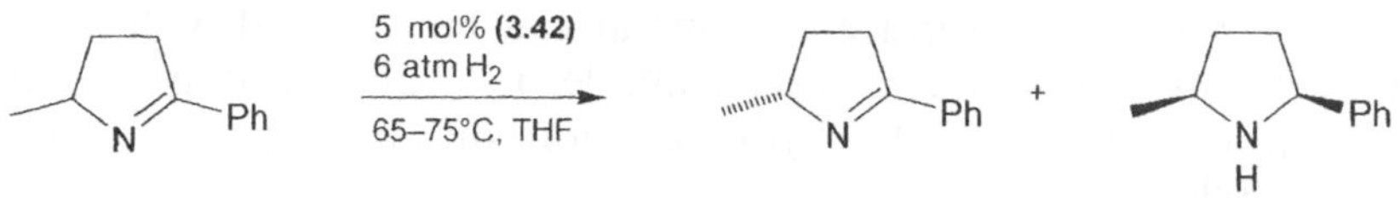

(3.45)   R-(3.45) 37% isolated   (3.46) 34% isolated
99% ee   99% ee

and further details of the scope of the reaction have been published previously.[24, 25]

Iridium-catalysed imine reduction has also been achieved with asymmetric induction by several research groups,[26] including the use of the phosphino-oxazoline complex (3.47), which provided up to 89% ee in the reduction of imine (3.48) into the corresponding amine (3.49).[27] Morimoto and co-workers employed an iridium/BINAP complex in the hydrogenation of the dihydroisoquinoline (3.50).[28] Additionally, Ru(OAc)$_2$BINAP catalysts have been used in the reduction of N-tosylimines, with up to 84% ee.[29]

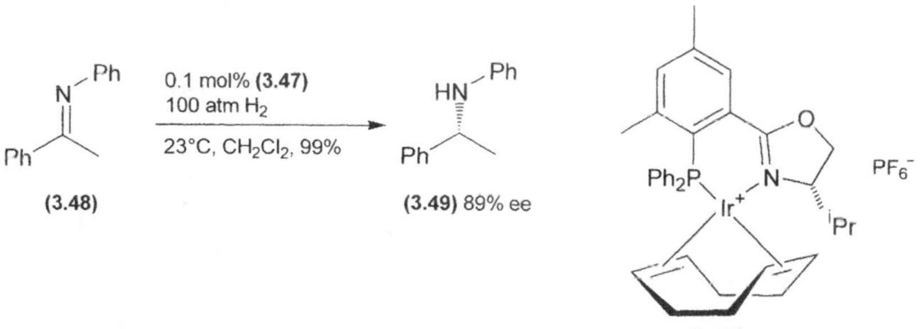

(3.48)   (3.49) 89% ee

(3.47)

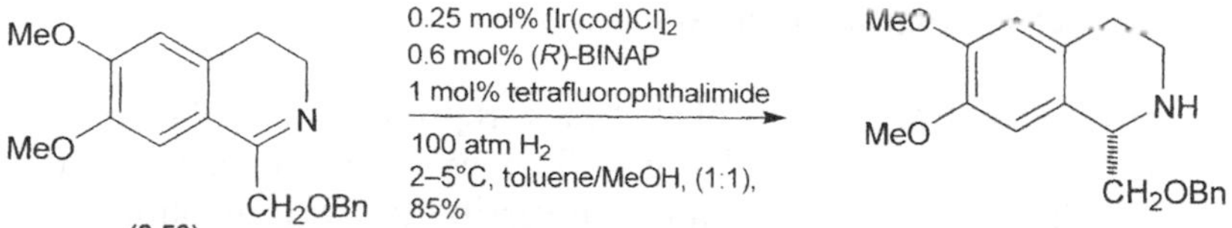

(3.50)   (3.51) 86% ee

3.3 Heterogeneous hydrogenation

The attachment of enantiomerically pure ligands to polymers provides one way of achieving asymmetric catalysis in a heterogeneous fashion, as well as the use of two (immiscible) liquids. These methods are considered alongside their fully homogeneous counterparts, although selectivities are usually somewhat compromised.

Alternatively, the surface of a metal can be modified with an enantiomerically pure additive. For example, in 1956 palladium metal/silk fibroin was used for the hydrogenation of alkenes, with moderate enantioselectivity.[30]

However, most success with modified metal surfaces has been achieved in the hydrogenation of ketones. The reduction of β-keto esters with a Raney nickel/tartaric acid/sodium bromide catalyst provides good enantioselectivities;[31, 32] for example, β-keto ester (3.52) gives rise to the β-hydroxy ester (3.53). Platinum metal modified with cinchona alkaloids has been successfully used with α-keto esters. The reaction yields product with up to 90% ee in the reduction of α-keto ester (3.54).[33, 34]

The use of polymer-stabilised colloidal platinum clusters (containing cinchonidine) in acetic acid provides up to 97.6% ee in the reduction of methyl pyruvate ($MeCOCO_2Me$).[35] Polyvinylpyrrolidine is used as the polymer and the particle size is small.[36]

(3.52) Raney Ni, 100 atm H_2 / (R,R)-tartaric acid NaBr, THF/ 0.9% CH_3CO_2H → (3.53)

(3.54) 50 atm H_2 / Pt/C/cinchonidine benzene, 87% → (3.55)

3.4   Transfer hydrogenation of ketones

The transfer hydrogenation of ketones represents an alternative to conventional methods using H_2 gas. The hydride source is usually a donor alcohol or formate. Although reversibility (with alcohol donors) may seem problematic to a highly selective reaction, several successful enantioselective transfer hydrogenation reactions have been reported. Amongst these reactions, iridium,[37] rhodium[38] and samarium[39] have all provided highly enantioselective catalysts for asymmetric transfer hydrogenation. However, ruthenium catalysts have attracted more attention.

Noyori and co-workers reported that ruthenium complex (3.52) gives very high enantioselectivity for many arylalkyl ketones, including various acetophenones (3.53) and cyclic ketones (3.54) and (3.55).[40] The same catalyst has been used in the reduction of α,β-acetylenic ketones.[41] Again, very high enantioselectivities have been recorded in this reaction, as shown by the reduction of acetylenic ketones **(3.56)** and **(3.57)**; 2-propanol was employed as the hydride source.

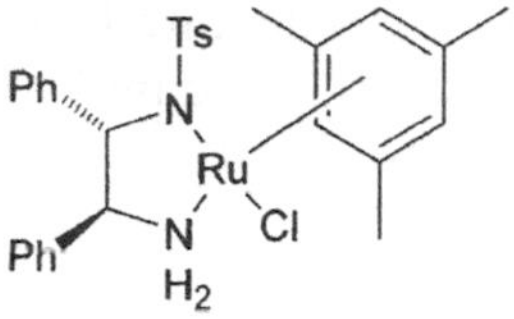
Ts
Ph
N
Ru
N
Cl
Ph
H2

(3.52)

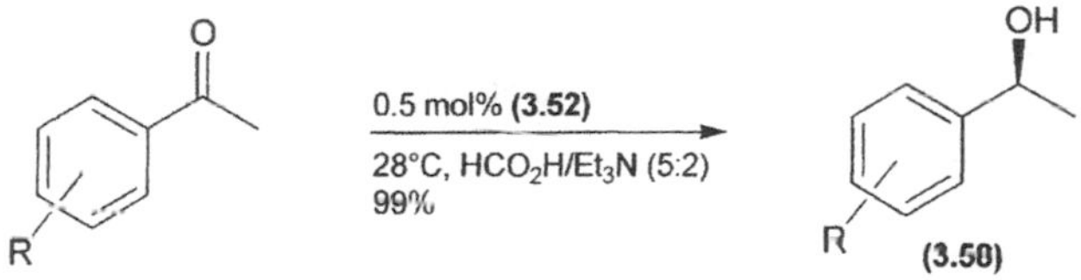
O
R
0.5 mol% (3.52)
28°C, HCO2H/Et3N (5:2)
99%
OH
R

(3.53) **(3.58)**
H, 98% ee, 20 h
p-Cl, 95% ee, 24 h
m-OMe, 98% ee, 50 h

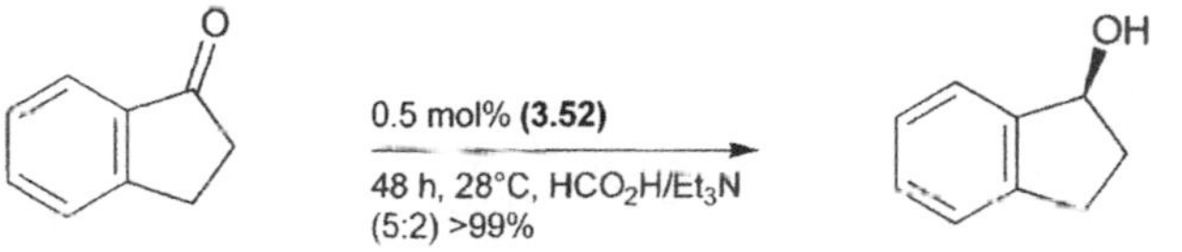
O
0.5 mol% (3.52)
48 h, 28°C, HCO2H/Et3N
(5:2) >99%
OH

(3.54) **(3.59)** 99% ee

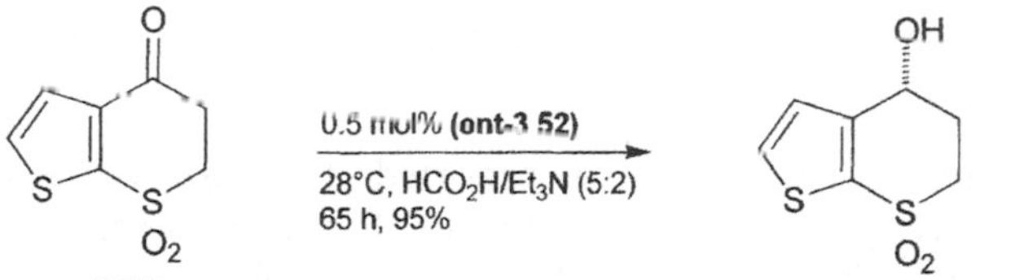
O
S
S
O2
0.5 mol% (ent-3.52)
28°C, HCO2H/Et3N (5:2)
65 h, 95%
OH
S
S
O2

(3.55) **(3.60)** 98% ee

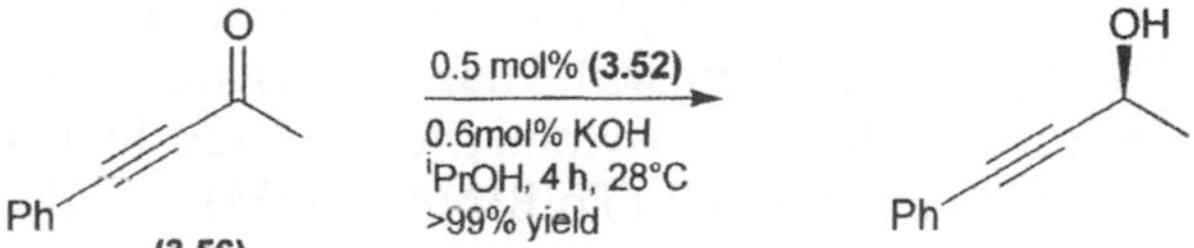
O
Ph
0.5 mol% (3.52)
0.6mol% KOH
iPrOH, 4 h, 28°C
>99% yield
OH
Ph

(3.56) **(3.61)** 97% ee

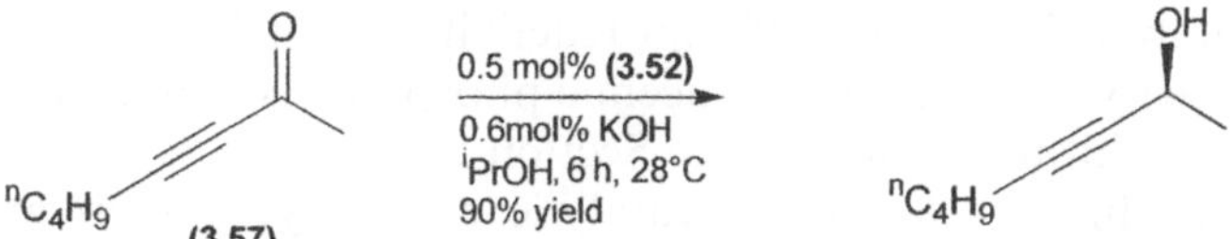
O
nC4H9
0.5 mol% (3.52)
0.6mol% KOH
iPrOH, 6 h, 28°C
90% yield
OH
nC4H9

(3.57) **(3.62)** >99% ee

Catalyst (**3.52**), whilst particularly efficient for asymmetric transfer hydrogenations,[42] is not the only ruthenium catalyst that has displayed high enantioselectivity. The reduction of acetophenone into enantio-merically-enriched phenethyl alcohol has also been achieved using the ruthenium complexes of the tetradentate ligand (**3.63**),[43] the alternative amino-sulfonamide ligand (**3.64**),[44] indanol (**3.65**),[45] oxazolines (**3.66**)[46] and (**3.67**),[47] and amino alcohol (**3.68**).[48]

(3.63) up to 97% ee (R)   **(3.64)** up to 96% ee (R)   **(3.65)** up to 91% ee (S)

(3.66) up to 94% ee (R)   **(3.67)** up to 98% ee (S)   **(3.68)** up to 95% ee (S)

Noyori and co-workers also reported the use of Ru(BINAP)/ enantiomerically pure diamine catalysts for enantioselective transfer hydrogenation.[49] Whilst Ru(II)/phosphine complexes are not normally very active catalysts for transfer hydrogenation,[50] the addition of diamines dramatically improves catalytic activity. Thus, the combination of RuCl$_2$[(S)-BINAP](DMF) (**3.70**) with enantiomerically pure diamine (**3.69**) provides a highly effective asymmetric transfer hydrogenation catalyst. The use of racemic RuCl$_2$(TolBINAP)(DMF) with enantio-merically pure diamine (**3.69**) also provides an active, asymmetric catalyst, which reduces ketones with up to 95% ee.[51] The diastereomeric ruthenium catalysts have very different reactivities, which is the basis for the high selectivity, even though a racemic phosphine is being used. (For a discussion of 'chiral poisoning', see Section 2.2).

In an extraordinary example of catalytic efficiency, Noyori and co-workers employed the RuCl$_2$[(R)-TolBINAP](R,R)-dpen catalyst in the reduction of acetophenone (**3.22**).[52] A turnover number in excess of two million was achieved, with a turnover frequency of $63\,s^{-1}$ at 30% conversion! The product (**3.24**) was isolated with excellent yield, despite the incredibly low catalyst loading (0.00004 mol%). Related catalysts achieved higher enantioselectivity (99% ee) with 0.001 mol% of catalyst.

The source of hydrogen in the product (i.e. from ^{i}PrOH or H_2) was not reported; however, these reactions have been included under the transfer hydrogenation section, where they appear to fit more comfortably. The catalyst was used as a preformed species, which provided the exceptional efficiency.

Asymmetric transfer hydrogenation offers one of the most selective methods for the conversion of ketones into enantiomerically-enriched alcohols, and seems likely to be the focus of considerable attention in the future.

3.5 Reduction of ketones using oxazaborolidine catalysts

Enantiomerically pure boranes have a long history in the reduction of prochiral ketones.[53] Amongst the early results using stoichiometric oxazaborolidines, the work of Itsuno and co-workers is of particular interest. Acetophenone could be reduced with the oxazaborolidines (**3.71**)[54] or (**3.72**),[55] where the ratio of amino alcohol to borane was 1:2, implying that one equivalent of oxazaborolidine and one equivalent of borane were present in the transition state. Itsuno and co-workers also reported that the oxazaborolidine reagent (**3.72**) could be used catalytically in the reduction of prochiral ketoxime ethers.[56]

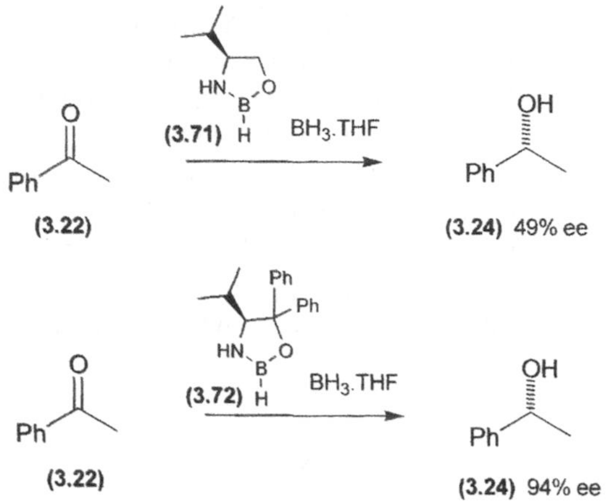

In 1987, Corey and co-workers demonstrated that the enantioselective reduction of ketones could be catalysed by oxazaborolidines. They showed that reduction of acetophenone was slow using BH$_3$-THF alone and that oxazaborolidine (3.72) alone did not cause reduction. However, in combination they reduced acetophenone in 1 min at room temperature. Using just 2.5 mol% of oxazaborolidine and stoichiometric BH$_3$-THF still provided excellent enantioselectivity.[57] In the same communication, proline-derived oxazaborolidine (3.73) was identified as a catalyst that was suitable for the reduction of a range of ketones, including acetophenone (3.22) and ketones (3.74 and 3.75).

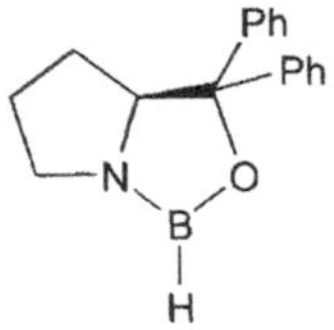

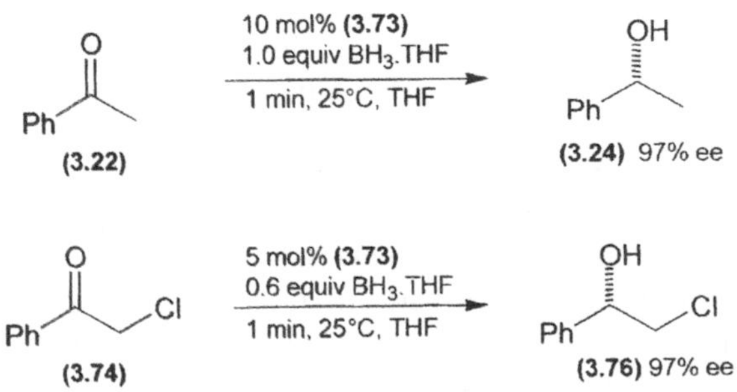

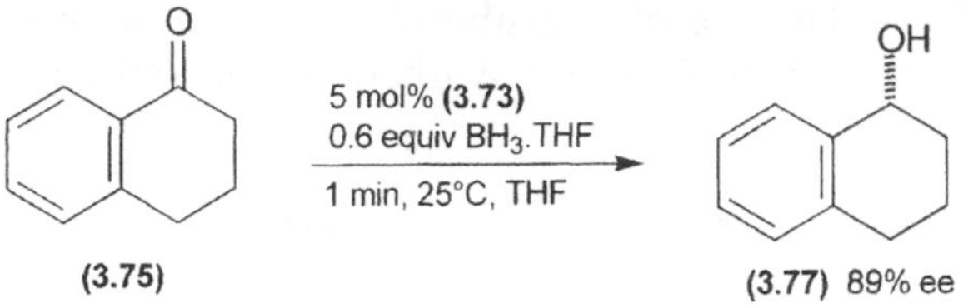

(3.75)   **(3.77)** 89% ee

Catalyst (3.73) is often referred to as the CBS catalyst after the names of the original authors (Corey, Bakshi and Shibata). A catalytic cycle was proposed, which explains the experimental observations. The oxazaborolidine interacts reversibly with borane, which then allows complexation of the ketone to give the key intermediate (3.77). In this intermediate, the borane is activated towards delivering hydride and the ketone is activated by the Lewis acidity of the boron in the oxazaborolidine. The geometry of the whole assembly leads to a highly enantioselective carbonyl reduction, in which the catalyst recycles.

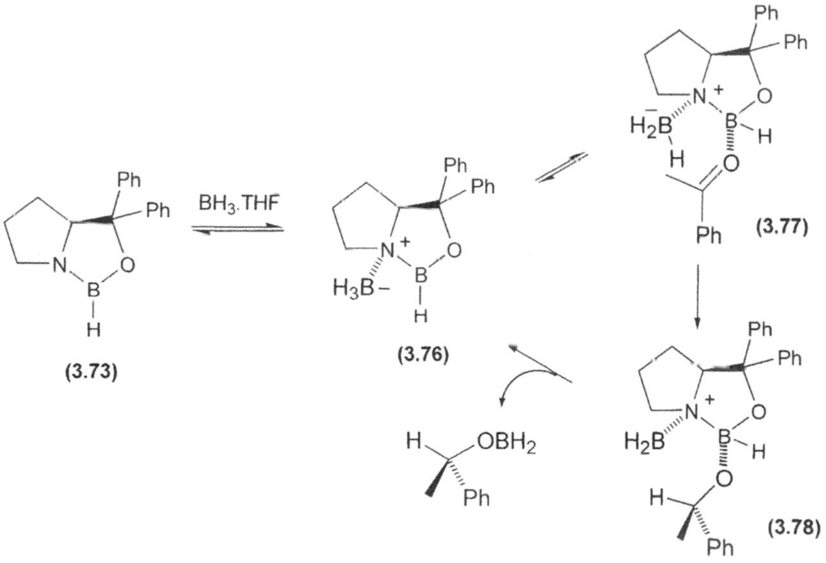

In further developments, catecholborane has been used as the stoichiometric reductant, and B-alkylated catalysts, such as compounds (3.79) and (3.80), which are more stable.[58, 59, 60, 61, 62] These catalysts have been used in the reduction of α,β-unsaturated ketones (3.81) and (3.82), and halogenated ketones (3.83) and (3.84). The reduction product (3.87) serves as a precursor for α-amino acids (3.89) and α-hydroxy esters (3.90). The reduction product (3.88) is a precursor to the antidepressant drug, fluoxetine (**3.91**). The use of oxazaborolidine catalysts was reviewed in 1998, with particular emphasis on synthetic scope and applications.[63]

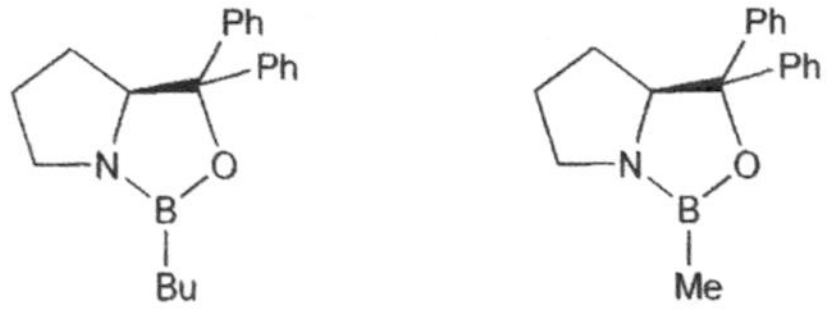

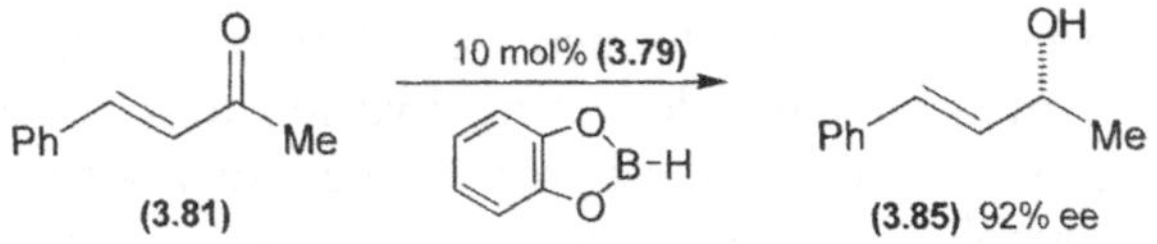

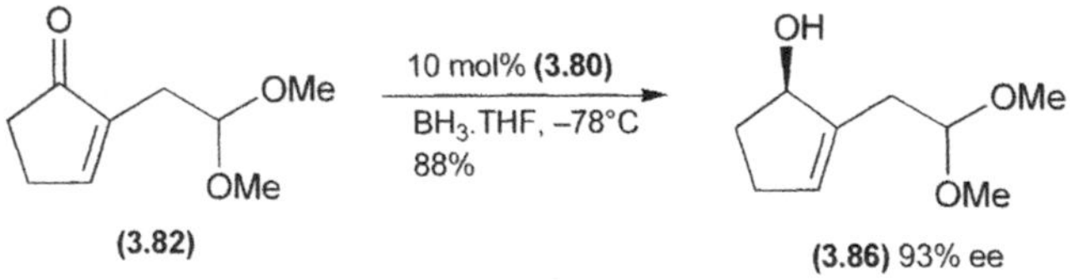

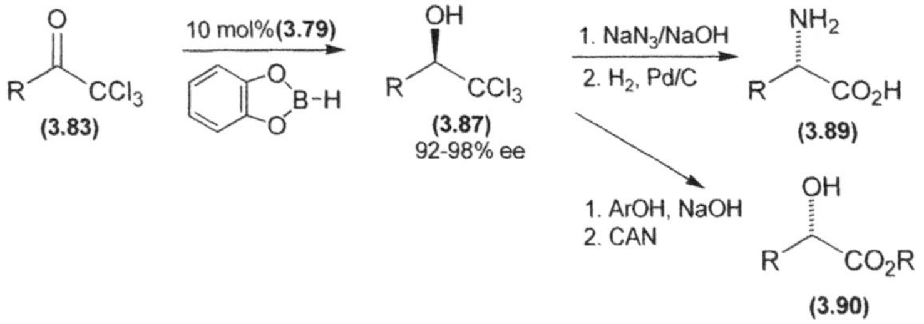

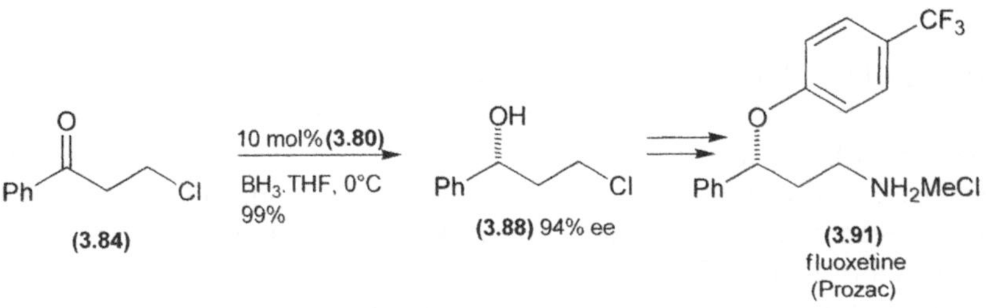

Many other amino alcohols have been examined for their ability to provide good oxazaborolidine catalysts, and these were reviewed in 1992.[64] It is difficult to know which amino alcohol provides the best enantioselectivity over a range of ketone reductions, since not all amino alcohols have been tried with all ketones. The four-membered ring analogue (3.92) is notable, giving high enantioselectivity in the catalytic reduction of acetophenone (3.22).[65]

The asymmetric reduction of ketones by catecholboranes has also been achieved using enantiomerically pure titanium complexes as catalysts.[66] Sodium borohydride has been used as an alternative stoichiometric reducing agent in conjunction with a cobalt catalyst, providing up to 92% ee.[67]

3.6 Hydrosilylation of ketones

The hydrosilylation of ketones provides an alternative to direct reduction with H_2. In terms of synthesis, the silyl ethers formed initially are almost invariably hydrolysed into the corresponding alcohols.[68] The rhodium-catalysed hydrosilylation of ketones has received more attention than other metals. One of the earliest ligands to give good enantioselectivity in this reaction was Brunner's Pythia ligand (3.93), which provides the asymmetric environment for the hydrosilylation of acetophenone (3.22).[69] This paved the way for the examination of other nitrogen-based ligands in this reaction. In 1989, three groups reported the use of oxazoline ligands (3.94),[70] (3.95)[71] and (3.96)[72] for rhodium-catalysed hydrosilylation. These were amongst the first reports of the now ubiquitous oxazoline moiety in asymmetric catalysis.[73,74] The pybox ligands, in particular, have been extended to a wide range of ketone substrates, with most products having high (generally 63–99%) ee.[75] The sense of asymmetric induction provided by these oxazoline ligands is very sensitive to the details of the ligand structure. Introduction of a methyl group into the 6-position of the pyridine ring in the pymox ligand also inverts the stereochemistry of the phenethyl alcohol subsequently formed.[76]

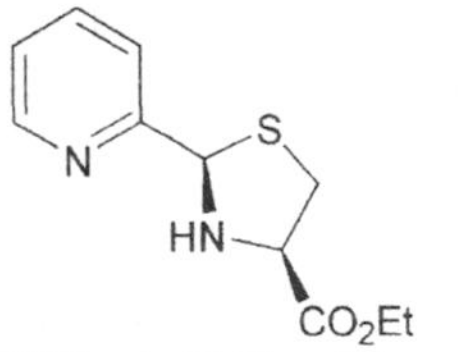

(3.93) up to 97.6% ee (S)

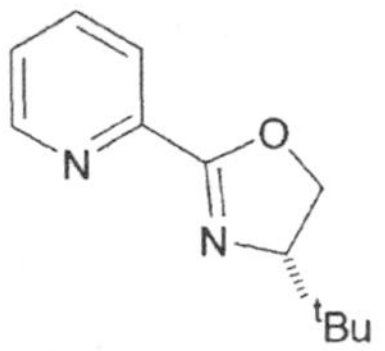

(3.94) up to 91% ee (R)

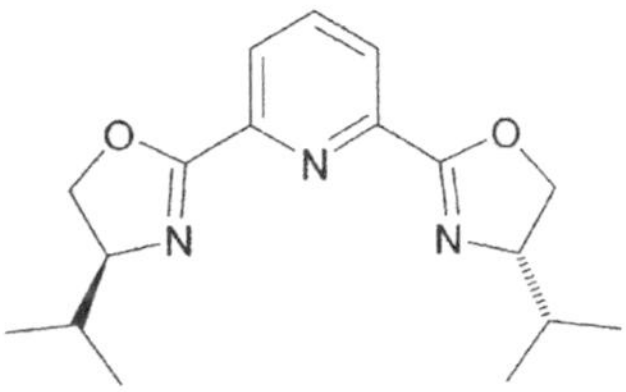

(3.95) up to 95% ee (S)

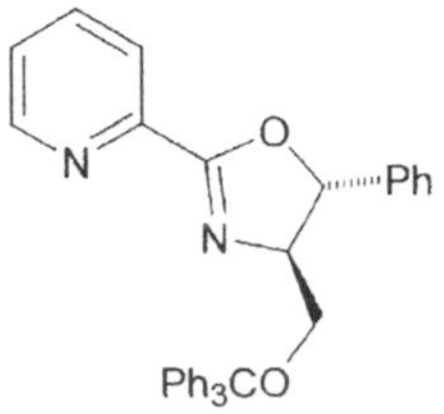

(3.96) up to 80% ee (R)

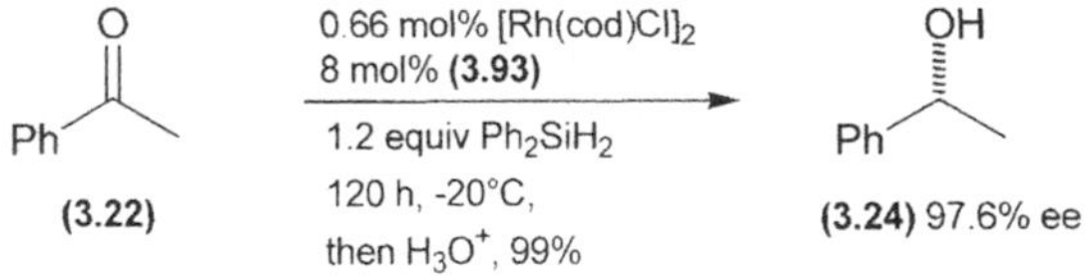

Alternative bis-oxazolines,[77, 78] as well as phosphino-oxazolines (see Section 10.2)[79, 80] have also been used, although these ligands are generally less selective. However, Uemura and co-workers reported interesting results with the oxazolinylferrocene-phosphine ligand (3.97). Whilst the rhodium-catalysed reaction of acetophenone (3.22) provides mainly the (R)-alcohol (3.24) as product, the corresponding reaction with an iridium catalyst provides the (S)-alcohol.[81] The iridium-catalysed system works well on many other ketones, including arylalkyl ketone (3.98) and α,β-unsaturated ketone (3.99) but very low selectivity is observed with 2-octanone as the substrate (19% ee in the product).

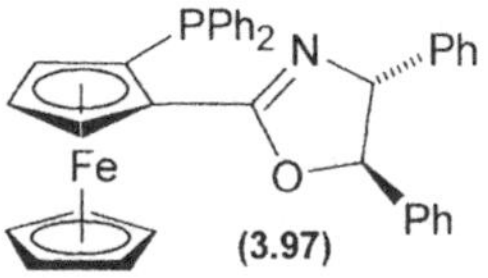

(3.97)

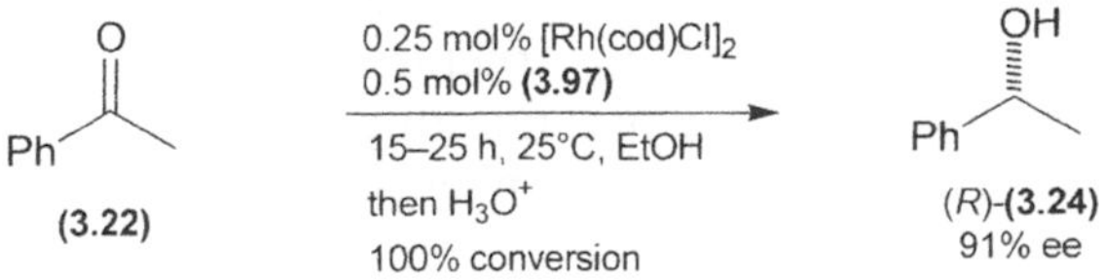

(3.22) (*S*)-(3.24) 96% ee

(3.98) (3.100) 83% ee

(3.99) (3.101) 84% ee absolute configuration not determined

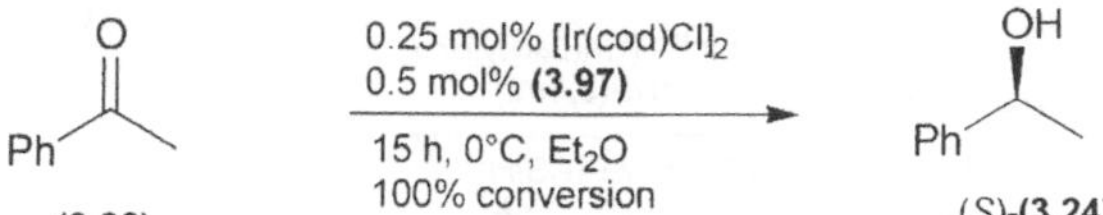

Other ligands have been used successfully in rhodium-catalysed hydrosilylation, including the *trans*-chelating diphosphine EtTRAP (**3.102**).[82, 83] For example, the δ-keto ester (**3.103**) undergoes enantioselective hydrosilylation followed by cyclisation to give the lactone (**3.104**).

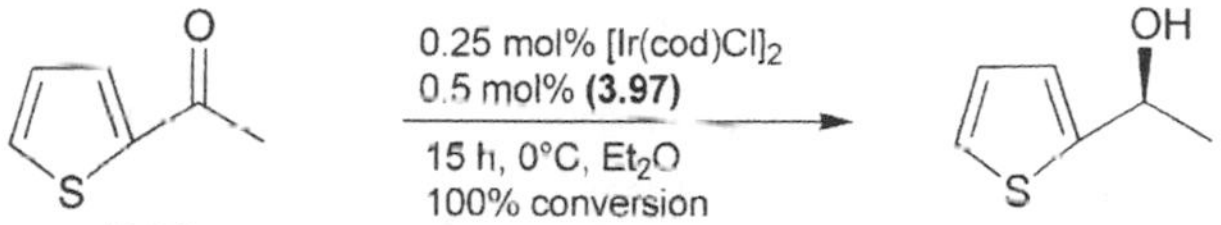

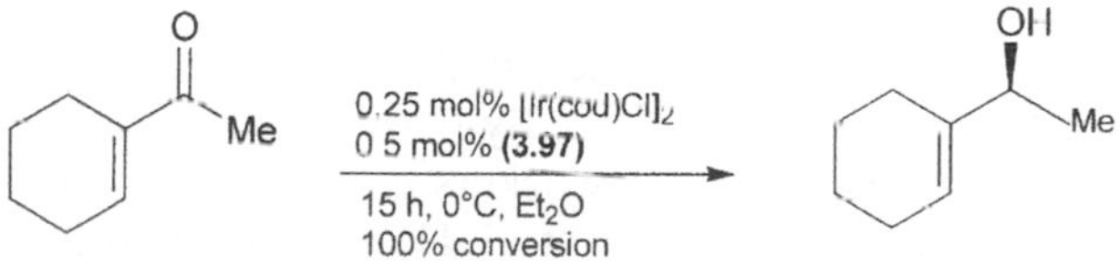

The hydrosilylation of a symmetrical ketone with a prochiral silane leads to the possibility of asymmetric induction in the newly formed silicon stereocentre. In their best example, Takaya and co-workers reported essentially complete control of asymmetric induction using the CyBINAP ligand (**3.105**) and a rhodium catalyst.[84] Pentan-3-one (**3.106**) and prochiral silane (**3.107**) are converted into the alkoxysilane (**3.108**), where the asymmetry is associated with the silicon atom.

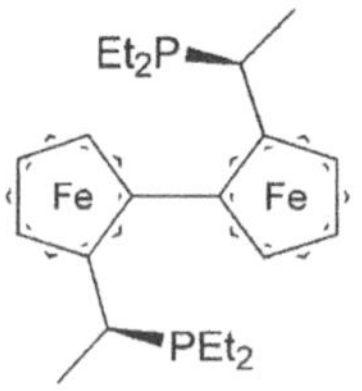

(3.102)

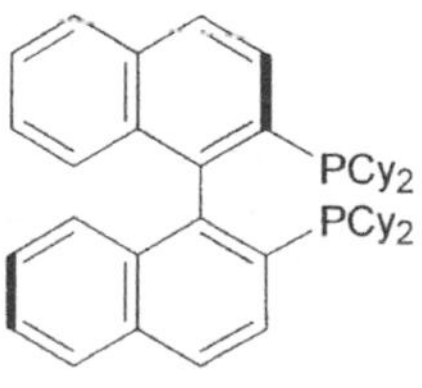

(3.105)

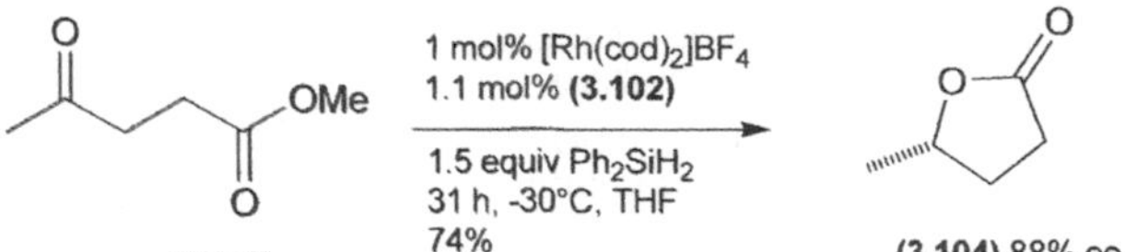

(3.103)

1 mol% [Rh(cod)₂]BF₄
1.1 mol% (3.102)

1.5 equiv Ph₂SiH₂
31 h, -30°C, THF
74%

(3.104) 88% ee

(3.106)

Ph
|
Si‴H
1-Np   H
H
(3.107)

2.5 mol% [Rh(cod)Cl]₂
5 mol% (3.105)
18 h, -20°C, THF
97%

(3.108) >99% ee

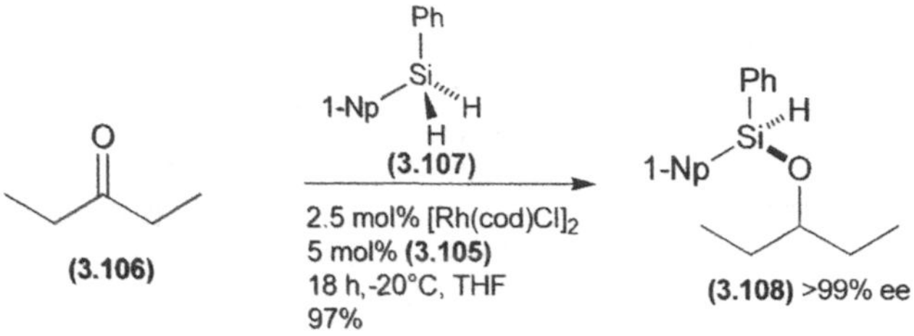

In addition to rhodium-catalysed hydrosilylation, asymmetric ruthe-nium-[85] and titanium-catalysed[86, 87] hydrosilylation have been reported. The report by Buchwald and co-workers concerning the hydrosilylation of ketones using titanocene catalyst (3.41) and inexpensive polymethyl-hydrosiloxane (3.109) appears to be the most general.[88] Reactions are run by activation of the titanocene catalyst (3.41) with two equivalents of butyllithium, followed by addition of polymethylhydrosiloxane and then ketone, and finally work-up with fluoride or acid to remove the silyl groups. It is suggested that the reactions take place *via* a titanium (III) hydride, although there is uncertainty over the details of the mechanism. Ketones (3.22) and (3.75) are reduced, with good enantioselectivity, following this procedure.

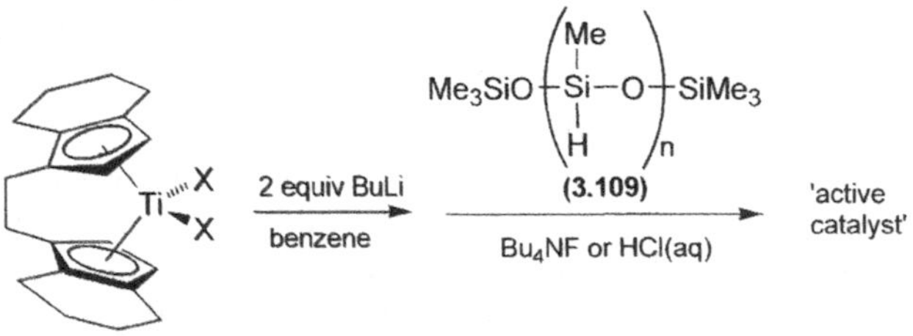

$$Me_3SiO\left(\begin{matrix} Me \\ | \\ Si-O \\ | \\ H \end{matrix}\right)_n SiMe_3$$

(3.109)

2 equiv BuLi
benzene

Bu₄NF or HCl(aq)

'active catalyst'

X,X = 1,1'-binaphth-2,2'-diolate
(3.41)

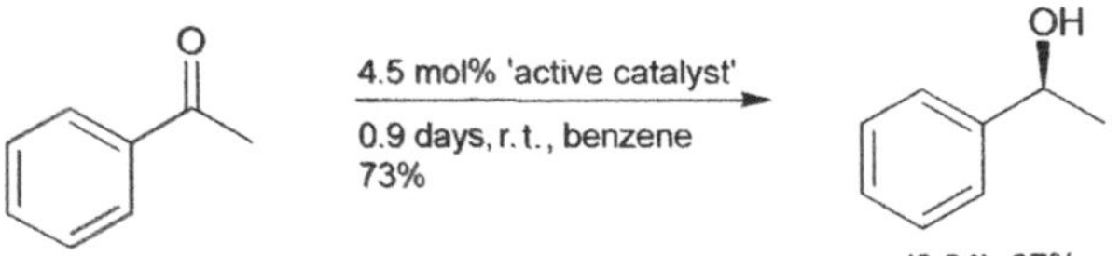

(3.22)

4.5 mol% 'active catalyst'

0.9 days, r. t., benzene
73%

(3.24) 97% ee

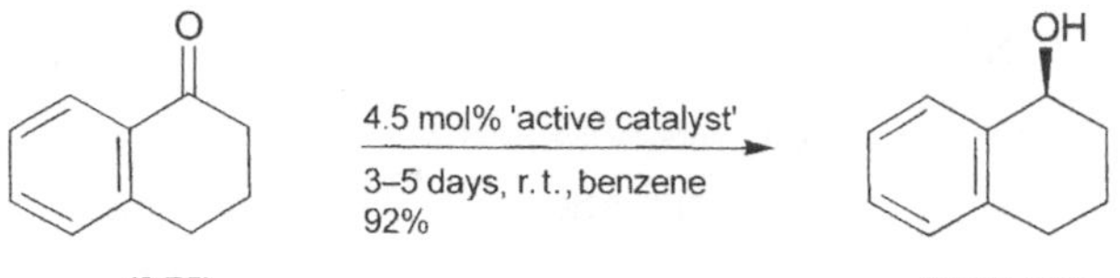

(3.75) → **(3.77)** 91% ee

4.5 mol% 'active catalyst'
3–5 days, r.t., benzene
92%

3.7 Hydrosilylation of imines and nitrones

Although little research has been carried out concerning the hydrosilylation of functionality other than carbonyls (and alkenes), the work published so far provides some exceptionally good examples of enantioselective catalysis.

Buchwald and co-workers have published a superb example of an asymmetric catalytic reaction using the difluorotitanocene catalyst (**3.110**);[89, 90] with catalyst loadings as low as 0.02 mol% (a substrate catalyst ratio of 5,000:1), very high ee's are obtained for the hydrosilylation of a wide range of imine substrates, including the cyclic imine (**3.111**) and the acyclic imines (**3.112**) and (**3.113**).

The hydrosilylation of nitrones provides a route to N,N-disubstituted hydroxylamines. The use of an Ru/TolBINAP complex was reported to give the best results in the asymmetric hydrosilylation of nitrones.[91] Nitrone (**3.117**) was converted into hydroxylamine (**3.118**), with reasonable yield and good enantioselectivity.

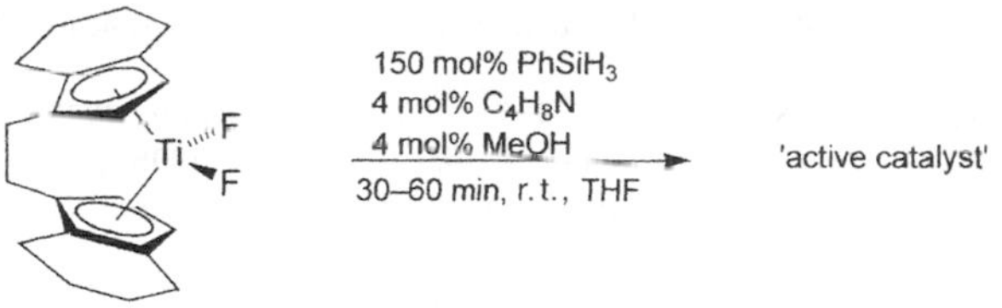

(3.110)

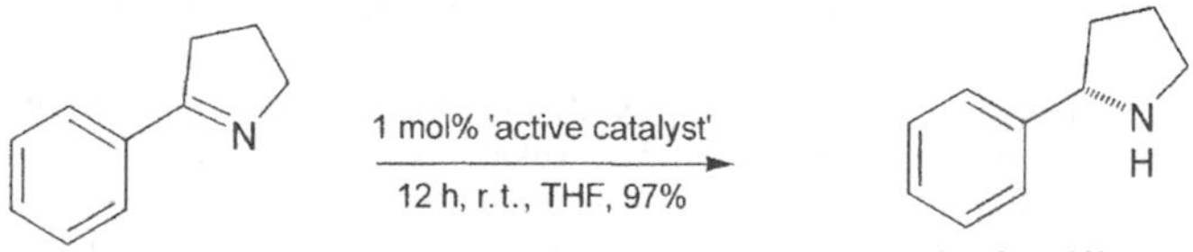

150 mol% PhSiH$_3$
4 mol% C$_4$H$_8$N
4 mol% MeOH
30–60 min, r.t., THF
→ 'active catalyst'

(3.111)

1 mol% 'active catalyst'
12 h, r.t., THF, 97%

(3.114) 99% ee

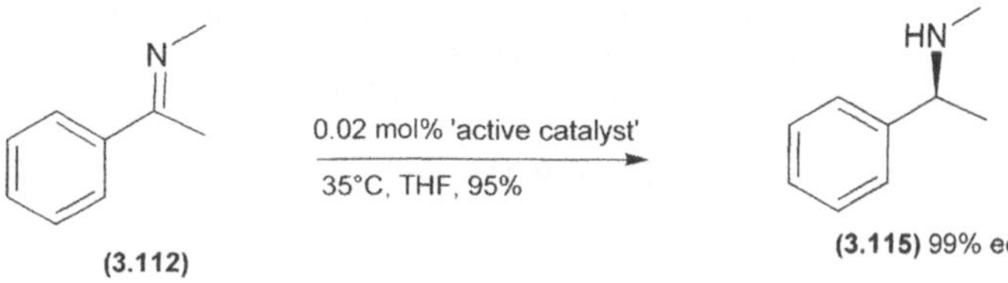

(3.112)

0.02 mol% 'active catalyst'
35°C, THF, 95%

(3.115) 99% ee

(3.113)

0.02 mol% 'active catalyst'

35°C, THF, 96%

(3.116) 92% ee

(3.117)

0.5 mol% Ru$_2$Cl$_4$[(S)-TolBINAP]$_2$

2 h, 0°C, dioxane, 64%

(3.118) 83% ee

References

1. Rhodium and iridium catalysts have also been used to give good enantioselectivities. For a review, see: R. Noyori, *Asymmetric Catalysis in Organic Synthesis*, John Wiley and Sons, New York, **1994**, 56.
2. M. Kitamura, T. Okuma, S. Inoue, N. Sayo, H. Kumobayashi, S. Akutagawa, T. Ohta, H. Takaya and R. Noyori, *J. Am. Chem. Soc.*, **1988**, *110*, 629.
3. R. Noyori, T. Okuma, M. Kitamura, H. Takaya, N. Sayo, H. Kumobayashi and S. Akutagawa, *J. Am. Chem. Soc.*, **1987**, *109*, 5856.
4. M. Kitamura, M. Tokunaga and R. Noyori, *J. Am. Chem. Soc.*, **1995**, *117*, 2931.
5. I. Gauter, V. Ratovelomanana-Vidal, P. Savignac, J.-P. Genêt, *Tetrahedron Lett.*, **1996**, *37*, 7721.
6. P. J. Pye, K. Rossen, R.A. Reamer, R.P. Volante and P.J. Reider, *Tetrahedron Lett.*, **1998**, *39*, 4441.
7. T. Chiba, A. Miyashita, H. Nohira and H. Takaya, *Tetrahedron Lett.*, **1993**, *34*, 2351.
8. M.J. Burk, T.G.P. Harper and C.S. Kalberg, *J. Am. Chem. Soc.*, **1995**, *117*, 4423.
9. D. Blanc, J.-C. Henry, V. Ratovelomanana-Vidal and J.-P. Genet, *Tetrahedron Lett.*, **1997**, *38*, 6603.
10. A. Roucoux, F. Agbossou, A. Mortreux and F. Petit, *Tetrahedron: Asymmetry*, **1993**, *4*, 2279.
11. C. Pasquier, S. Naili, L. Pelinski, J. Brocard, A. Mortreux and F. Agbossou, *Tetrahedron: Asymmetry*, **1998**, *9*, 193.
12. Q. Jiang, Y. Jiang, D. Xiao, P. Cao and X. Zhang, *Angew. Chem., Int. Ed.*, **1998**, *37*, 1100.
13. T. Doi, M. Kokubo, K. Yamamoto and T. Takahashi, *J. Org. Chem.*, **1998**, *63*, 428.
14. R. Noyori and H. Takaya, *Acc. Chem. Res.*, **1990**, *23*, 245.
15. For a mathematical treatment of dynamic resolution reactions, see: M. Kitamura, M. Tokunga and R. Noyori, *J. Am. Chem. Soc.*, **1993**, *115*, 144.
16. K. Mashima, K.-H. Kusano, N. Sato, Y.-I. Matsumura, K. Nozaki, H. Kumobayashi, N. Sayo, Y. Hori, T. Ishizaki, S. Akutagawa and H. Takaya, *J. Org. Chem.*, **1994**, *59*, 3064.
17. J.-P. Genêt, M.C. Caño de Andrade and V. Ratovelomanana-Vidal, *Tetrahedron Lett.*, **1995**, *36*, 2063.
18. M. Kitamura, M. Tokunaga, T. Pham, W.D. Lubell and R. Noyori, *Tetrahedron Lett.*, **1995**, *36*, 5769.

19. C. Bolm, *Angew. Chem., Int. Ed. Engl.*, **1993**, *32*, 232.
20. M.J. Burk and J.E. Feaster, *J. Am. Chem. Soc.*, **1992**, *114*, 6266.
21. C.A. Willoughby and S.L. Buchwald, *J. Am. Chem. Soc.*, **1992**, *114*, 7562.
22. C.A. Willoughby and S.L. Buchwald, *J. Am. Chem. Soc.*, **1993**, *58*, 7627.
23. A. Viso, N.E. Lee and S.L. Buchwald, *J. Am. Chem. Soc.*, **1994**, *116*, 9373.
24. C.A. Willoughby and S.L. Buchwald, *J. Am. Chem. Soc.*, **1994**, *116*, 11703.
25. C.A. Willoughby and S.L. Buchwald, *J. Am. Chem. Soc.*, **1994**, *116*, 8952.
26. R. Sablong and J.A. Osborn, *Tetrahedron Lett.*, **1996**, *37*, 4937.
27. P. Schnider, G. Koch, R. Prêtôt, G. Wang, F.M. Bohnen, C. Krüger and A. Pfaltz, *Chem. Eur. J.*, **1997**, *3*, 887.
28. T. Morimoto, N. Suzuki and K. Achiwa, *Tetrahedron: Asymmetry*, **1998**, *9*, 183.
29. A.B. Charette and A. Giroux, *Tetrahedron Lett.*, **1996**, *37*, 6669.
30. S. Akabori, S. Sakurai, Y. Izumi and Y. Fujii, *Nature*, **1956**, *178*, 323.
31. Y. Izumi, *Adv. Catal.*, **1983**, *32*, 215.
32. M. Nakahata, M. Imaida, H. Ozaki, T. Harada and A. Tai, *Bull. Chem. Soc. Jpn.*, **1982**, *55*, 2186.
33. Y. Orito, S. Imai and S. Niwa, *J. Chem. Soc. Jpn.*, **1980**, 670.
34. M. Garland and H.-U. Blaser, *J. Am. Chem. Soc.*, **1990**, *112*, 7048.
35. X. Zuo, H. Liu and M. Liu, *Tetrahedron Lett.*, **1998**, *39*, 1941.
36. H.-U. Blaser, *Tetrahedron: Asymmetry*, **1991**, *2*, 843.
37. D. Müller, G. Umbricht, B. Weber and A. Pfaltz, *Helv. Chim. Acta*, **1991**, *74*, 232.
38. P. Gamez, F. Fache, P. Mangeney and M. Lemaire, *Tetrahedron Lett.*, **1993**, *34*, 6897.
39. D.A. Evans, S.G. Nelson, M.R. Gagné, A.R. Muci, *J. Am. Chem. Soc.*, **1993**, *115*, 9800.
40. A. Fujii, S. Hashiguchi, N. Uematsu, T. Ikariya and R. Noyori, *J. Am. Chem. Soc.*, **1996**, *118*, 2521.
41. K. Matsumura, S. Hashiguchi, T. Ikariya and R. Noyori, *J. Am. Chem. Soc.*, **1997**, *119*, 8738.
42. (a) Noyori and S. Hashiguchi, *Acc. Chem. Res.*, **1997**, *30*, 97. (b) K.J. Haack, S. Hashiguchi, A. Fujii, T. Ikariya and R. Noyori, *Angew. Chem., Int. Ed. Engl.*, **1997**, *36*, 285.
43. J.-X. Gao, T. Ikariya and R. Noyori, *Organometallics*, **1996**, *15*, 1087.
44. K. Püntener, L. Schwink and P. Knochel, *Tetrahedron Lett.*, **1996**, *37*, 8165.
45. M. Palmer, T. Walsgrove and M. Wills, *J. Org. Chem.*, **1997**, *62*, 5226.
46. T. Sammakia and E.L. Stangeland, *J. Org. Chem.*, **1997**, *62*, 6104.
47. Y. Jiang, Q. Jiang and X. Zhang, *J. Am. Chem. Soc.*, **1998**, *120*, 3817.
48. [1] D.A. Alonso, D. Guijarro, P. Pinho, O. Temme and P.G. Andersson, *J. Org. Chem.*, **1998**, *63*, 2749.
49. T. Ohkuma, H. Ooka, S. Hashiguchi, T. Ikariya and R. Noyori, *J. Am. Chem. Soc.*, **1995**, *117*, 2675.
50. But P,N ligands have been shown to be effective. For example, see: T. Langer and G. Helmchen, *Tetrahedron Lett.*, **1996**, *37*, 1381.
51. T. Ohkuma, H. Doucet, T. Pham, K. Mikami, T. Korenaga, M. Terada and R. Noyori, *J. Am. Chem. Soc.*, **1998**, *120*, 1086.
52. H. Doucet, T. Ohkuma, K. Murata, T. Yokozawa, M. Kozawa, E. Katayama, A.F. England, T. Ikariya and R. Noyori, *Angew. Chem., Int. Ed. Engl.*, **1998**, *37*, 1703.
53. For an overview, see: G. Procter, *Asymmetric Synthesis*, Oxford University Press, Oxford, **1996**, 161.
54. A. Hirao, S. Itsuno, S. Nakahama and N. Yamazaki, *J. Chem. Soc., Chem. Commun.*, **1981**, 315.
55. S. Itsuno, M. Nakano, K. Miyazaki, H. Masuda and K. Ito, *J. Chem. Soc. Perkin Trans. 1*, **1983**, 1673.

56. S. Itsuno, Y. Sakurai, K. Ito, A. Hirao and S. Nakahama, *Bull. Chem. Soc. Jpn.*, **1987**, 395.
57. E.J. Corey, R.K. Bakshi and S. Shibata, *J. Am. Chem. Soc.*, **1987**, *109*, 5551.
58. E.J. Corey and J.O. Link, *Tetrahedron Lett.*, **1989**, *30*, 6275.
59. E.J. Corey and R.K. Bakshi, *Tetrahedron Lett.*, **1990**, *31*, 611.
60. E.J. Corey and A.V. Gavai, *Tetrahedron Lett.*, **1988**, *29*, 3201.
61. E.J. Corey and J.O. Link, *Tetrahedron Lett.*, **1992**, *33*, 3431.
62. E.J. Corey and G.A. Reichard, *Tetrahedron Lett.*, **1989**, *30*, 5207.
63. E.J. Corey and C.J. Helal, *Angew. Chem.*, *Int. Ed. Engl.*, **1998**, *37*, 1986.
64. S. Wallbaum and J. Martens, *Tetrahedron: Asymmetry*, **1992**, *3*, 1475.
65. W. Behnen, C. Dauelsberg, S. Wallbaum and J. Martens, *Synth. Commun.*, **1992**, *22*, 2143.
66. G. Giffels, C. Dreisbach, U. Kragl, M. Weigerding, H. Waldmann and C. Wandrey, *Angew. Chem.*, *Int. Ed. Engl.*, **1995**, *34*, 2005.
67. T. Nagata, K. Yorozu, T. Yamada and T. Mukaiyama, *Angew. Chem.*, *Int. Ed. Engl.*, **1995**, *34*, 2145.
68. H. Brunner, H. Nishiyama and K. Itoh, in *Catalytic Asymmetric Synthesis*, VCH, New York, **1993**, 303.
69. H. Brunner, R. Becker and G. Riepl, *Organometallics*, **1984**, *3*, 1354.
70. H. Brunner and U. Obermann, *Chem. Ber.*, **1989**, *122*, 499.
71. H. Nishiyama, H. Sakaguchi, T. Nakamura, M. Horihata, M. Kondo and K. Itoh, *Organometallics*, **1989**, *8*, 847.
72. G. Balavoine, J.C. Clinet and I. Lellouche, *Tetrahedron Lett.*, **1989**, *38*, 5141.
73. The first report of an oxazoline ligand in asymmetric catalysis was from Brunner's group: H. Brunner, U. Obermann and P. Wimmer, *J. Organomet. Chem.*, **1986**, *316*, C1.
74. For a review on the use of bis-oxazolines in asymmetric catalysis, see: A.K. Ghosh, P. Mathivanan and J. Cappiello, *Tetrahedron: Asymmetry*, **1998**, *9*, 1.
75. H. Nishiyama, M. Kondo, T. Nakamura and K. Itoh, *Organometallics*, **1991**, *10*, 500.
76. H. Brunner and P. Brandl, *J. Organomet. Chem.*, **1990**, *390*, C81.
77. G. Helmchen, A. Krotz, K.-T. Ganz and D. Hansen, *Synlett*, **1991**, 257.
78. Y. Imai, W. Zhang, T. Kida, Y. Nakatsuji and I. Ikedan, *Tetrahedron: Asymmetry*, **1996**, *7*, 2453.
79. L.M. Newman, J.M.J. Williams, R. McCague and G.A. Potter *Tetrahedron: Asymmetry*, **1996**, *7*, 1597.
80. T. Langer, J. Janssen and G. Helmchen, *Tetrahedron: Asymmetry*, **1996**, *7*, 1599.
81. Y. Nishibayashi, K. Segawa, H. Takada, K. Ohe and S. Uemura, *J. Chem. Soc.*, *Chem. Commun.*, **1996**, 847.
82. M. Sawamura, R. Kuwano, J. Shirai and Y. Ito, *Synlett*, **1995**, 347.
83. M. Sawamura, R. Kuwano and Y. Ito, *Angew. Chem.*, *Int. Ed. Engl.*, **1994**, *33*, 111.
84. T. Ohta, M. Ito, A. Tsuneto and H. Takaya, *J. Chem. Soc.*, *Chem. Commun.*, **1994**, 2525.
85. G. Zhu, M. Terry, X. Zhang, *J. Organometallic Chem.*, **1997**, *547*, 97–101.
86. J.-I. Sakaki, W.B. Schweizer and D. Seebach, *Helv. Chim. Acta*, **1993**, *76*, 2654.
87. H. Imma, M. Mori and T. Nakai, *Synlett*, **1996**, 1229.
88. M.B. Carter, B. Schiøtt, A. Gutiérrez and S.L. Buchwald, *J. Am. Chem. Soc.*, **1994**, *116*, 11667.
89. X. Verdaguer, U.E.W. Lange, M.T. Reding and S.L. Buchwald, *J. Am. Chem. Soc.*, **1996**, *188*, 6784.
90. X. Verdaguer, U.E.W. Lange and S.L. Buchwald, *Angew. Chem.*, *Int. Ed. Engl.*, **1998**, *37*, 1103.
91. S.-I. Murahashi, S. Watanabe and T. Shiota, *J. Chem. Soc.*, *Chem. Commun.*, **1994**, 725.

4 Epoxidation of alkenes

This chapter considers the enantioselective catalytic epoxidation of alkenes. The individual sections are ordered by reaction type, rather than by substrate type. Despite the very high selectivities obtained with many substrates, there is still no universal catalyst which is effective for all classes of alkene structure.

4.1 Epoxidation of allylic alcohols

The asymmetric epoxidation of allylic alcohols using titanium/tartrate ester/t-butyl hydroperoxide was developed by Sharpless during the 1980s and has become one of the most important methods of asymmetric catalysis. Whilst there has been little further significant development of this reaction in recent years, this is simply a testament to the fact that the reaction already works well. There are several reviews available detailing the 'Sharpless asymmetric epoxidation' reaction.[1]

The first reports from Sharpless and co-workers described the stoichiometric use of the 'catalyst';[2] however, the truly catalytic variant of the reaction was found to be more general in the presence of activated molecular sieves.[3] The benefits of using catalytic amounts of the titanium/tartrate combination include not only reduced cost but also an easier work-up procedure. This is particularly true for water soluble epoxide products, such as glycidol.

The sense of asymmetric induction in Sharpless asymmetric epoxidation reactions can be reliably predicted using the model presented in Figure 4.1. In order for the model to predict the stereochemical outcome

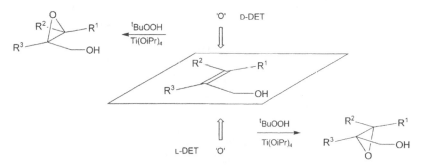

Figure 4.1 Schematic representation of the Sharpless asymmetric epoxidation reaction. Abbreviation: DET = diethyl tartrate.

correctly, only two points need to be remembered. The allylic hydroxy group resides in the bottom right corner and D-(-)-diethyltartrate (D-DET) (which has the (S,S)-configuration) attacks from above the plane. Throughout this section, allylic alcohols will be drawn in this way, such that it is clear how the stereochemistry can be related to the model. Thus, allyl alcohol (**4.01**) is converted into (R)-glycidol (**4.02**) using the D-(-)-diethyltartrate/titanium tetraisopropoxide combination. However, the use of L-(+)-diethyltartrate as the ligand affords the opposite enantiomer, (S)-glycidol (**4.03**).

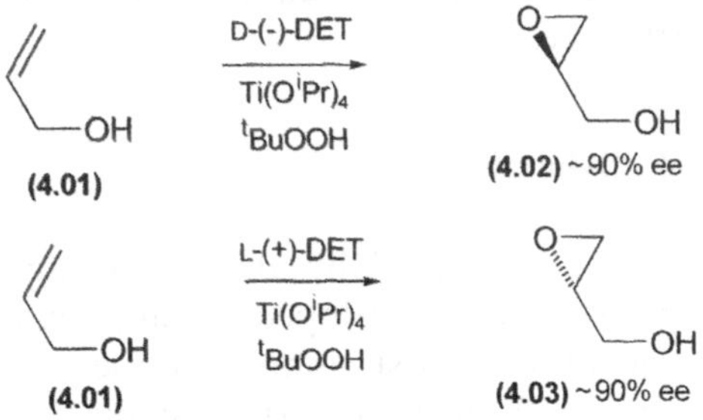

Variously substituted allylic alcohols all obey this model, and a range of examples is provided by the conversions of the allylic alcohols (**4.04–4.08**) into the corresponding epoxides (**4.09–4.13**). In the case of the diene (**4.08**), it is noteworthy that only the alkene which is part of the allylic alcohol unit undergoes epoxidation. Dimethyl tartrate (DMT), diethyltartrate (DET) and diisopropyl esters (DIPT) are all used fairly commonly and generally give similar levels of selectivity, although optimal selectivities can be obtained by screening the various esters. Polymer-linked tartrates have also been used successfully in Sharpless asymmetric epoxidation reactions.[4]

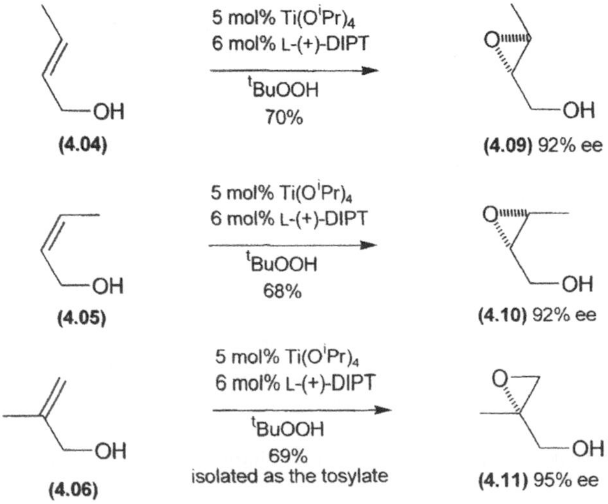

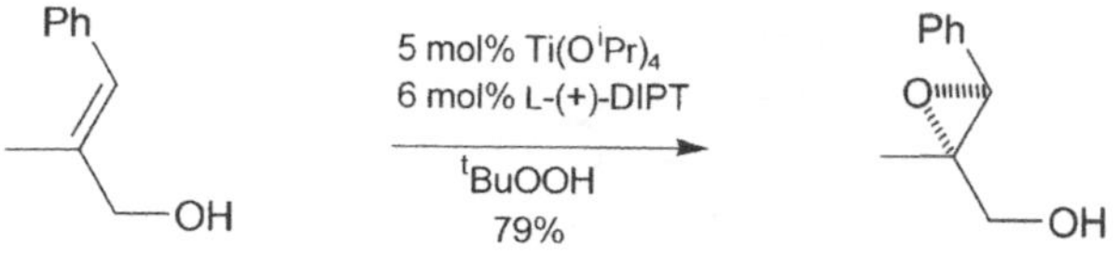
Ph
5 mol% Ti(O^{i}Pr)$_4$
6 mol% L-(+)-DIPT
^{t}BuOOH
79%
OH
Ph
O
OH

(4.07) **(4.12)** >98% ee

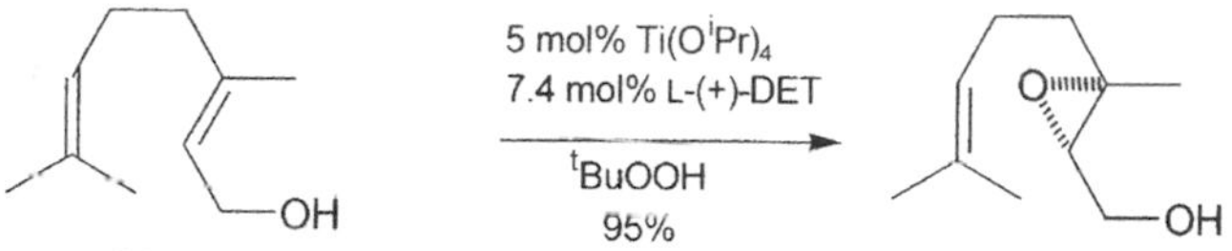
OH
5 mol% Ti(O^{i}Pr)$_4$
7.4 mol% L-(+)-DET
^{t}BuOOH
95%
O
OH

(4.08) **(4.13)** 91% ee

There can be no doubt that the reliability of the Sharpless reaction amongst many different classes of allyl alcohol contributes to its success as a synthetic tool in asymmetric synthesis. Another remarkable attribute of the Sharpless asymmetric epoxidation is the very high level of discrimination between substrate enantiomers (kinetic resolution). The kinetic resolution (Figure 4.2) was first reported using stoichiometric amounts of titanium/tartrate, and this is the procedure generally adopted,[5] although catalytic amounts of titanium may also be employed.[6,7]

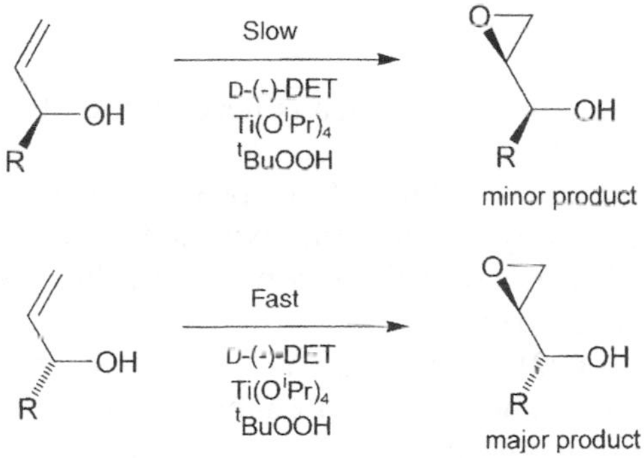
OH
R
Slow
D-(-)-DET
Ti(O^{i}Pr)$_4$
^{t}BuOOH
O
OH
R
minor product
OH
R
Fast
D-(-)-DET
Ti(O^{i}Pr)$_4$
^{t}BuOOH
O
OH
R
major product

Figure 4.2 Sharpless kinetic resolution of racemic substrates. Abbreviation: DET = diethyl tartrate.

Examples of the kinetic resolution include the reaction of the secondary alcohols **(4.14–4.16)**, which show large differences in the rates of the faster- and slower-reacting enantiomers. In the case of substrate **(4.16)**, the k_{rel} ($= k_{fast}/k_{slow}$) is as high as 700! k_{rel} is also known as the selectivity factor, S (see Section 12.8). Whilst these results are obtained under stoichiometric conditions, kinetic resolutions are successful under catalytic conditions, especially when the alkene possesses a bulky group remote from the hydroxy-containing substituent.[3b] In general, as with other kinetic resolutions, the best results are obtained by stopping the reaction before 50% conversion, if it is the product which is needed. If it is

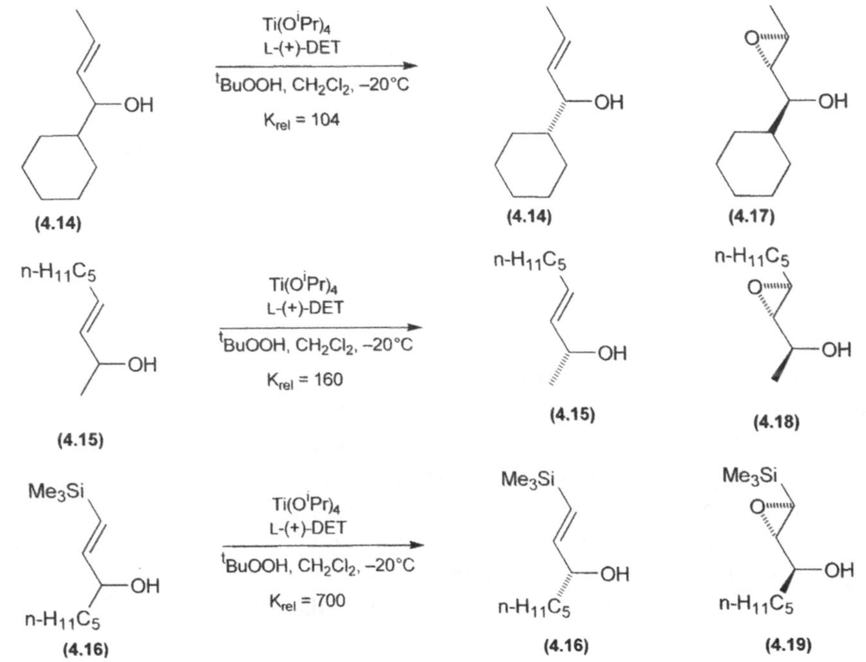

(4.14)   (4.14)   (4.17)

(4.15)   (4.15)   (4.18)

(4.16)   (4.16)   (4.19)

the recovered starting material which is needed, then the reaction should be allowed to proceed beyond 50% conversion.

The substrate (4.20) is achiral, but epoxidation of the enantiotopic alkenes occurs selectively, to give the expected product (4.21), where a new chiral centre is formed at the secondary alcohol, as well as the one associated with the epoxide.[8]

The substrate (4.22) represents a real challenge for a selective epoxidation reaction. There are two enantiomers of the substrate, each of which has two different alkenes which could react. Each alkene group has two diastereotopic faces. In all, there are eight possible outcomes, of which only one product, epoxide (4.23) is observed in practice.[9] This result was also obtained using stoichiometric titanium reagent, but the underlying principles would be the same for the catalytic variant.

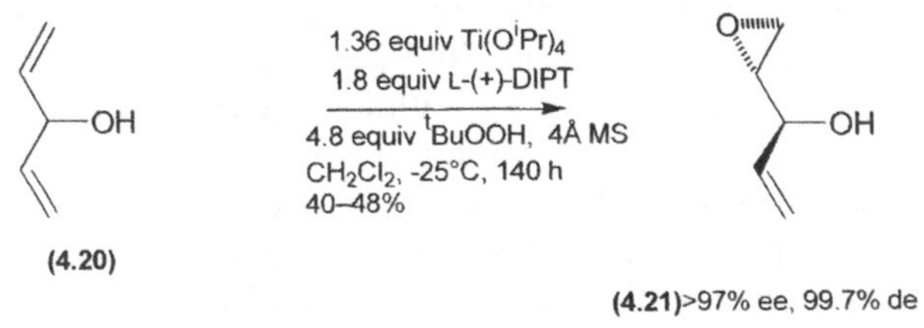

(4.20)

(4.21)>97% ee, 99.7% de

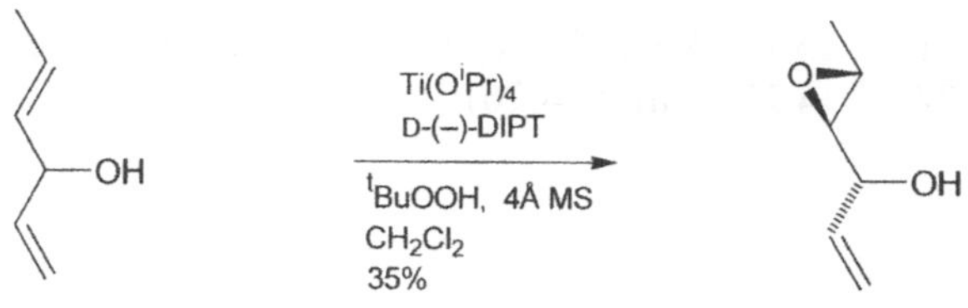

(4.22)

Ti(O^{i}Pr)$_4$
D-(−)-DIPT
^{t}BuOOH, 4Å MS
CH$_2$Cl$_2$
35%

(4.23) 95% ee

The mechanism of the Sharpless asymmetric epoxidation reaction has been a topic of some debate.[10] It is clear that the isopropoxy ligands can be displaced by the diol (tartrate), the hydroperoxide and the allylic alcohol substrate. It is reasonable that the titanium 'gathers together' the substrates and that the asymmetric environment provided by the tartrate fixes the stereochemistry of the products. In fact, at the point of epoxidation, a catalytic ensemble containing two titanium atoms seems likely. Figure 4.3 presents a likely assembly in the Sharpless epoxidation reaction. Despite all the mechanistic information available, it can be difficult to determine which complex in the reaction mixture is the most reactive, providing the pathway through which the process takes place. Nevertheless, the reaction is predictable from a synthetic point of view.

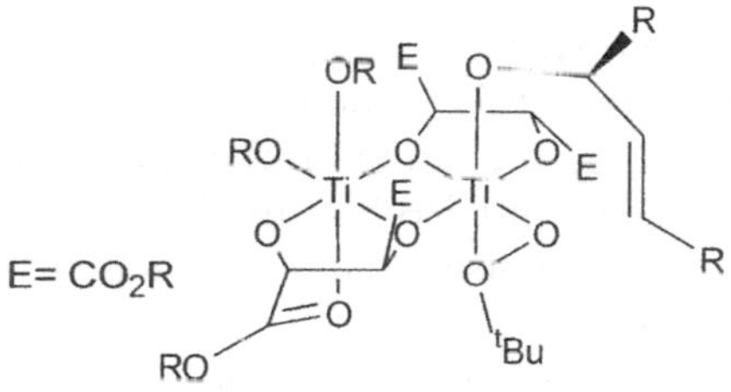

E= CO$_2$R

Figure 4.3 Possible assembly in the Sharpless asymmetric epoxidation reaction.

In conclusion, the Sharpless asymmetric epoxidation reaction should be used to epoxidise an allylic alcohol; other alkenes require alternative methods.

4.2   Epoxidation with manganese(salen) complexes

Enantiomerically pure manganese(salen) complexes currently provide the most selective method for the catalytic asymmetric epoxidation of unfunctionalised alkenes, although they are not general for all substrates.[11] In 1990, the use of C$_2$-symmetric salen ligands in manganese-catalysed epoxidation was reported, independently, by Jacobsen and co-workers[12] and Katsuki and co-workers.[13] Typical catalysts are represented by complexes (**4.24**),[14] (**4.25**)[15] and (**4.26**).[16]

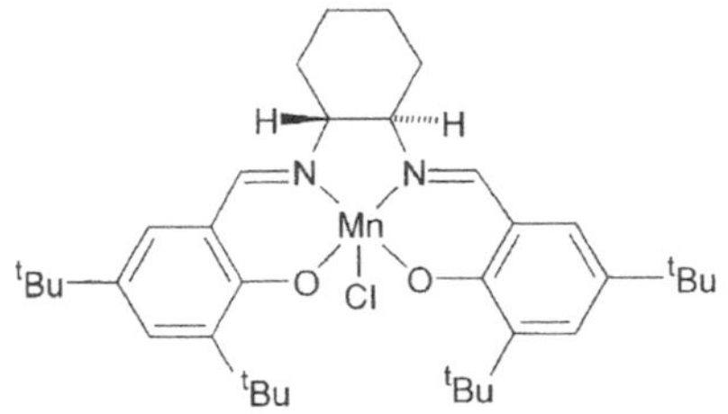

H
H
N
N
Mn
O
Cl
O
tBu
tBu
tBu
tBu

(4.24)

H
H
N
N
Mn
O
Cl
O
iPr3SiO
OSiiPr3
tBu
tBu

(4.25)

Ar
Ar
N
N
Mn+
AcO−
O
O
Ph
Ph

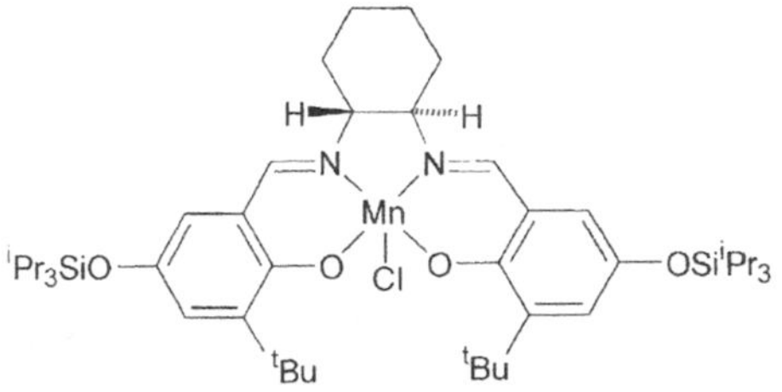

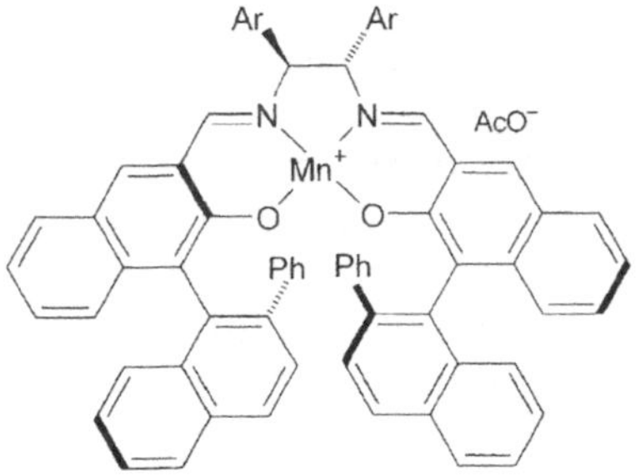

(4.26)

Most frequently, iodosylbenzene (PhIO) or sodium hypochlorite (NaOCl) are used as the stoichiometric oxidant, although alternative reagents have been used, including hydrogen peroxide,[17] periodate,[18] dimethyldioxirane[19] and an m-CPBA/N-methylmorpholine-N-oxide combination, which allows the use of a lower temperature and provides higher enantioselectivities.[20]

Catalyst (**4.24**) is effective in the epoxidation of many (*Z*)-alkenes, including the acyclic alkenes (**4.27**) and (**4.28**), as well as the cyclic alkenes (**4.29**) and (**4.30**).[21] Acyclic (*Z*)-alkenes usually yield mainly *cis* epoxides with a small amount of *trans* product as well. The product of the epoxidation reaction of (*Z*)-ethylcinnamate (**4.28**) has been converted into the taxol side chain (**4.35**).[22]

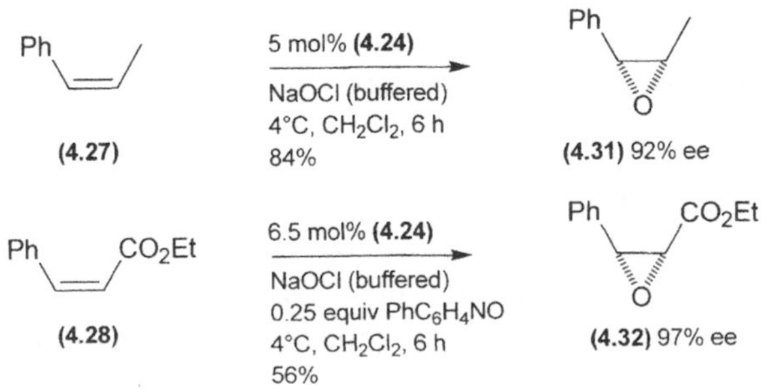

Ph
(4.27)
5 mol% (4.24)
NaOCl (buffered)
4°C, CH2Cl2, 6 h
84%
Ph
O
(4.31) 92% ee
Ph
CO2Et
(4.28)
6.5 mol% (4.24)
NaOCl (buffered)
0.25 equiv PhC6H4NO
4°C, CH2Cl2, 6 h
56%
Ph
CO2Et
O
(4.32) 97% ee

(4.29)   (4.33) 86% ee

4 mol% **(4.26)**
PhIO, CH$_2$Cl$_2$
96%

(4.30)   (4.34) 98% ee

5 mol% **(4.24)**
NaOCl (buffered)
4°C, CH$_2$Cl$_2$, 6 h
87%

(4.35)

Trisubstituted alkenes have been successfully epoxidised with the same catalyst **(4.24)**.[23] The addition of 4-phenyl pyridine N-oxide was found to be beneficial in using lower catalytic loadings in industrial applications.[24,25] This is exemplified by the epoxidation of alkenes **(4.36)** and **(4.37)**. Cyclic tetrasubstituted alkenes have also been epoxidised, with high enantiomeric excess. The best results seem to be with chromene substrates, such as compound **(4.40)**.[26] Di- and trisubstituted alkenes on a chromene scaffold also undergo epoxidation, with high enantioselectivity (94 to >98% ee).[27] Enynes generally give a mixture of cis and trans epoxides. Katsuki and co-workers converted a mixture of epoxides **(4.43)** and **(4.44)** into an alcohol **(4.45)**, containing one stereocentre, by reaction with lithium aluminium hydride; the products were used in synthetic procedures to give enantiomerically-enriched insect pheromones.[28]

Jacobsen and co-workers used enantiomerically pure quaternary ammonium salts to enhance the formation of trans epoxides.[29] It is not clear why these salts reverse the normal diastereoselectivity of the reaction; however, the epoxidation of (Z)-stilbene to trans stilbene oxide can be achieved with >96:4 trans selectivity and 90% ee using catalyst **(4.25)**.

(4.36)   (4.38) 95% ee

3 mol% **(4.24)**
20 mol% 4-PhC$_5$H$_4$NO
NaOCl, 0°C, CH$_2$Cl$_2$
91%

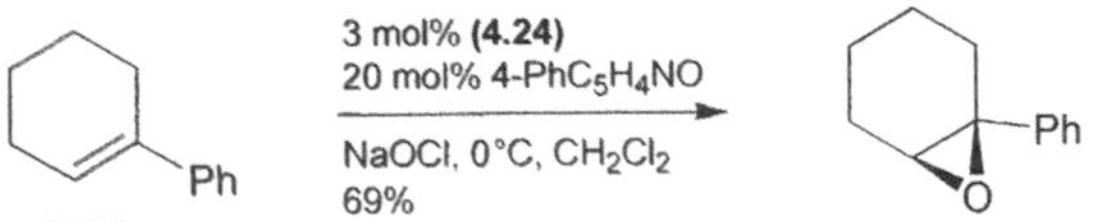

(4.37) → **(4.39)** 93% ee

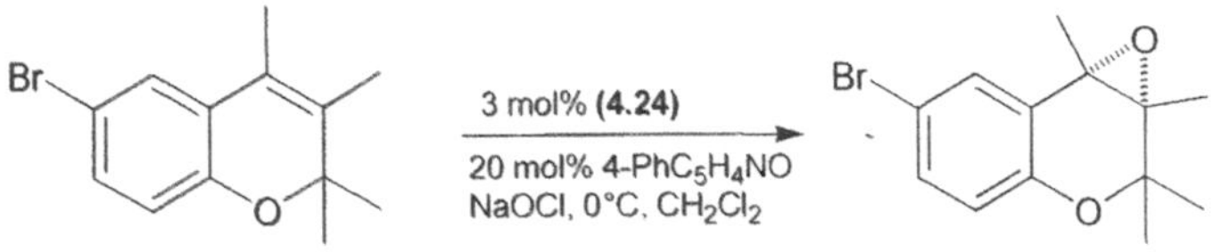

3 mol% **(4.24)**
20 mol% 4-PhC5H4NO
NaOCl, 0°C, CH2Cl2
69%

(4.40) → **(4.41)** 85% ee

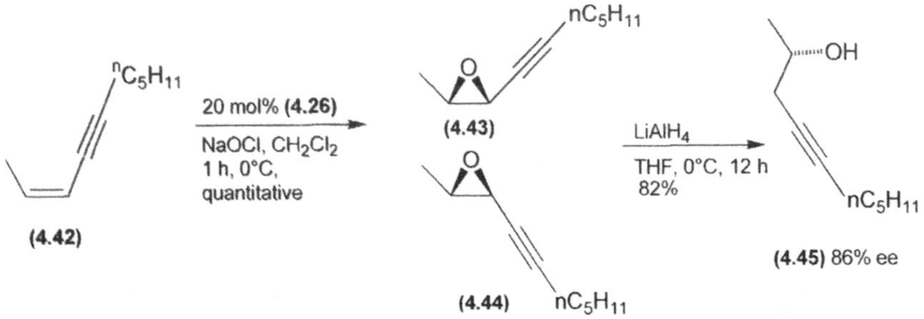

3 mol% **(4.24)**
20 mol% 4-PhC5H4NO
NaOCl, 0°C. CH2Cl2

(4.42)

20 mol% **(4.26)**
NaOCl, CH2Cl2
1 h, 0°C,
quantitative

(4.43)

(4.44)   nC5H11

LiAlH4
THF, 0°C, 12 h
82%

(4.45) 86% ee

Dienes[30] and cyclic dienes[31] have been used as substrates to give high enantioselectivity in the product mono-epoxides. Hentemann and Fuchs have shown that dienylsulfones are especially good substrates, at least in cyclic cases.[32] Thus, dienes **(4.46)** and **(4.47)** are converted into the mono-epoxides **(4.48)** and **(4.49)** as single regioisomers, with excellent enantioselectivity.

MCPBA (*m*-chloroperbenzoic acid) can be used as the stoichiometric oxidant at lower temperatures, with good enantioselectivities in the epoxidation of monosubstituted alkenes, such as styrenes **(4.50)**. The latter give fairly good enantioselectivities in the formation of mono-substituted epoxides **(4.51)**, although the enantioselectivity is less impressive than it is for many other classes of alkene epoxidation. The MCPBA must be added once the solution has been cooled, since no epoxidation occurs if the NMO and MCPBA are premixed.

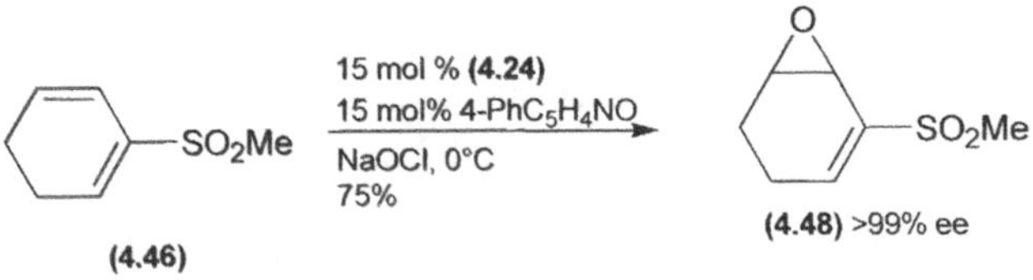

(4.46) → **(4.48)** >99% ee

15 mol % **(4.24)**
15 mol% 4-PhC5H4NO
NaOCl, 0°C
75%

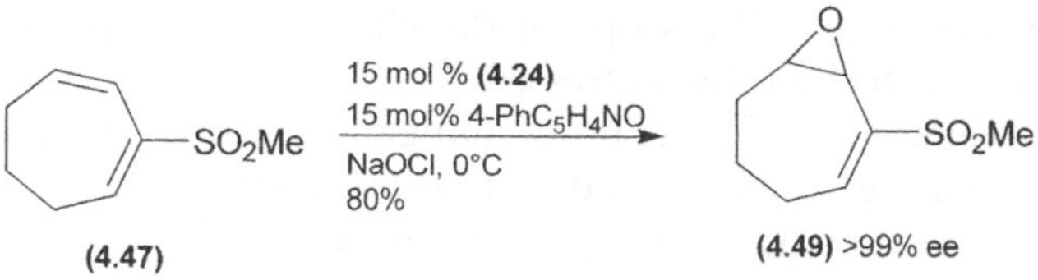

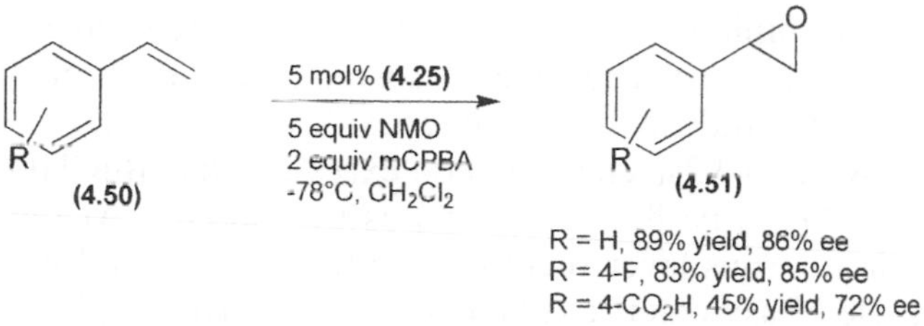

SO2Me
15 mol % (4.24)
15 mol% 4-PhC5H4NO
NaOCl, 0°C
80%
O
SO2Me
(4.47)
(4.49) >99% ee

R
(4.50)
5 mol% (4.25)
5 equiv NMO
2 equiv mCPBA
-78°C, CH2Cl2
O
R
(4.51)
R = H, 89% yield, 86% ee
R = 4-F, 83% yield, 85% ee
R = 4-CO2H, 45% yield, 72% ee

Racemic substrates have been employed in Mn(salen) catalysed epoxidation reactions. Reasonable to good kinetic resolution of racemic chromanes (selectivity factor, S=2.7–9.3)[33] and racemic allenes[34] have been achieved using enantiomerically pure manganese(salen) complexes to catalyse epoxidation.

Enol ethers are interesting substrates for 'epoxidations', since α-hydroxy ketones or the corresponding acetals are isolated, depending on the choice of solvent. Fukuda and Katsuki used enol ethers as substrates, including the cyclic enol ether (**4.52**), which gives rise to the hydroxy acetal product (**4.53**).[35] Adam and co-workers have used silyl enol ethers and silyl ketene acetals as substrates.[36] A typical example is provided by the asymmetric oxidation of silyl enol ether (**4.54**), generating the α-hydroxy ketone (**4.55**) after a suitable work-up.

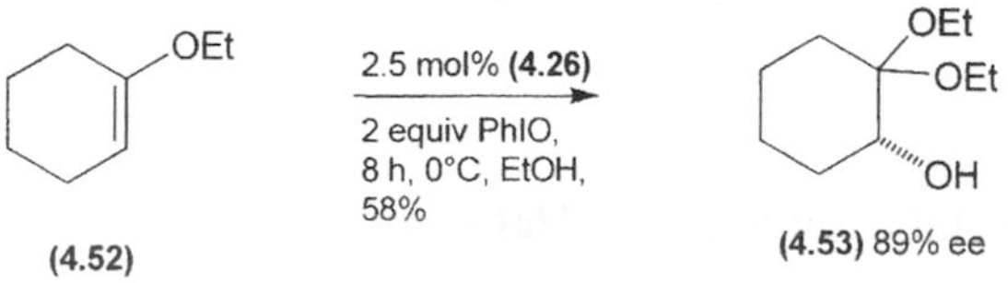

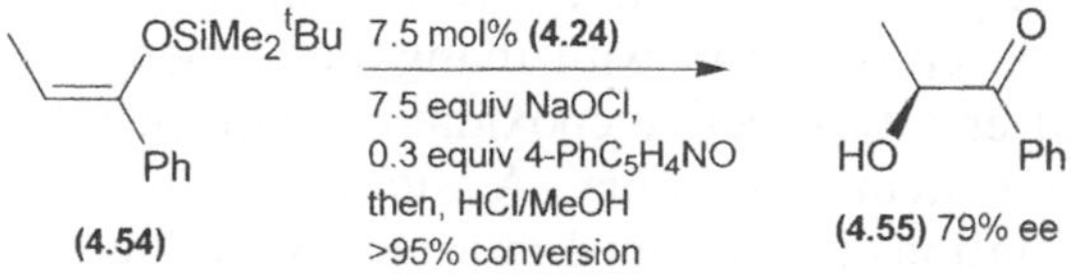

OEt
2.5 mol% (4.26)
2 equiv PhIO,
8 h, 0°C, EtOH,
58%
OEt
OEt
OH
(4.52)
(4.53) 89% ee

OSiMe2tBu
Ph
7.5 mol% (4.24)
7.5 equiv NaOCl,
0.3 equiv 4-PhC5H4NO
then, HCl/MeOH
>95% conversion
O
HO
Ph
(4.54)
(4.55) 79% ee

Bousquet and Gilheany reported the use of chromium(salen) complexes, where the epoxidation of (E)-alkenes leads to higher enantioselectivity than for (Z)-alkenes.[37] The scope of the chromium(salen) complexes as catalysts remains to be fully addressed.[38]

Polymeric enantiomerically pure manganese(salen) complexes have given disappointing results, so far.[39] However, salen ligands possessing perfluoroalkyl groups (C_8F_{17}), replacing tBu in Jacobsen's ligand (4.24), have been used successfully in the first example of asymmetric catalysis using fluorous biphase chemistry.[40] The manganese(salen) complexes derived are soluble in perfluorinated solvents and insoluble in common organic solvents. Indene has been epoxidised under fluorous biphase conditions, with 90–92% enantiomeric excess, using $tBuCHO/O_2$ (which is readily soluble in perfluorocarbons) as the oxidant. At the end of the reaction, the fluorous layer retains the catalyst and could be reused. No doubt the use of asymmetric catalysis under fluorous biphase conditions will be widely examined in other reactions in the future.

The details of manganese(salen) catalysed epoxidation are a topic of current debate. What seems clear is the shuttle of manganese between oxidation state (III) to oxidation state (V) by whatever oxidant is used.[41] The orientation of the alkene as it approaches the metal-oxo complex is not totally clear.[42] The oxygen transfer cannot always be completely concerted, since this does not explain how isomerisation from *cis* to *trans* occurs during the epoxidation cycle. The involvement of a radical intermediate provides a neat explanation of the isomerisation, and also explains why it is the better radical-stabilising group which becomes stereochemically scrambled.[43] Thus, oxidation of an Mn(III) complex (4.56) to an Mn(V) complex (4.57) provides a catalytic species capable of oxidising an alkene, e.g. alkene (4.27). In this case, a benzylic radical intermediate (4.58) either undergoes direct collapse to the *cis* epoxide (4.31) or is able to rotate before collapse, give rise to the *trans* epoxide (4.59), as shown in Figure 4.4. The involvement of manganaoxetane complexes by [2 + 2] addition has been proposed[44] and disputed.[45]

4.3   Other metal-catalysed epoxidations

Whilst the Sharpless epoxidation with titanium catalysts and the Jacobsen-Katsuki epoxidation with manganese(salen) complexes are at the forefront of enantioselective epoxidation with metal catalysts, there are alternative systems available. A summary of alternative metal-catalysed epoxidations is contained in a review by Jacobsen.[11c] Many of the earlier systems involved mimics of cytochrome P-450, employing enantiomerically pure metalloporphyrin catalysts. In 1983, Groves and

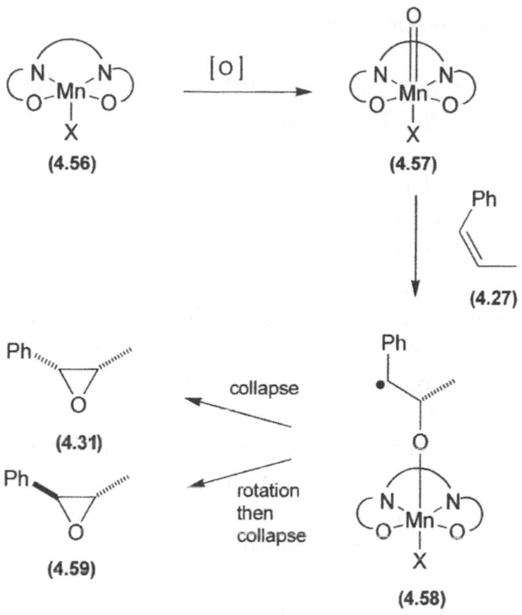

Figure 4.4 Pathways for the Mn(salen)-catalysed epoxidations.

Myers achieved up to 51% ee using an Fe(III) porphyrin catalyst.[46] There have been many other asymmetric epoxidations catalysed by metallo-porphyrin sytems, which have met with various degrees of success.[47] One of the problems with porphyrin ligands is that systematic variation of the ligand is difficult, hampering the search for the very best systems.

Ruthenium bis-oxazoline complexes have been reported, independently, by two groups to give moderately good enantioselectivity in the epoxidation of stilbene.[48,49] Strukul and co-workers reported the use of enantiomerically pure platinum complexes for the epoxidation of alkenes.[50] Although very high enantioselectivities have so far been elusive, the reaction is interesting since the mechanism involves co-ordination of the alkene to the metal catalyst and attack by a Pt-OOH species. In other metal-catalysed epoxidation reactions, the mechanism involves attack of an oxo/peroxo or similar complex onto an non-complexed alkene.

Enantiomerically pure manganese complexes using ligands other than the salen structure have been reported,[51,52] but so far with lower enantioselectivities. At present, the titanium/tartrate and manganese(salen) complexes provide the best metal-catalysed epoxidation methods; however, the prospects for alternative systems are good.

4.4   Julia-Colonna epoxidation

The epoxidation of α,β-unsaturated ketones catalysed by polyamino acids is known as the Julia-Colonna epoxidation.[53] The reaction is especially effective using poly-L-leucine as a catalyst in the epoxidation of chalcone (4.60) to chalcone epoxide (4.61).

There have been several modifications of the Julia-Colonna procedure, most particularly the work of Roberts and co-workers, who performed the reaction in the absence of water.[54] Replacement of aqueous sodium hydroxide/hydrogen peroxide by urea hydrogen peroxide (UHP) and 1,8-diazabicyclo[5.4.0]undec-7-ene (DBU) provides a reactive mixture, where poly-L-leucine is the preferred catalyst. The chalcone epoxide product (4.61) is formed in just 30 min, with excellent yield and enantioselectivity.

Roberts and co-workers have shown how the chalcone epoxides can be used in the synthesis of natural products. [55] Mechanistically, the reaction involves nucleophilic conjugate addition of the deprotonated hydrogen peroxide onto the chalcone, followed by cyclisation with loss of hydroxide. The asymmetric induction may be imparted by hydrogen-bonding between the chalcone and the polyamino acid. Polyamino acids are also able to catalyse reactions other than epoxidation.[56]

4.5   Phase-transfer-catalysed epoxidation

Asymmetric phase-transfer-catalysed reactions in epoxidation and in other processes are attracting increasing interest, and have recently shown their ability to impart high levels of enantioselectivity.[57] Like the related use of polyamino acid catalysts, the use of phase-transfer catalysts works well with the epoxidation of electron deficient alkenes. Quinine- and cinchonine-derived quaternary ammonium salts have been most widely used since the pioneering work of Wynberg and Greijdanus.[58, 59] Although fairly high selectivities have been achieved, there have, so far, been no selectivities to rival the best results from other epoxidation methods.

Recent examples include the epoxidation of chalcone (4.60), with high enantioselectivity, using an anthracenyl quaternary ammonium salt (4.62);[60] however, the reaction awaits development into a more general procedure.

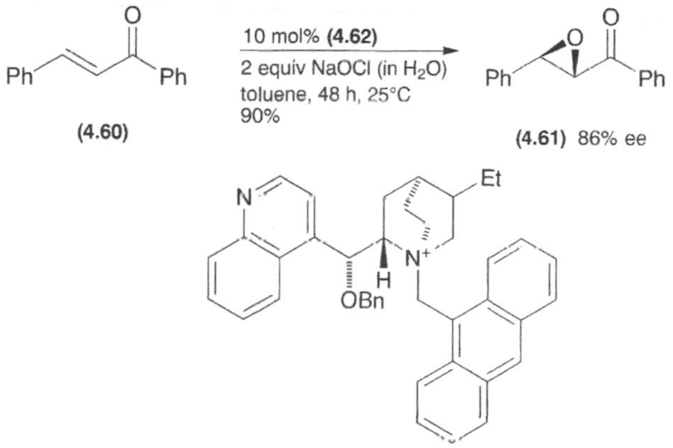

(4.60)

10 mol% (4.62)

2 equiv NaOCl (in H₂O)
toluene, 48 h, 25°C
90%

(4.61) 86% ee

(4.62)

4.6 Epoxidation with iminium salts

A French group demonstrated that iminium salts (4.63) are able to catalyse the epoxidation of alkenes (4.64) with oxone (KHSO₅) under basic conditions. The reaction proceeds *via* formation of an intermediate oxaziridinium salt (4.65), which converts the alkene (4.64) into an epoxide (4.66).[61] The same group prepared enantiomerically pure iminium salts (4.67), and achieved epoxidation of stilbene (4.68) to stilbene oxide (4.69), with 33% ee.[62]

Aggarwal and Wang used the iminium catalyst (4.70) to provide up to 71% ee in the epoxidation of 1-phenylcyclohexene (4.71) to the corresponding epoxide (4.72).[63] Bulman Page and co-workers designed a straightforward preparation of iminium salts, including catalyst (4.73), which provides reasonable selectivity in the epoxidation of stilbene (4.68).[64]

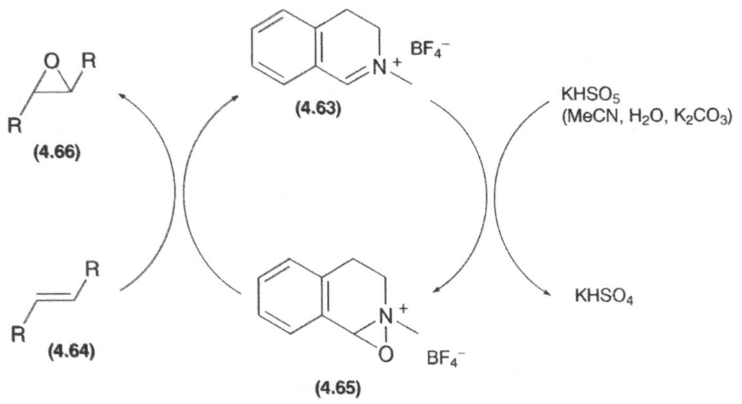
BF4⁻
(4.63)
KHSO5
(MeCN, H2O, K2CO3)
(4.66)
(4.64)
KHSO4
BF4⁻
(4.65)

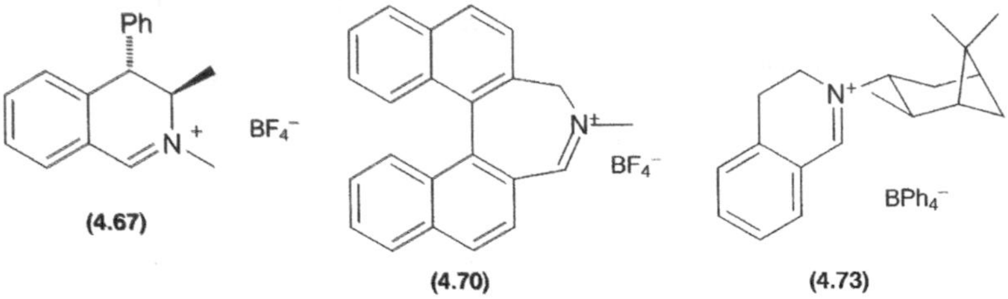
Ph
BF4⁻
(4.67)
BF4⁻
(4.70)
BPh4⁻
(4.73)

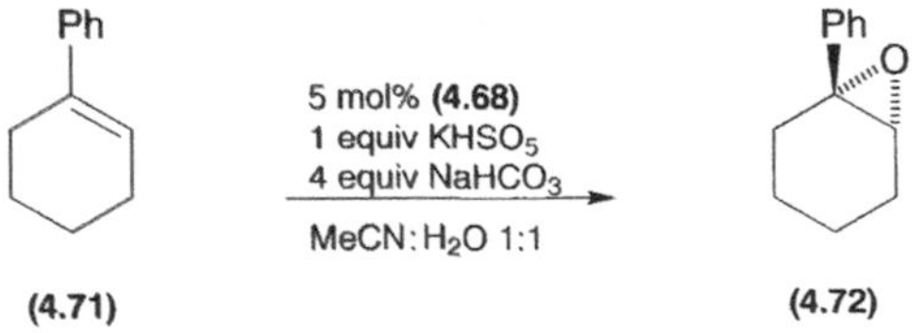
Ph
5 mol% (4.68)
1 equiv KHSO5
4 equiv NaHCO3
MeCN:H2O 1:1
(4.71)
(4.72)

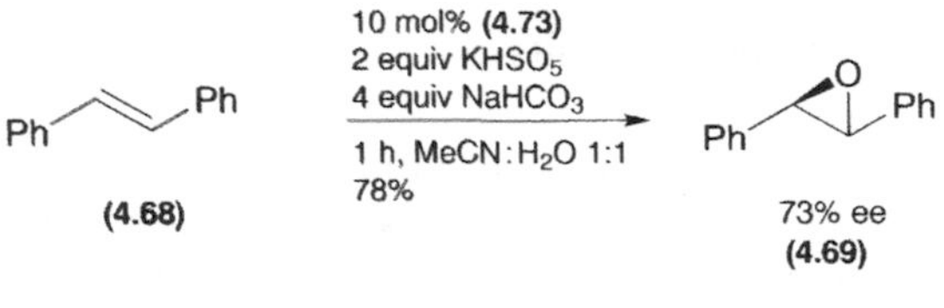
10 mol% (4.73)
2 equiv KHSO5
4 equiv NaHCO3
1 h, MeCN:H2O 1:1
78%
(4.68)
73% ee
(4.69)

4.7 Epoxidation with ketone catalysts

Alkenes can be epoxidised by dioxiranes; this process can be achieved enantioselectively if an enantiomerically pure dioxirane is employed. Since ketones can be oxidised to dioxiranes with oxone under conditions where alkenes are not directly epoxidised, this allows the following catalytic cycle to be established, where a ketone (**4.74**) is converted into a dioxirane (**4.75**). This oxidises the alkene (**4.64**) into an epoxide (**4.66**). There are clear parallels between this mechanism and the iminium-catalysed epoxidation described in Section 4.6. This catalytic cycle was first demonstrated by Curci and co-workers in 1984, using an enantio-merically pure ketone.[65] However, it was not until 1996 that high levels of enantioselectivity were reported.

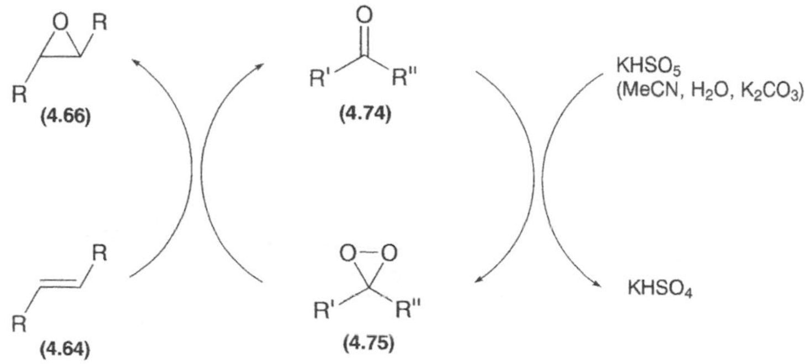

The choice of ketone is governed by its ability to form a dioxirane rapidly (an electron withdrawing group in the α-position is helpful) and not to be prone to either racemisation or Baeyer-Villiger oxidation. The intermediate dioxirane must also be willing to donate an oxygen to the alkene substrate.

Yang and co-workers used the catalyst (4.76) with great effect in the epoxidation of stilbenes (4.77).[66, 67] Armstrong and Hayter used the α-fluorinated tropinone-derived ketone (4.79), which provides good selec-tivity for the epoxidation of phenylstilbene (4.80).[68]

The highest selectivities have been achieved using the fructose-derived ketone catalyst (4.82), although this is typically used in 20–30 mol% catalyst loading. For example, stilbene (4.68) and the cinnamyl ether (4.83) are epoxidised with high selectivity.[69] The reaction has also been successfully applied to the mono-epoxidation of many conjugated dienes, all with excellent enantioselectivity.[70] A typical example is provided by the epoxidation of diene (4.84) into the mono-epoxide (4.86).

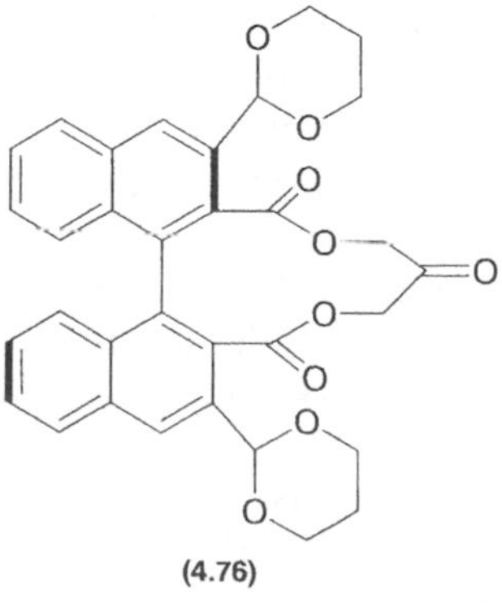

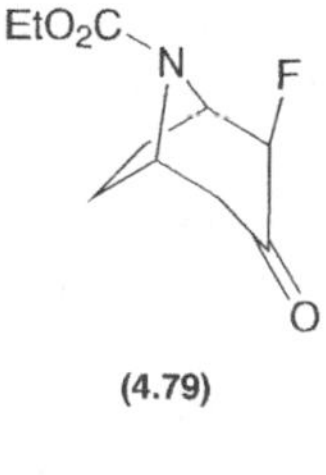

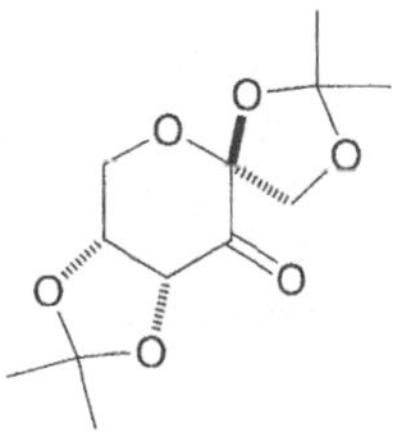

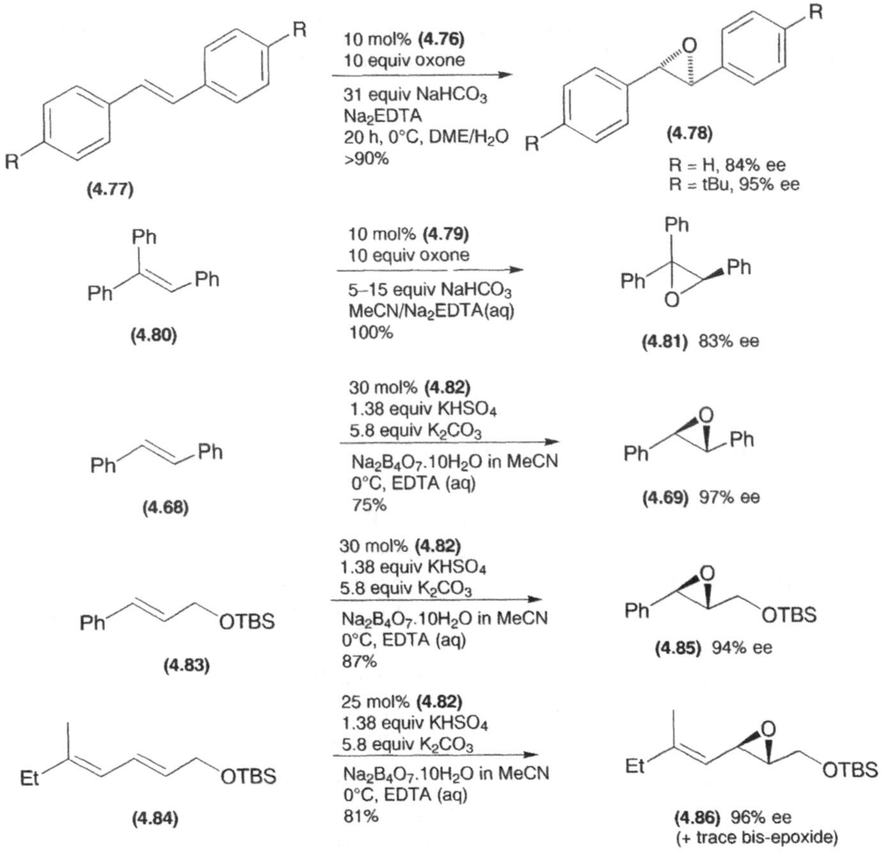

4.8   Aziridination

So far, aziridination reactions have, in some ways, had more in common with cyclopropanation reactions (see Section 9.1) than with epoxidation reactions. Nevertheless, the aziridination reaction is more synthetically akin to epoxidation and, on that basis, is included in the present chapter.

Copper complexes of bis-oxazoline ligands (4.87), which are very efficient cyclopropanation catalysts, are also competent in aziridination reactions. Evans and co-workers have shown that high selectivities are obtained using cinnamates as the substrate, and the 'nitrene' source PhI = NTs.[71] The substrates (4.88) and (4.89) both undergo enantio-selective aziridination, but it is the cinnamate ester (4.88) which provides the more selective reaction.

Jacobsen and co-workers employed diimine ligands (4.92), which worked especially well on substrate (4.93), although in other examples lower enantioselectivities (30–87%) were obtained.[72, 73]

Most (salen)manganese (III) complexes have been found to be less effective for aziridination than for epoxidation, both in terms of enantioselectivity and yield.[74] However, the salen complex (**4.95**) has been shown to give high enantioselectivity for some styrenes.[75] The best substrate is styrene (**4.89**), which is converted into the tosylaziridine (**4.96**).

An asymmetric rhodium-catalysed aziridination reaction has shown encouraging levels of enantioselectivity (up to 73% ee).[76] An alternative approach to aziridine synthesis, involving carbenoid transfer to imine, has so far proceeded with only moderate yield and selectivity.[77]

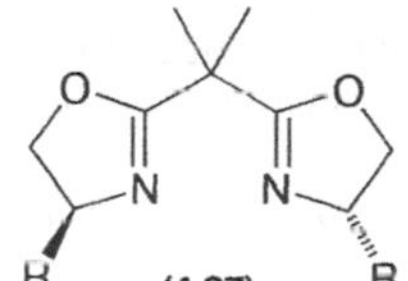
O
O
N
N
R
R

(4.87)

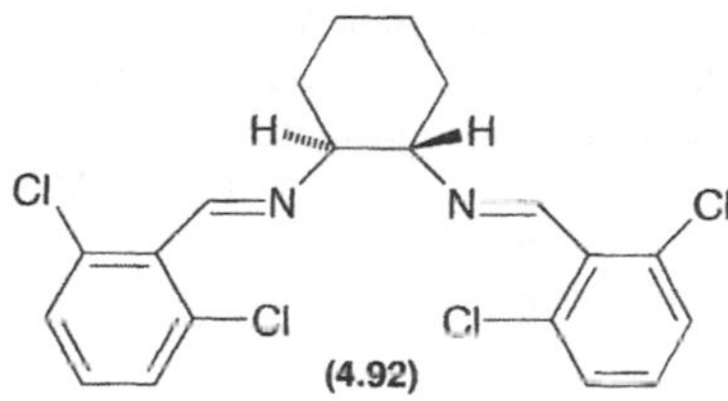
H
H
Cl
N
N
Cl
Cl
Cl

(4.92)

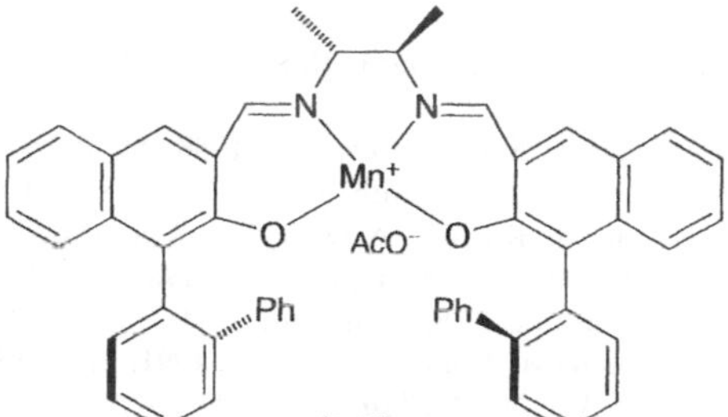
N
N
Mn+
O
AcO-
O
Ph
Ph

(4.95)

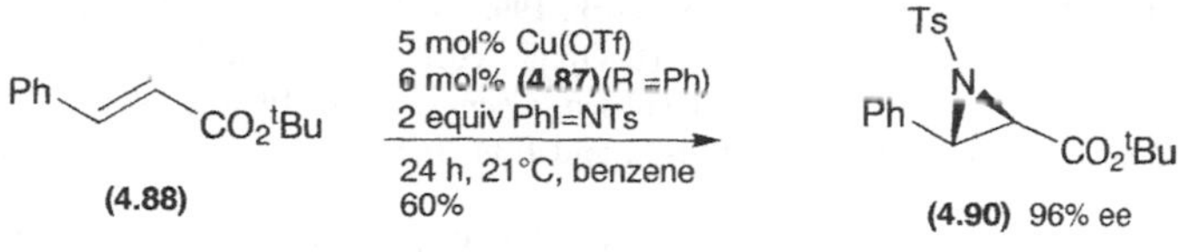
Ph
CO2tBu
(4.88)
5 mol% Cu(OTf)
6 mol% (4.87)(R =Ph)
2 equiv PhI=NTs
24 h, 21°C, benzene
60%
Ts
N
Ph
CO2tBu
(4.90) 96% ee

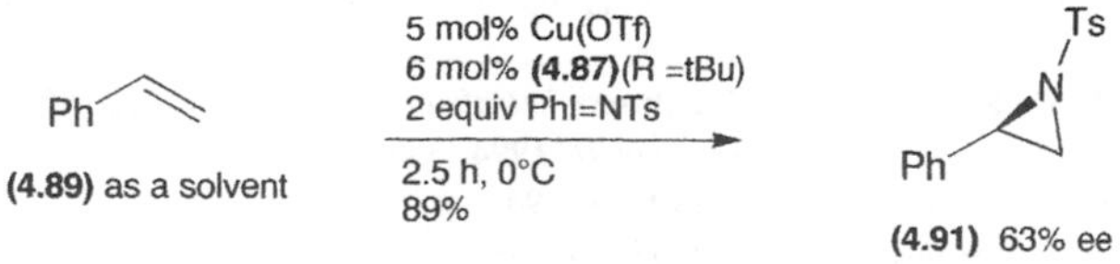
Ph
(4.89) as a solvent
5 mol% Cu(OTf)
6 mol% (4.87)(R =tBu)
2 equiv PhI=NTs
2.5 h, 0°C
89%
Ts
N
Ph
(4.91) 63% ee

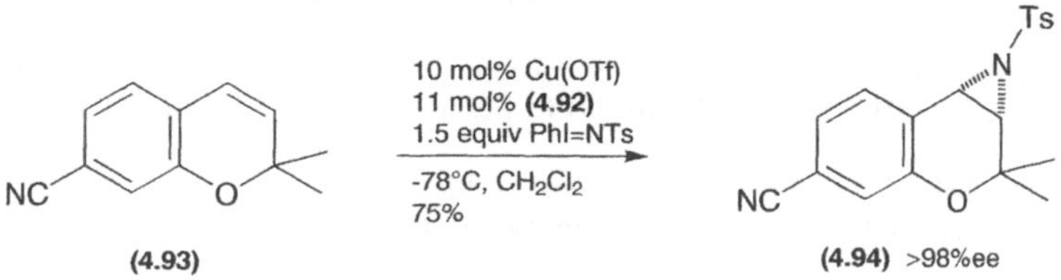
NC
O
(4.93)
10 mol% Cu(OTf)
11 mol% (4.92)
1.5 equiv PhI=NTs
-78°C, CH2Cl2
75%
Ts
N
NC
O
(4.94) >98%ee

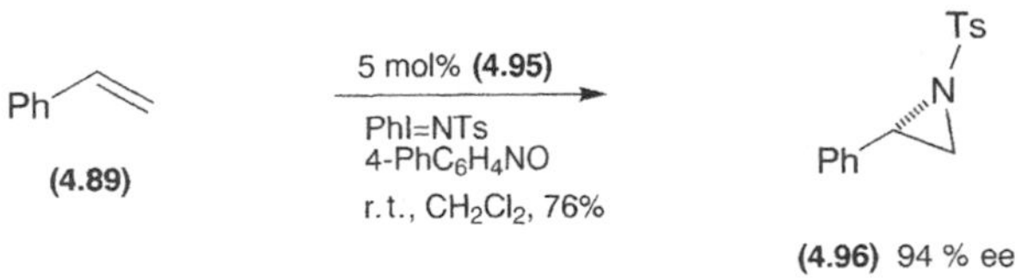
Ph
(4.89)
5 mol% (4.95)
PhI=NTs
4-PhC6H4NO
r.t., CH2Cl2, 76%
Ts
N
Ph
(4.96) 94 % ee

References

1. R.A. Johnson and K.B. Sharpless, in *Catalytic Asymmetric Synthesis*, (I. Ojima, ed.) VCH, New York, **1993**, 103. (b) T. Katsuki and V.S. Martin, *Organic Reactions*, **1996**, *48*, 1. (c) R.A. Johnson and K.B. Sharpless, *Comprehensive Organic Synthesis*, (B.M. Trost, ed.) Pergamon, Oxford **1991**, Vol. 7, 389.
2. T. Katsuki and K.B. Sharpless, *J. Am. Chem. Soc.*, **1980**, *102*, 5974.
3. (a) R.M. Hanson and K.B. Sharpless, *J. Org. Chem.*, **1986**, *51*, 1922. (b) Y. Gao, R.M. Hanson, J.M. Klunder, S.Y. Ko, H. Masamune and K.B. Sharpless, *J. Am. Chem. Soc.*, **1987**, *109*, 5765.
4. J.K. Karjalainen, O.E.O. Hormi and D.C. Sherrington, *Tetrahedron: Asymmetry*, **1998**, *9*, 1563.
5. V.S. Martin, S.S. Woodard, T. Katsuki, Y. Yamada, M. Ikeda and K.B. Sharpless, *J. Am. Chem. Soc.*, **1981**, *103*, 6237.
6. P.R. Carlier, W.S. Mungall, G. Schröder and K.B. Sharpless, *J. Am. Chem. Soc.*, **1988**, *110*, 2978.
7. Y. Kitano, T. Matsumoto and F. Sato, *Tetrahedron*, **1988**, *44*, 4072.
8. S.L. Schreiber, T.S. Schreiber and D.B. Smith, *J. Am. Chem. Soc.*, **1987**, *109*, 1525.
9. K.B. Sharpless, C.H. Behrens, T. Katsuki, A.W.M. Lee, V.S. Martin, M. Takatani, S.M. Viri, F.J. Walker and S.S. Woodard, *Pure Appl. Chem.*, **1983**, *55*, 589.
10. (a) S.S. Woodard, M.G. Finn and K.B. Sharpless, *J. Am. Chem. Soc.*, **1991**, *113*, 106. (b) M.G. Finn and K.B. Sharpless, *J. Am. Chem. Soc.*, **1991**, *113*, 113.
11. For reviews, see: (a) T. Katsuki, *Coord. Chem. Rev.*, **1995**, *140*, 189. (b) E.N. Jacobsen, in *Comprehensive Organometallic Chemistry II*, (G. Wilkinson, F.G.A. Stone, E.W. Abel and L.S. Hegedus, eds.) Pergamon, New York, **1995**, Vol. 12, Chapter 11.1. (c) E.N. Jacobsen, in *Catalytic Asymmetric Synthesis*, (I. Ojima, ed.) VCH, New York, **1993**, 159. (d) P.J. Pospisil, D.H. Carsten and E.N. Jacobsen, *Chem. Eur. J.*, **1996**, *2*, 974.
12. W. Zhang, J.L. Loebach, S.R. Wilson and E.N. Jacobsen, *J. Am. Chem. Soc.*, **1990**, *112*, 2801.
13. R. Irie, K. Noda, Y. Ito, N. Matsumoto and T. Katsuki, *Tetrahedron Lett.*, **1990**, *31*, 7345.
14. J.F. Larrow, E.N. Jacobsen, Y. Gao, Y. Hong, X. Nie and C.M. Zepp, *J. Org. Chem.*, **1994**, *59*, 1939.
15. S. Chang, R.M. Heid and E.N. Jacobsen, *Tetrahedron Lett.*, **1994**, *35*, 669.
16. H. Sasaki, R. Irie and T. Katsuki, *Synlett*, **1994**, 356.
17. P. Pietikäinen, *Tetrahedron Lett.*, **1994**, *35*, 941.
18. P. Pietikäinen, *Tetrahedron Lett.*, **1995**, *36*, 319.
19. W. Adam, J. Jekö, A. Lévai, C. Nemes, T. Patonay and P. Sebök, *Tetrahedron Lett.*, **1995**, *36*, 3669.
20. M. Palucki, G.J. McCormick and E.N. Jacobsen, *Tetrahedron Lett.*, **1995**, *36*, 5457.
21. E.N. Jacobsen, W. Zhang, A.R. Muci, J.R. Ecker and L. Deng, *J. Am. Chem. Soc.*, **1991**, *113*, 7063.
22. L. Deng and E.N. Jacobsen, *J. Org. Chem.*, **1992**, *57*, 4320.
23. B.D. Brandes and E.N. Jacobsen, *J. Org. Chem.*, **1994**, *59*, 4378.

24. D. Bell, M.R. Davies, F.J.L. Finney, G.R. Green, P.M. Kincey and I.S. Mann, *Tetrahedron Lett.*, **1996**, *37*, 3895.
25. C.H. Senanayake, G.B. Smith, K.M. Ryan, L.E. Fredenburgh, J. Liu, F.E. Roberts, D.L. Hughes, R.D. Larsen, T.R. Verhoeven and P.J. Reider, *Tetrahedron Lett.*, **1996**, *37*, 3271.
26. B.D. Brandes and E.N. Jacobsen, *Tetrahedron Lett.*, **1995**, *36*, 5123.
27. N.H. Lee, A.R. Muci and E.N. Jacobsen, *Tetrahedron Lett.*, **1991**, *32*, 5055.
28. (a) T. Hamada, K. Daikai, R. Irie and T. Katsuki, *Synlett*, **1995**, 407. (b) T. Hamada, K. Daikai, R. Irie and T. Katsuki, *Tetrahedron: Asymmetry*, **1995**, *6*, 2441.
29. S. Chang, J.M. Galvin and E.N. Jacobsen, *J. Am. Chem. Soc.*, **1994**, *116*, 6937.
30. S. Chang, N.H. Lee and E.N. Jacobsen, *J. Org. Chem.*, **1993**, *58*, 6939.
31. D. Mikame, T. Hamda, R. Irie and T. Katsuki, *Synlett*, **1995**, 827.
32. M.F. Hentemann and P.L. Fuchs, *Tetrahedron Lett.*, **1997**, *38*, 5615.
33. S.L. Vander Velde and E.N. Jacobsen, *J. Org. Chem.*, **1995**, *60*, 5380.
34. Y. Noguchi, H. Takiyama and T. Katsuki, *Synlett*, **1998**, 543.
35. T. Fukuda and T. Katsuki, *Tetrahedron Lett.*, **1996**, *37*, 4389.
36. (a) W. Adam, R.T. Fell, C. Mock-Knoblauch, C.R. Saha-Möller, *Tetrahedron Lett.*, **1996**, *37*, 6531. (b) W. Adam, R.T. Fell, V.R. Stegmann and C.R. Saha-Möller, *J. Am. Chem. Soc.*, **1998**, *120*, 708.
37. C. Bousquet and D.G. Gilheany, *Tetrahedron Lett.*, **1995**, *36*, 7739.
38. H. Imanishi and T. Katsuki, *Tetrahedron Lett.*, **1997**, *38*, 251.
39. B.B. De, B.B. Lohray, S. Sivaram and P.K. Dhal, *Tetrahedron: Asymmetry*, **1995**, *6*, 2105.
40. G. Pozzi, F. Cinato, F. Montanari and S. Quici, *J. Chem. Soc., Chem. Commun.*, **1998**, 877.
41. D. Feichtinger and D.A. Plattner, *Angew. Chem., Int. Ed. Engl.*, **1997**, *36*, 1718.
42. T. Hamada, R. Irie and T. Katsuki, *Synlett*, **1994**, 479.
43. (a) N. Hosoya, A. Hatayama, K. Yanai, H. Fujii, R. Irie and T. Katsuki, *Synlett*, **1993**, 641. (b) M. Palucki, N.S. Finney, P.J. Pospisil, M.L. Güler, T. Ishida and E.N. Jacobsen, *J. Am. Chem. Soc.*, **1998**, *120*, 948, and references therein.
44. C. Linde, M. Arnold, P.-O. Norrby and B. Åkermark, *Angew. Chem., Int. Ed. Engl.*, **1997**, *36*, 1723.
45. N.S. Finney, P.J. Pospisil, S. Chang, M. Palucki, R.G. Konsler, K.B. Hansen and E.N. Jacobsen, *Angew. Chem., Int. Ed. Engl.*, **1997**, *36*, 1720.
46. J.T. Groves and R.S. Myers, *J. Am. Chem. Soc.*, **1983**, *105*, 5791.
47. For a few examples, see: (a) S.O'Malley and T. Kodadek, *J. Am. Chem. Soc.*, **1989**, *111*, 9176. (b) R.L. Halterman and S.-T. Jan, *J. Org. Chem.*, **1991**, *56*, 5253. (c) Y. Naruta, F. Tani, N. Ishihara and K. Maruyama, *J. Am. Chem. Soc.*, **1991**, *113*, 6865. (d) D. Mansuy, P. Battoni, J.-P. Renaud and P. Guerin, *J. Chem. Soc., Chem. Commun.*, **1985**, 155. (e) Z. Gross and S. Ini, *J. Org. Chem.*, **1997**, *62*, 5514. (f) J.P. Collman, V.J. Lee, X. Zhang, J.A. Ibers and J.I. Brauman, *J. Am. Chem. Soc.*, **1993**, *115*, 3834. (g) A. Berkessel and M. Frauenkron, *J. Chem. Soc., Perkin Trans. 1*, **1997**, 2265.
48. H. Nishiyama, T. Shimada, H. Itoh, H. Sugiyama and Y. Motoyama, *J. Chem. Soc., Chem. Commun.*, **1997**, 1863.
49. N. End and A. Pfaltz, *J. Chem. Soc., Chem. Commun.*, **1998**, 589.
50. C. Baccin, A. Gusso, F. Pinna and G. Strukul, *Organometallics*, **1995**, *14*, 1161.
51. T. Mukaiyama, T. Yamada, T. Nagata and K. Imagawa, *Chem. Lett.*, **1993**, 327.
52. C. Bolm, D. Kadereit and M. Valacchi, *Synlett*, **1997**, 687.
53. S. Banfi, S. Colonna, H. Molinari, S. Julia and J. Quixer, *Tetrahedron*, **1984**, *40*, 5207.
54. P.A. Bentley, S. Bergeron, M.W. Cappi, D.E. Hibbs, M.B. Hursthouse, T.C. Nugent, R. Pulido, S.M. Roberts and L.E. Wu, *J. Chem. Soc., Chem. Commun.*, **1997**, 739.
55. For example, see: M.W. Cappi, W.-P. Chen, R.W. Flood, Y.-W. Liao, S.M. Roberts, J. Skidmore, J.A. Smith and N.M. Williamson, *J. Chem. Soc., Chem. Commun.*, **1998**, 1159.
56. S. Ebrahim and M. Wills, *Tetrahedron: Asymmetry*, **1997**, *8*, 3163.

57. M.J. O'Donnell, in *Catalytic Asymmetric Synthesis*, (I. Ojima, ed.) VCH, New York, **1993**, 89.
58. For example, see: H. Wynberg and B. Greijdanus, *J. Chem. Soc., Chem. Commun.*, **1978**, 427.
59. For a review, see: H. Wynberg, *Topics in Stereochemistry*, **1986**, *16*, 87.
60. B. Lygo and P.G. Wainwright, *Tetrahedron Lett.*, **1998**, *39*, 1599.
61. See: X. Lusinchi and G. Hanquet, *Tetrahedron*, **1997**, *53*, 13727, and references therein.
62. L. Bohé, G. Hanquet, M. Lusinchi and X. Lusinchi, *Tetrahedron Lett.*, **1993**, *34*, 7271.
63. V.K. Aggarwal and M.F. Wang, *J. Chem. Soc., Chem. Commun.*, **1996**, 191.
64. P.C. Bulman Page, G.A. Rassias, D. Bethell and M.B. Schilling, *J. Org. Chem.*, **1998**, *62*, 2774.
65. R. Curci, M. Fiorentino and M.R. Serio, *J. Chem. Soc., Chem. Commun.*, **1984**, 155.
66. D. Yang, X.-C. Wang, M.-K. Wong, Y.-C. Yip and M.-W. Tang, *J. Am. Chem. Soc.*, **1996**, *118*, 11311.
67. D. Yang, Y.-C. Yip, M.-W. Tang, M.-K. Wong, J.-H. Zheng and K.-K. Cheung, *J. Am. Chem. Soc.*, **1996**, *118*, 491.
68. A. Armstrong and B.R. Hayter, *J. Chem. Soc., Chem. Commun.*, **1998**, 621.
69. Z.-X. Wang, Y. Tu, M. Frohn and Y. Shi, *J. Org. Chem.*, **1997**, *62*, 2328.
70. M. Frohn, M. Dalkiewicz, Y. Tu, Z.-X. Wang and Y. Shi, *J. Org. Chem.*, **1998**, *62*, 2948.
71. D.A. Evans, M.M. Faul, M.T. Bilodeau, B.A. Anderson and D.M. Barnes, *J. Am. Chem. Soc.*, **1993**, *115*, 5328.
72. Z. Li, K.R. Conser and E.N. Jacobsen, *J. Am. Chem. Soc.*, **1993**, *115*, 5326.
73. Z. Li, R.W. Quan and E.N. Jacobsen, *J. Am. Chem. Soc.*, **1995**, *117*, 5889.
74. K. Noda, N. Hosoya, R. Irie, Y. Ito, T. Katsuki, *Synlett*, **1993**, 469.
75. H. Nishikori and T. Katsuki, *Tetrahedron Lett.*, **1996**, *37*, 9245.
76. P. Müller, C. Baud and Y. Jacquier, *Tetrahedron*, **1996**, *52*, 1543.
77. K.B. Hansen, N.S. Finney and E.N. Jacobsen, *Angew. Chem., Int. Ed. Engl.*, **1995**, *34*, 676.

5 Further oxidation reactions

There are many oxidation reactions apart from epoxidation. Alkenes can be oxidised to diols or amino alcohols directly; but, of course, oxidation reactions are not restricted to alkene substrates. Other sections in this chapter consider oxidation of C—H bonds, ketones and sulfides.

5.1 Dihydroxylation

The asymmetric dihydroxylation of alkenes (the AD reaction) using osmium catalysts was discovered and developed by Sharpless and now represents one of the most impressive achievements of asymmetric catalysis. The majority of early results did not use catalytic systems; however, breakthrough in the catalytic asymmetric dihydroxylation reaction was reported by Sharpless and co-workers in 1988.[1] Major developments in the reaction occurred during the early 1990s as reported in several reviews.[2]

The ligands used by Sharpless and co-workers are based on dihydroquinidine (DHQD) (**5.01**) and dihydroquinine (DHQ) (**5.02**). Dihydroquinidine and dihydroquinine are diastereomers, although their derivatives behave as 'pseudo-enantiomers' in osmium-catalysed dihydroxylation reactions, providing opposite and approximately equal selectivity.

Early ligands were simple esters of dihydroquinidine and dihydroquinine, such as acetates and *p*-chlorobenzoates, and gave very good enantioselectivity.[3] Nevertheless, the C_2-symmetric ligands, which use a phthalazine spacer unit, have proved to be the most generally applicable ligands, working well with several classes of alkene.[4] Dihydroquinidine and dihydroquinine can both be attached to the phthalazine spacer, as illustrated by (DHQD)$_2$-PHAL (**5.03**). The ligands accelerate the rate of the osmium-catalysed dihydroxylation, which is particularly helpful in asymmetric catalysis. This ligand and its pseudo-enantiomeric partner (DHQ)$_2$-PHAL (**5.04**) provide excellent enantioselectivity in the dihydroxylation reaction. The ligands used in the asymmetric dihydroxylation reactions have been modified and attached in various ways to silica, polyethylene glycol and other polymers. The use of polymer-supported AD reactions was reviewed in 1998.[5]

The usual stoichiometric oxidant used in these reactions is now potassium ferricyanide ($K_3Fe(CN)_6$), and this is used in the commercially-available AD-mix. AD-mix-α contains $K_3Fe(CN)_6$, K_2CO_3 and

(DHQ)$_2$-PHAL, together with involatile potassium-osmate (VI) dihy-drate. The alternative AD-mix-β contains the (DHQD)$_2$-PHAL ligand.

Using the AD-mix reagents, (E)-disubstituted alkenes often give particularly good selectivities, as represented by the conversion of an alkene (5.05) into either enantiomer of the diol (5.06).

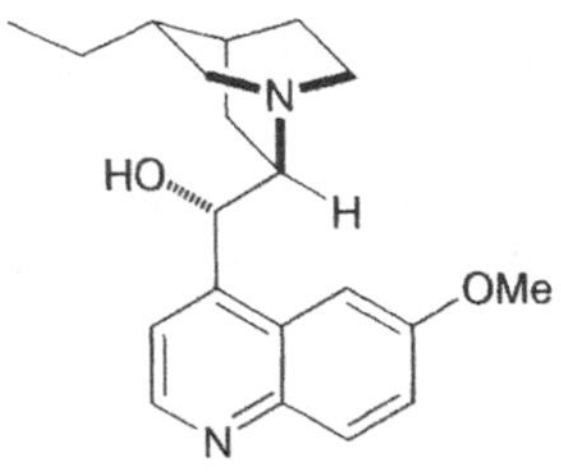

(5.01) DHQD

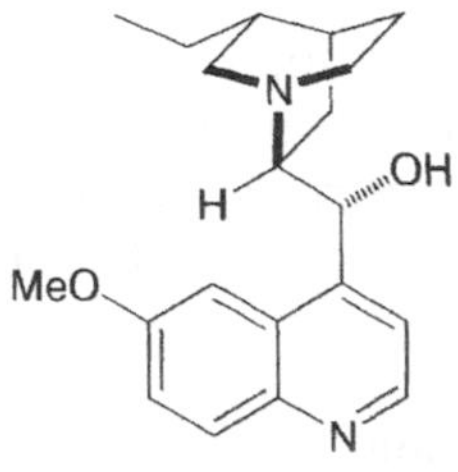

(5.02) DHQ

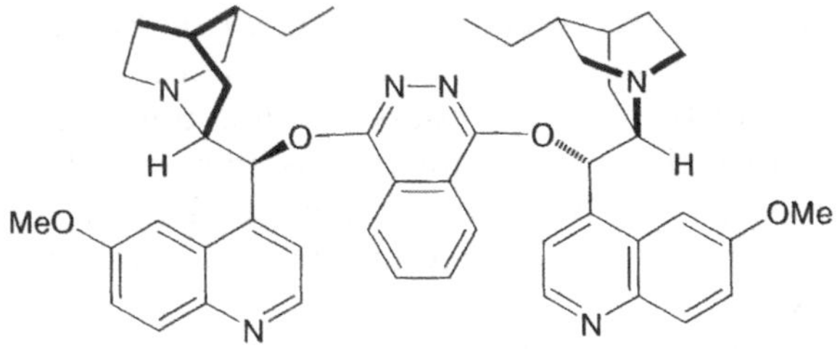

(5.03) (DHQD)$_2$PHAL

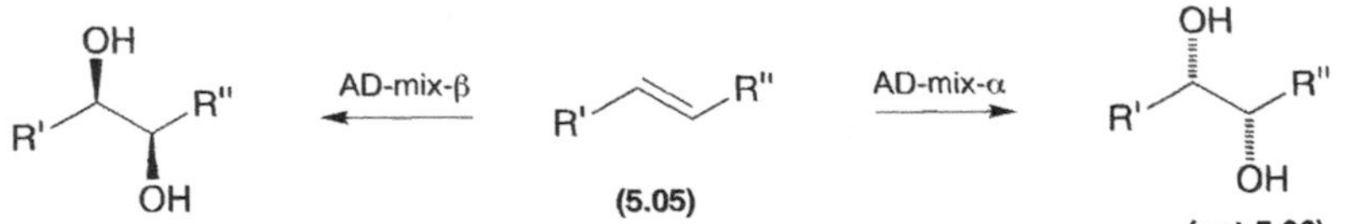

(5.06)

(5.05)

(ent-5.06)

Specific examples are given by the dihydroxylation of the *trans*-disubstituted alkenes (5.07–5.09) using AD-mix-β, which yields diols (5.10–5.12) in very high enantiomeric excess. Dihydroxylation of stilbene (5.07) has been developed, such that 1 kg of substrate can undergo dihydroxylation in a 5 litre vessel.[6] The corresponding AD-mix-α con-taining (DHQ)$_2$PHAL often gives slightly lower selectivities. In the last three cases, 99.5% ee, 93% ee and 95% ee, respectively, are obtained. Methylsulfonamide is used to increase the rate of reaction, and this generally allows lower temperatures to be used.

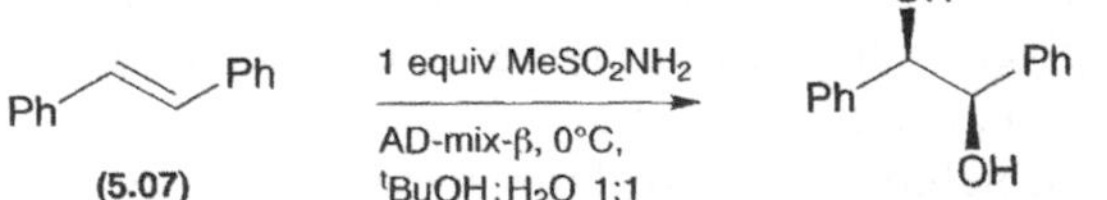

(5.07)

1 equiv MeSO$_2$NH$_2$

AD-mix-β, 0°C,
^{t}BuOH:H$_2$O 1:1

(5.10) 99.8% ee

(5.08) → (5.11) 97% ee

1 equiv MeSO$_2$NH$_2$, AD-mix-β, 0°C, ^{t}BuOH:H$_2$O 1:1

(5.09) → (5.12) 97% ee

1 equiv MeSO$_2$NH$_2$, AD-mix-β, r.t., ^{t}BuOH:H$_2$O 1:1

Asymmetric dihydroxylation reactions can also be carried out on substrates containing heteroatoms.[7,8] In the case of allylic halides, including cinnamyl chloride (5.13), NaHCO$_3$ is required in addition to K$_2$CO$_3$ in the AD-mix. This suppresses hydrolysis of the starting material and ring closure of the product. In the case of allylic sulfides, such as compound (5.14), the alkene is selectively oxidised in preference to the sulfur.

Allylic alcohols can also be dihydroxylated with good enantioselectivity, although the corresponding ethers or esters generally give higher enantioselectivity.[9] Thus, the allylic alcohol (5.15) undergoes dihydroxylation, with good enantioselectivity, but the corresponding benzoate ester yields a product with 99% ee. Tertiary allylic alcohols can also be used as substrates for asymmetric dihydroxylation.[10] Corey and co-workers used a ligand closely related to the (DHQP)$_2$-PHAL ligand, where the linker is a pyridazine unit (the benzene ring is not included).[11] In their examples, they found that p-methoxybenzoate derivatives of allylic alcohols and p-methoxyphenyl ether derivatives of homoallylic alcohols were the most suitable.

Dienes are interesting substrates for the dihydroxylation reaction. In cases where the diene is symmetrical, the enediols are isolated in good yield and enantioselectivity.[12] For example, diene (5.19) yields the enediol (5.22).

(5.13) → (5.16) 98% ee

1 equiv MeSO$_2$NH$_2$, 3 equiv NaHCO$_3$, AD-mix-β, r.t., ^{t}BuOH:H$_2$O 1:1, 80%

(5.14) → (5.17) 98% ee

1 equiv MeSO$_2$NH$_2$, AD-mix-β, 0°C, ^{t}BuOH:H$_2$O 1:1, 75%

(5.15) (5.18) 93% ee

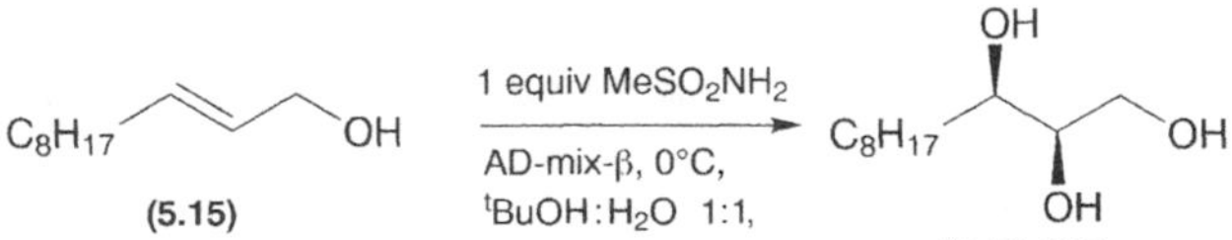

Similarly, enynes, including substrate **(5.20)**, are selectively oxidised in the alkene moiety to give products which have been shown to be synthetically useful.[13] However, for nonsymmetrical dienes, there is the added issue of regioselectivity. The reactions tend to proceed to retain any conjugation (as in product **5.24**), although other electronic effects are also operative in some cases.[14]

(5.19) (5.22) 99% ee

(5.20) (5.23) 97% ee

(5.21) (5.24) 92% ee

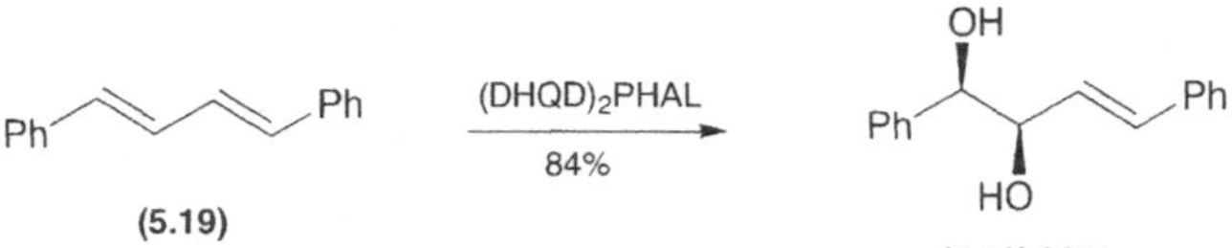

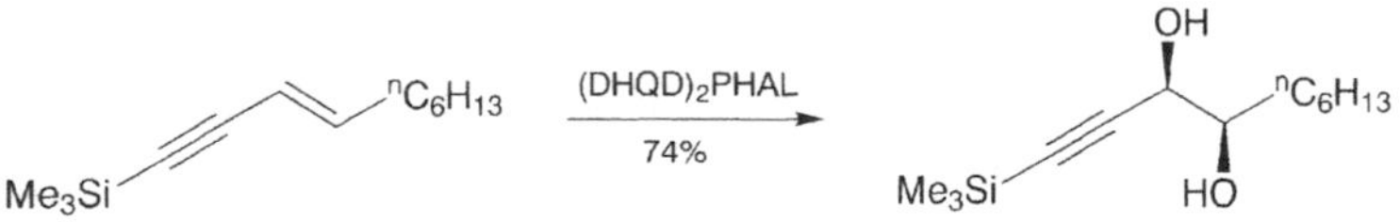

β,γ-and γ,δ-Unsaturated esters give rise to hydroxy lactones under the Sharpless AD reaction conditions, where the diol closes selectively to provide γ-lactones.[15] The enantiomeric excess can be excellent, as with the unsaturated esters **(5.25)** and **(5.26)**, which produce the γ-lactones **(5.27)** and **(5.28)**. A similar approach has also been used with vinyl silanes which possess a pendant ester group.[16]

(5.25) (5.27) 99% ee

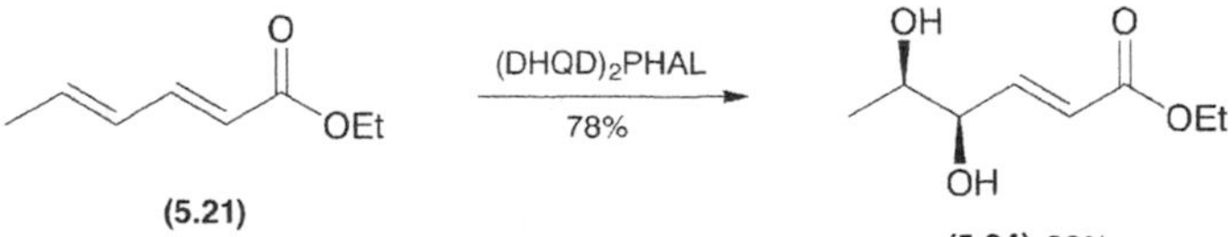

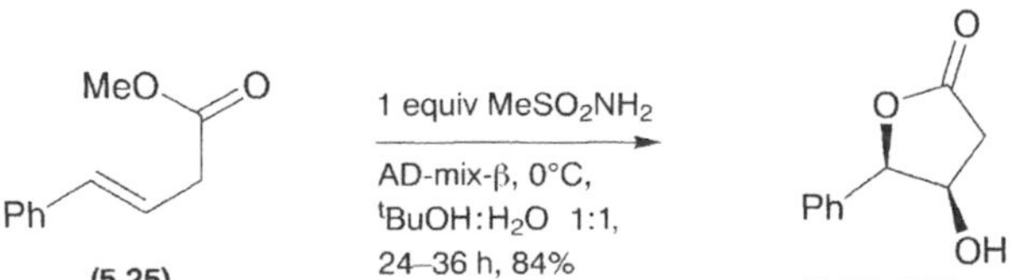

(5.26) → (5.28) 96% ee

1 equiv MeSO$_2$NH$_2$
AD-mix-β, 0°C,
^{t}BuOH:H$_2$O 1:1,
24–36 h, 84%

Like disubstituted alkenes, trisubstituted alkenes are good substrates for the AD reaction, and the (DHQD)$_2$-PHAL and (DHQ)$_2$PHAL ligands generally give the best selectivities.[17] The acyclic trisubstituted alkene (5.29) and the cyclic trisubstituted alkene (5.30) both yield the corresponding diols, with very high enantioselectivity. Enol ethers represent an interesting class of alkene, since the isolated products are α-hydroxy ketones.[18] (E)-Enol ethers, such as compound (5.31), tend not to give selectivities as high as do (Z)-enol ethers, such as compound (5.32). However, both geometries of a given enol ether substrate preferentially provide the same enantiomer of α-hydroxy ketone.

(5.29)

1 equiv MeSO$_2$NH$_2$
AD-mix-β, 24–36 h,
0°C, ^{t}BuOH:H$_2$O 1:1

(5.33) 98% ee

(5.30)

1 equiv MeSO$_2$NH$_2$
AD-mix-β, 24–36 h,
0°C, ^{t}BuOH:H$_2$O 1:1

(5.34) 97% ee

(5.31)

1 equiv MeSO$_2$NH$_2$
AD-mix-β, 24–36 h,
0°C, ^{t}BuOH:H$_2$O 1:1

(5.35) 90% ee

(E)/(Z) 4:96
(5.32)

1 equiv MeSO$_2$NH$_2$
AD-mix-β, 24–36 h,
0°C, ^{t}BuOH:H$_2$O 1:1

(5.36) 95% ee

For monosubstituted and tetrasubstituted alkenes, enantioselectivities are somewhat more variable. In some, but not all cases, the alternative pyrimidine ligands (DHQD)$_2$PYR (5.37) and (DHQ)$_2$PYR (5.38) provide superior results.[19] Some examples, where the pyrimidine ligands give better results then the phthalazine ligands, are provided by the dihydroxylation of monosubstituted alkenes (5.39) and (5.40), as well as the enol ether (5.41). For tetrasubstituted alkenes other than enol ethers, enantioselectivities tend to be reduced and isolated yields also suffer.[20] (Z)-Disubstituted alkenes are not usually good substrates for the Sharpless AD reaction, and although alternative ligands have been used,[21, 22] enantioselectivities are moderate, with only a few exceptions.

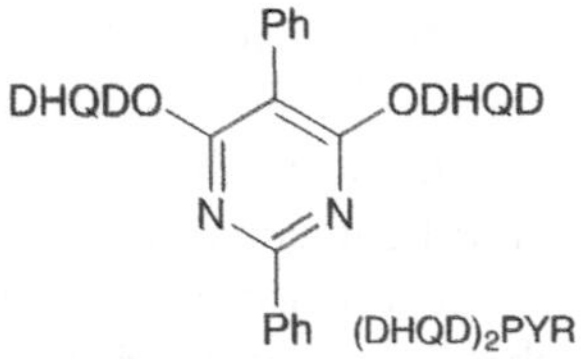

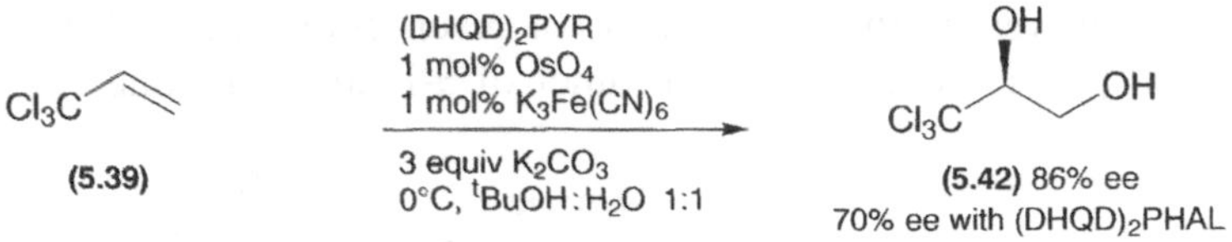

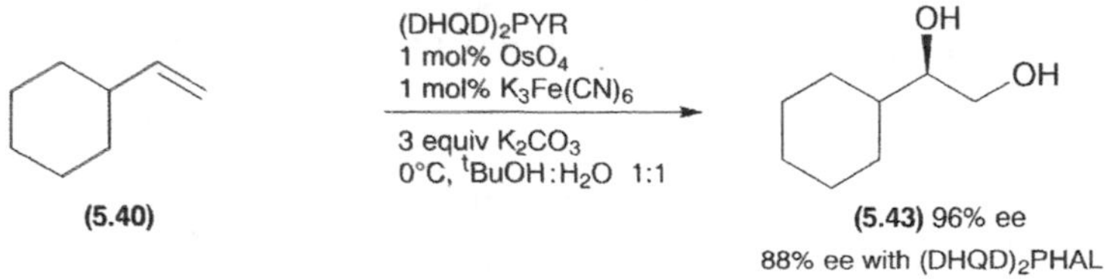

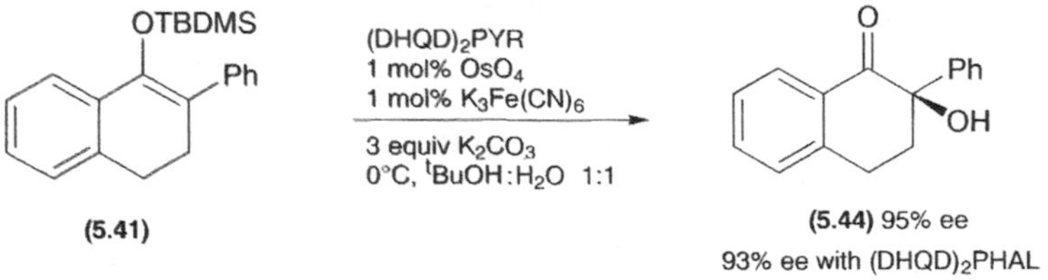

In the Sharpless asymmetric dihydroxylation of alkenes which already contain a stereocentre, the possibility of matched and mismatched selectivity arises. There have been several reports of this phenomenon, illustrated by the work of Cha and co-workers,[23] who showed that in the absence of a ligand, the alkene (5.45) has a slight preference for formation of diastereomer (5.46) in the dihydroxylation reaction. The use of (DHQD)₂PHAL (5.03) reinforces this preference (matched), whereas the (DHQ)₂PHAL (5.04) ligand represents a mismatched case and overturns the underlying selectivity, giving rise to diastereomer (5.47) as the major product.

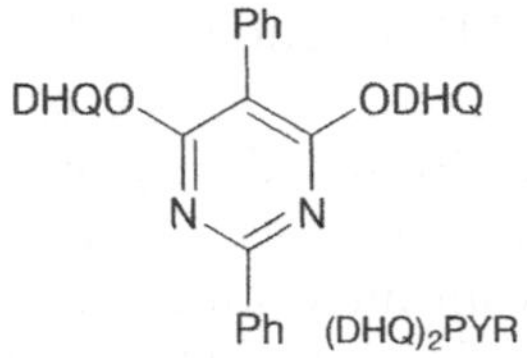

In fact, the rate at which one enantiomer of a substrate reacts may be different from the other, and therefore a kinetic resolution of racemic substrates is also possible with the AD reaction. The kinetic resolution reactions have generally given good, but not exceptional, discrimination. Some of the more selective examples of kinetic resolution include the axially chiral alkene (5.48)[24] and allylic acetate (5.49).[25] In an unusual example of a kinetic resolution, Hawkins and Meyer performed the AD reaction of the racemic buckyball, C_{76}.[26] This procedure yields an enantiomerically-enriched element!

	(5.46) : (5.47)
No ligand	2:1
(DHQD)$_2$	>20:1
(DHQ)$_2$PHAL	1:10

using (DHQD)$_2$-PHAL

(5.48) K$_{rel}$ = 32

using

(5.49) K$_{rel}$ = 25

In this section, many different asymmetric dihydroxylation reactions have been described. Sharpless uses an empirical mnemonic device to predict the sense of asymmetric induction in the product, as shown in Figure 5.1.[27] The south east (SE) quadrant presents a steric barrier, preferring the smallest substituent (usually a hydrogen atom). The north west (NW) quadrant also presents a steric barrier but a smaller one than in the south east. It is clear that (E)-disubstituted alkenes fit this mnemonic very well, since the substituents can fit interchangeably in the available south west (SW) and north east (NE) quadrants. The NE quadrant is available for moderately-sized substituents, whilst the SW quadrant is considered an attractive area, with a preference for flat aromatic substituents. Aliphatic substituents are also accepted in the SW quadrant but ROCH$_2$ and methyl groups do not favour this region.[28]

Not surprisingly, there has been considerable debate over the mechanism of a reaction as important as the AD process. The two main

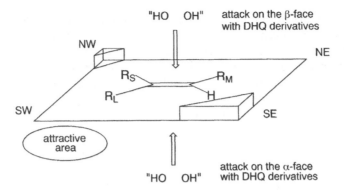

Figure 5.1 Predictive model for the asymmetric dihydroxylation reaction. Abbreviation: DHQ = dihydroquinine.

pathways are either concerted [3 + 2] cycloaddition of the osmium tetroxide with alkene, or [2 + 2] cycloaddition followed by subsequent conversion into the osmium (VI) glycolate (**5.51**), as identified in Figure 5.2. It seems reasonable that the ligand adds before the cycloaddition, and Lohray and co-workers have indicated that this is the case.[29] Sharpless and Gobel presented evidence that a nonlinear relationship between ee and temperature is not consistent with a one-step mechanism, suggesting that the [2 + 2] mechanism is more likely.[30] Once the osmium (VI) glycolate (**5.51**) is formed, the catalytic cycle is completed by reoxidation with the stoichiometric oxidant. Water is required to liberate the diol from the osmium.

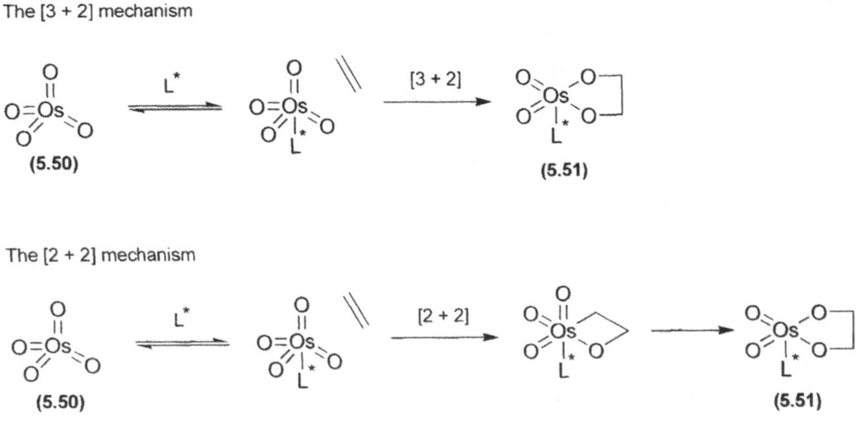

The [3 + 2] mechanism

O=Os (5.50) ⇌ L* → O=Os L ‖ [3 + 2] → Os (5.51)

The [2 + 2] mechanism

O=Os (5.50) ⇌ L* → O=Os L ‖ [2 + 2] → Os L → Os (5.51)

Figure 5.2 Possible mechanisms of dihydroxylation.

Molecular modelling,[31] as well as construction of modified ligands,[32, 33] have been used in an effort to understand the origins of enantioselectivity in AD reactions.

5.2 Aminohydroxylation

In 1996, Sharpless and co-workers reported an exciting development of the AD reaction, which was a modification to give asymmetric aminohydroxylation.[34, 35] The first modification reported by Sharpless and co-workers employed chloramine-T-trihydrate (TsNClNa.$3H_2O$) as the amino alcohol source. Subsequently, alternative reagents have been used, giving rise to amino alcohol derivatives with high enantioselectivity. N-halocarbonates,[36] chloramine-M (MeSO$_2$NClNa)[37] and N-bromo-acetamide (+base),[38] have all been used to give high enantioselectivity in the asymmetric aminohydroxylation (AA) reaction.

Typical examples include the asymmetric aminohydroxylation (AA) of alkenes (5.52–5.54), all with excellent enantioselectivity. The AA reaction has also been applied to the dienylsilane (5.55) by Landais and co-workers.[39] Whilst the enantioselectivity is not perfect, the reaction is still remarkably regio- and diastereoselective.

The aminohydroxylation of styrenes gives primarily the benzylic amino derivative as the major regioisomer of product. Thus, p-bromostyrene

4 mol% K$_2$OsO$_2$(OH)$_4$
5 mol% (DHQ)$_2$PHAL
3 equiv TsNClNa.$3H_2O$
3 h, r.t., CH$_3$CN:H$_2$O 1:1,
66%

(5.52)

(5.56) 81% ee
71% ee with (DHQD)$_2$PHAL

MeSO$_2$NClNa
4 mol% K$_2$OsO$_2$(OH)$_4$
5 mol% (DHQD)$_2$PHAL
3 h, 0°C
n-PrOH/H$_2$O,

(5.52)

(5.57) 95% ee

1.1 equiv Me—C(O)—NHBr
1.02 equiv LiOH.H$_2$O
4 mol% K$_2$OsO$_2$(OH)$_4$
5 mol% (DHQD)$_2$PHAL
3 h, 4°C
tBuOH:H$_2$O 1:1,
64%

(5.53)

(5.58) 99% ee

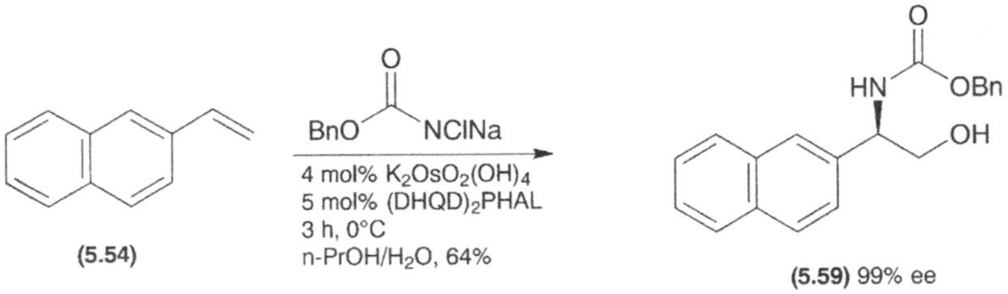

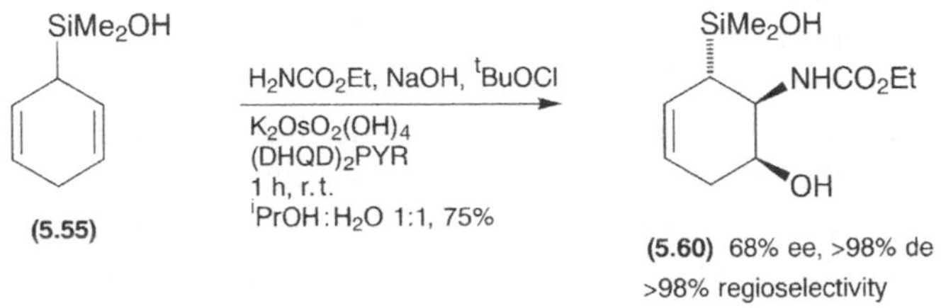

(5.60) yields the regioisomeric products (5.61) and (5.62) in an 80:20 ratio. These products can be converted into diamines[40] and amino acids.[41] Interestingly, the amino acid synthesis is effective on the mixture of regioisomers, since only the primary alcohols can be converted into the acid, which is readily separated during work-up. In this way, the protected *p*-bromophenylglycine (5.63) can be prepared in a straightforward manner. The use of the alternative ligand, (DHQD)₂AQN (5.64), can overturn the normal regiochemistry of these reactions. The use of this ligand in the aminohydroxylation of cinnamates has also been reported.[42] For example, cinnamate (5.52) gives rise to the amino alcohols (5.65) and (5.66), with the benzylic alcohol regioisomer (5.65) predominating when (DHQD)₂AQN (5.64) is employed as the ligand.

Thus, the opposite regiochemistry but the same sense of enantioselectivity is obtained using the AQN ligands rather than the PHAL ligands. No doubt, further developments in the asymmetric aminohydroxylation reaction will be forthcoming.

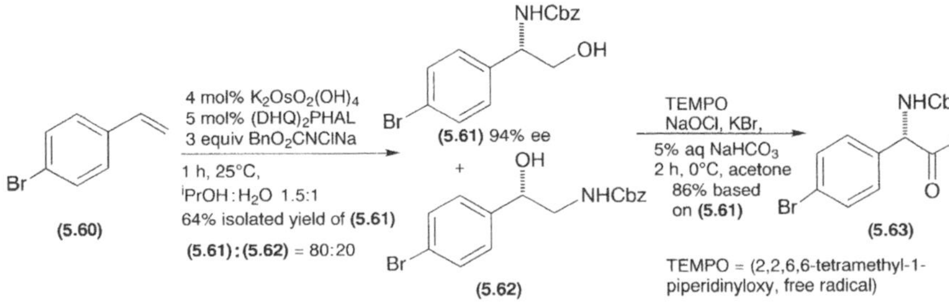

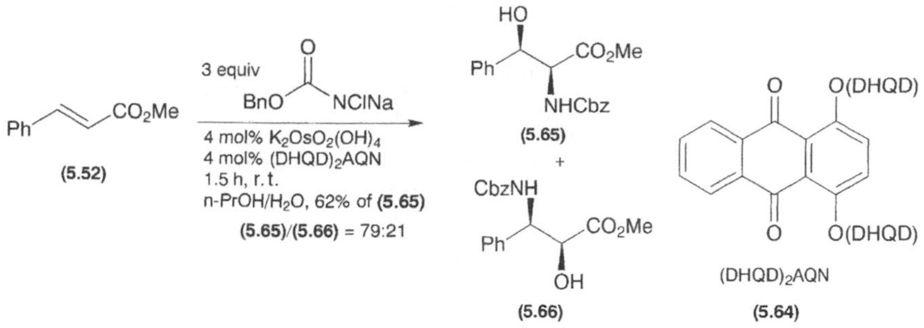

5.3 Oxidation of C—H

The direct oxidation of unfunctionalised alkanes in an asymmetric fashion is a formidable challenge. However, oxidation of C—H adjacent to suitable functional groups provides a handle on which to operate. In particular, the allylic oxidation of cyclic alkenes has received considerable attention since 1995, when several groups reported an asymmetric variant of the Kharasch reaction.[43, 44, 45, 46, 47, 48] The reaction is catalysed by copper salts and requires a perester to give the allylic ester as product. The ligands employed, so far, have fallen into two main categories: those based on proline;[49] and those based on oxazolines.[50,51]

A few specific examples are given here, using ligands (5.67–5.70). Cycloalkenes (5.71–5.73) are particularly good substrates, although the enantioselectivities have been good but not excellent. Propargylic oxidation of the acyclic alkyne (5.74) can be achieved, with moderate enantioselectivity. It is noteworthy that, in this case, the quoted yield is based on the starting material (rather than being based on the perester).[52]

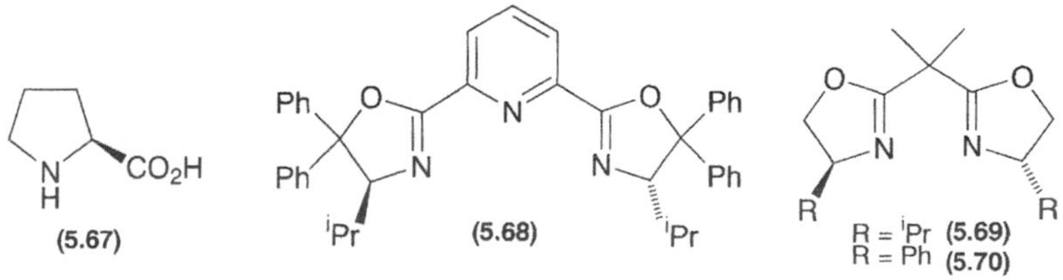

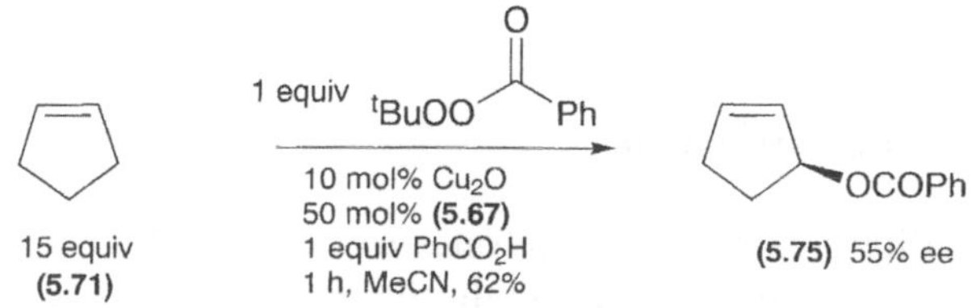

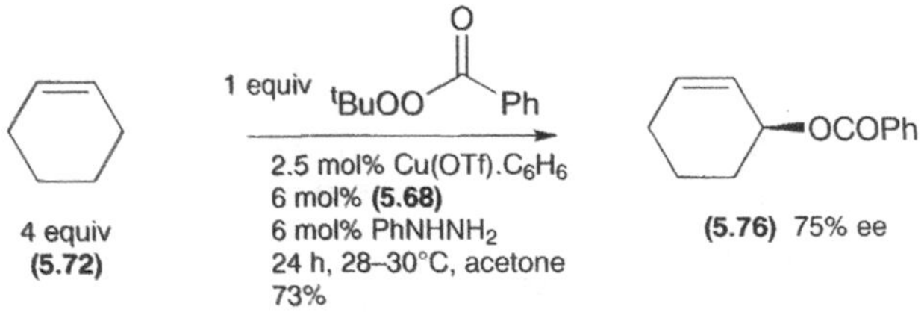

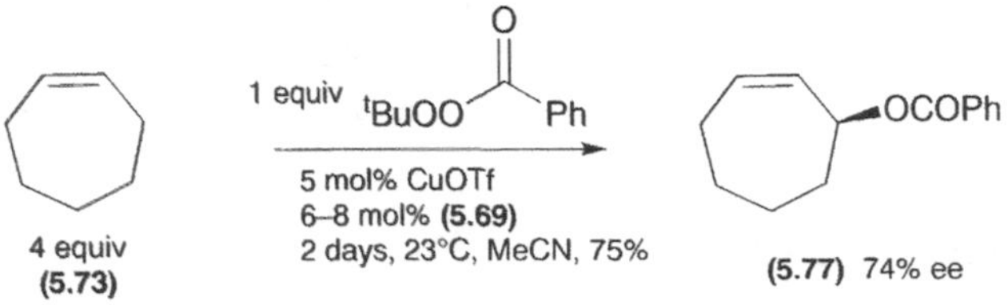

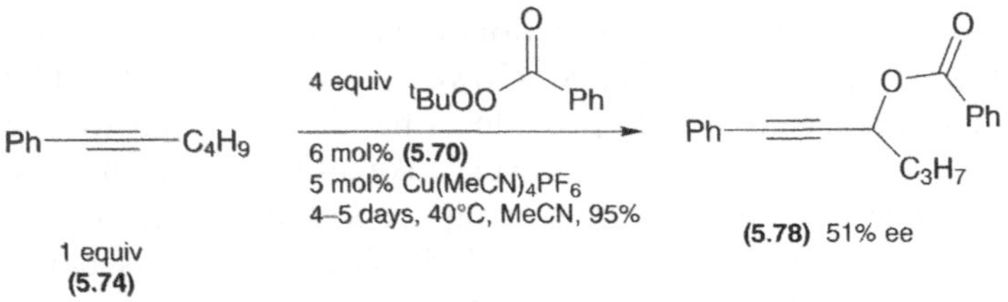

Benzylic oxidations using vaulted binaphthyl iron(III) porphyrins,[53] and (salen)manganese (III) complexes[54] have suffered from moderate yields and/or enantioselectivies. However, Miyafuji and Katsuki obtained reasonable yield and good enantioselectivity for C$-$H oxidation adjacent to an ether in substrate (5.79), using manganese(salen) complex (5.80).[55]

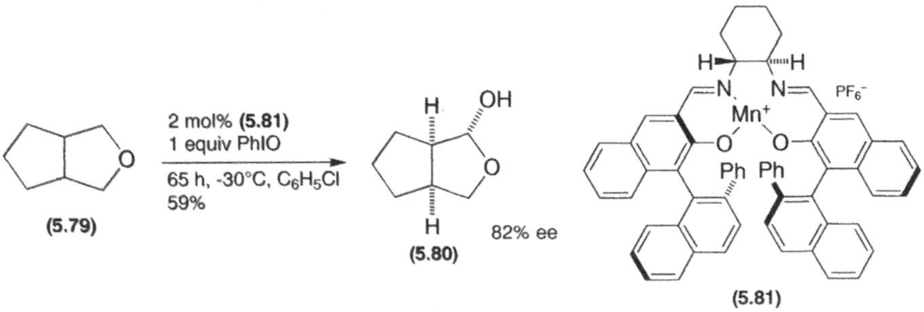

5.4 Baeyer-Villiger oxidation

The Baeyer-Villiger oxidation of ketones to esters can be achieved using transition metal catalysts,[56] although the enantioselective variant has not been widely studied. Strukul and co-workers reported the use of enantiomerically pure platinum complexes in the conversion of cyclic ketones into lactones by kinetic resolution (up to 58% ee).[57] The copper complex (5.81) has been reported to effect similar kinetic resolutions (up to 69% ee),[58] but in the case of bicyclic cyclobutanones higher selectivities are observed.[59] One enantiomer of ketone substrate (5.82) is converted into the 'normal' lactone (5.83), whereas the other enantiomer is converted in the 'abnormal' lactone (5.84). Baeyer-Villiger oxidations by biocatalysis can also be particularly effective for asymmetric induction.[60]

(5.81)

5.5 Oxidation of sulfides

The asymmetric oxidation of sulfides has been successfully achieved using biotransformations.[61, 62] However, a detailed discussion of these reactions is beyond the scope of the present book.

Kagan and Pitchen[63] and Modena and co-workers[64] independently reported the oxidation of sulfides to sulfoxides using modified Sharpless epoxidation catalyst (titanium/diethyl tartrate). By 1987,[65] Kagan had already reported a catalytic variation of the reaction. However, a more recent publication describes an improved catalytic system, which allows for the use of lower (10 mol%) loading of catalyst.[66] Sulfides (5.85–5.87) undergo sulfoxidation, with fair to very good enantioselectivity. An alternative catalyst based on $Ti(OiPr)_4$ and BINOL is also effective for sulfoxidation, providing up to 96% ee.[67, 68] Jacobsen and co-workers [69] and Katsuki and co-workers[70, 71] have both reported the use of (salen)manganese(III) catalysts for sulfide oxidation. These catalysts can be effective for the enantioselective oxidation of several arylmethyl-sulfides, using iodosylbenzene as the stoichiometric oxidant. The vana-dium-catalysed oxidation of disulfides has also been reported recently.[72]

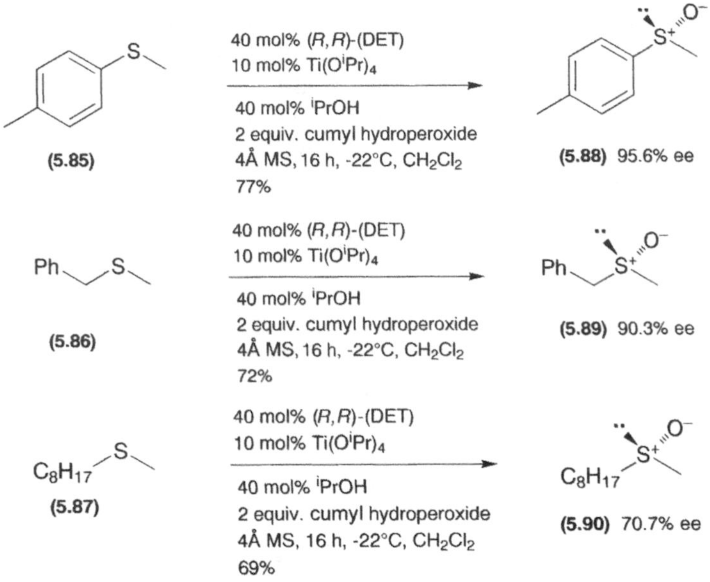

Sulfides can usually be selectively oxidised to sulfoxides without (too much) over-oxidation to the corresponding sulfones. However, the conversion of sulfoxides into sulfones can be achieved under relatively mild catalytic conditions, if required.

Uemura and co-workers demonstrated a kinetic resolution of sulfoxides by catalysted oxidation.[73] For example, the oxidation of racemic phenylmethylsulfoxide (5.91) yields the sulfone (5.92), but the unreacted starting material is recovered with very high enantiomeric excess.

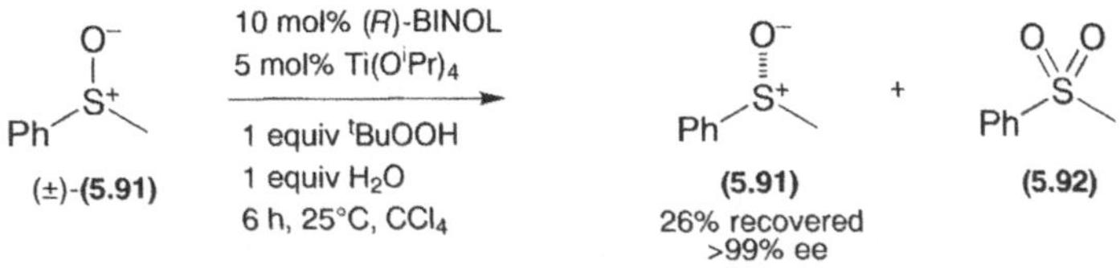

References

1. E.N. Jacobsen, I. Markó, W.S. Mungall, G. Schröder and K.B. Sharpless, *J. Am. Chem. Soc.*, **1988**, *110*, 1968.
2. B.B. Lohray, *Tetrahedron: Asymmetry*, **1992**, *3*, 1317. (b) R.A. Johnson and K.B. Sharpless, in *Catalytic Asymmetric Synthesis*, (I. Ojima, ed.) VCH, New York, **1993**, 227. (c) H.C.

Kolb, M.S. VanNieuwenhze and K.B. Sharpless, *Chem. Rev.*, **1994**, *94*, 2483. (d) D.J. Berrisford, C. Bolm and K.B. Sharpless, *Angew. Chem., Int. Ed. Engl.*, **1995**, *34*, 1059.

3. K.B. Sharpless, W. Amberg, M. Beller, H. Chen, J. Hartung, Y. Kawanami, D. Lübben, E. Manoury, Y. Ogino, T. Shibata and T. Ukita, *J. Org. Chem.*, **1991**, *56*, 4585.
4. K.B. Sharpless, W. Amberg, Y.L. Bennani, G.A. Crispino, J. Hartung, K.-S. Jeong, H.-L. Kwong, K. Morikawa, Z.-M. Wang, D. Xu and X.-L. Zhang, *J. Org. Chem.*, **1992**, *57*, 2768.
5. C. Bolm and A. Gerlach, *Eur. J. Org. Chem.*, **1998**, 21.
6. Z.-M. Wang and K.B. Sharpless, *J. Org. Chem.*, **1994**, *59*, 8302.
7. K.P.M. Vanhessche, Z-M. Wang and K.B. Sharpless, *Tetrahedron Lett.*, **1994**, *35*, 3469.
8. P.J. Walsh, P.T. Ho, S.B. King and K.B. Sharpless, *Tetrahedron Lett.*, **1994**, *35*, 5129.
9. D. Xu, C.Y. Park and K.B. Sharpless, *Tetrahedron Lett.*, **1994**, *35*, 2495.
10. Z.-M. Wang and K.B. Sharpless, *Tetrahedron Lett.*, **1993**, *34*, 8225.
11. E.J. Corey, A. Guzman-Perez and M.C. Noe, *Tetrahedron Lett.*, **1995**, *36*, 3481.
12. D. Xu, G.A. Crispino and K.B. Sharpless, *J. Am. Chem. Soc.*, **1992**, *114*, 7570.
13. K. Tani, Y. Sato, S. Okamoto and F. Sato, *Tetrahedron Lett.*, **1993**, *34*, 4975.
14. H. Becker, M.A. Soler and K.B. Sharpless, *Tetrahedron*, **1995**, *51*, 1345.
15. Z.-M. Wang, X.-L. Zhang, K.B. Sharpless, S.C. Sinha, A. Sinha-Bagchi and E. Keinan, *Tetrahedron Lett.*, **1992**, *33*, 6407.
16. Y. Miyazaki, H. Hotta and F. Sato, *Tetrahedron Lett.*, **1994**, *35*, 4389.
17. H. Becker, P.T. Ho, H.C. Kolb, S. Loren, P.-O. Norrby and K.B. Sharpless, *Tetrahedron Lett.*, **1994**, *35*, 7315.
18. T. Hashiyama, K. Morikawa and K.B. Sharpless, *J. Org. Chem.*, **1992**, *57*, 5067.
19. G.A. Crispino, K.-S. Jeong, H.C. Kolb, Z.-M. Wang, D. Xu and K.B. Sharpless, *J. Org. Chem.*, **1993**, *58*, 3785.
20. K. Morikawa, J. Park, P.G. Andersson, T. Hashiyama and K.B. Sharpless, *J. Am. Chem. Soc.*, **1993**, *115*, 8463.
21. L. Wang and K.B. Sharpless, *J. Am. Chem. Soc.*, **1992**, *114*, 7568.
22. L. Wang, K. Kakiuchi and K.B. Sharpless, *J. Org. Chem.*, **1994**, *59*, 6895.
23. N.-S. Kim, J.-R. Choi and J.K. Cha, *J. Org. Chem.*, **1993**, *58*, 7096.
24. M.S. VanNieuwenhze and K.B. Sharpless, *J. Am. Chem. Soc.*, **1993**, *115*, 7864.
25. B.B. Lohray and V. Bhushan, *Tetrahedron Lett.*, **1993**, *34*, 3911.
26. J.M. Hawkins and A. Meyer, *Science*, **1993**, *260*, 1918.
27. H.C. Kolb, P.G. Andersson and K.B. Sharpless, *J. Am. Chem. Soc.*, **1994**, *116*, 1278.
28. K.J. Hale, S. Manaviazar and S.A. Peak, *Tetrahedron Lett.*, **1994**, *35*, 425.
29. B.B. Lohray, V. Bhushan and E. Nandanan, *Tetrahedron Lett.*, **1994**, *35*, 4209.
30. T. Gobel and K.B. Sharpless, *Angew. Chem., Int. Ed. Engl.*, **1993**, *32*, 1329.
31. P.-O. Norrby, H.C. Kolb and K.B. Sharpless, *J. Am. Chem. Soc.*, **1994**, *116*, 8470.
32. E.J. Corey and M.C. Noe, *J. Am. Chem. Soc.*, **1993**, *15*, 12579.
33. E.J. Corey, M.C. Noe and M.J. Grogan, *Tetrahedron Lett.*, **1994**, *35*, 6427.
34. G. Li, H.-T. Chang and K.B. Sharpless, *Angew. Chem., Int. Ed. Engl.*, **1996**, *35*, 451.
35. O. Reiser, *Angew. Chem., Int. Ed. Engl.*, **1996**, *35*, 1308.
36. (a) G. Li, H.H. Angert and K.B. Sharpless, *Angew. Chem., Int. Ed. Engl.*, **1996**, *35*, 2813. (b) K.L. Reddy, K.R. Dress and K.B. Sharpless, *Tetrahedron Lett.*, **1998**, *39*, 3667.
37. J. Rudolph, P.C. Sennhenn, C.P. Vlaar and K.B. Sharpless, *Angew. Chem., Int. Ed. Engl.*, **1996**, *35*, 2810.
38. M. Bruncko, G. Schlingloff and K.B. Sharpless, *Angew. Chem., Int. Ed. Engl.*, **1997**, *36*, 1483.
39. R. Angelaud, Y. Landais and K. Schenk, *Tetrahedron Lett.*, **1997**, *38*, 1407.
40. P. O'Brien, S.A. Osborne and D.D. Parker, *Tetrahedron Lett.*, **1998**, *39*, 4099.
41. K.L. Reddy and K.B. Sharpless, *J. Am. Chem. Soc.*, **1998**, *120*, 1207.

42. B. Tao, G. Schlingloff and K.B. Sharpless, *Tetrahedron Lett.*, **1998**, *39*, 2507.
43. A. Levina and J. Muzart, *Tetrahedron: Asymmetry*, **1995**, *6*, 147.
44. A. Levina and J. Muzart, *Synth. Commun.*, **1995**, *25*, 1789.
45. M.T. Rispens, C. Zondervan and B.L. Feringa, *Tetrahedron: Asymmetry*, **1995**, *6*, 661.
46. A.S. Gokhale, A.B.E. Minidis and A. Pfaltz, *Tetrahedron Lett.*, **1995**, *36*, 1831.
47. M.B. Andrus, A.B. Argade, X. Chen and M.G. Pamment, *Tetrahedron Lett.*, **1995**, *36*, 2945.
48. K. Kawasaki, S. Tsumura and T. Katsuki, *Synlett*, **1995**, 1245.
49. M.K. Södergren and P.G. Andersson, *Tetrahedron Lett.*, **1996**, *37*, 7577.
50. G. Sekar, A. DattaGupta and V.K. Singh, *J. Org. Chem.*, **1998**, *63*, 2961.
51. J.S. Clark, K.F. Tolhurst, M. Taylor and S. Swallow, *J. Chem. Soc., Perkin Trans 1.*, **1998**, 1167.
52. J.S. Clark, K.F. Tolhurst, M. Taylor and S. Swallow, *Tetrahedron Lett.*, **1998**, *39*, 4913.
53. J.T. Groves and P. Viski, *J. Org. Chem.*, **1990**, *55*, 3628.
54. K. Hamachi, R. Irie and T. Katsuki, *Tetrahedron Lett.*, **1996**, *37*, 4979.
55. A. Miyafuji and T. Katsuki, *Synlett*, **1997**, 836.
56. For a review, see: G. Strukul, *Angew. Chem., Int. Ed. Engl.*, **1998**, *37*, 1198.
57. A. Gusso, C. Baccin, F. Pinna and G. Strukul, *Organometallics*, **1994**, *13*, 3442.
58. C. Bolm, G. Schlingloff and K. Weickhardt, *Angew. Chem., Int. Ed. Engl.*, **1994**, *33*, 1848–1849.
59. C. Bolm and G. Schlingloff, *J. Chem. Soc., Chem. Commun.*, **1995**, 1247.
60. C.-H. Wong and G.M. Whitesides, in *Enzymes in Synthetic Organic Chemistry*, Tetrahedron Organic Chemistry Series, Volume 12, Pergamon, Oxford, **1994**, p. 169.
61. H.L. Holland, *Chem. Rev.*, **1988**, *88*, 473.
62. G. Ottolina, P. Pasta, G. Carrea, S. Colonna, S. Dallavalle and H.L. Holland, *Tetrahedron: Asymmetry*, **1995**, *6*, 1375.
63. P. Pitchen and H.B. Kagan, *Tetrahedron Lett.*, **1984**, *25*, 1049. See also: J.M. Brunel, P. Diter, M. Duetsch and H.B. Kagan, *J. Org. Chem.*, **1995**, *60*, 8086.
64. S.H. Di Furia, G. Modena and G. Seraglia, *Synthesis*, **1984**, 325.
65. S. Zhao, O. Samuel and H.B. Kagan, *Tetrahedron*, **1987**, *43*, 5135.
66. J.M. Brunel and H.B. Kagan, *Synlett*, **1995**, 404.
67. N. Komatsu, Y. Nishibayashi, T. Sugita and S. Uemura, *Tetrahedron Lett.*, **1992**, *33*, 5391.
68. N. Komatsu, M. Hashizume, T. Sugita and S. Uemura, *J. Org. Chem.*, **1993**, *58*, 4529.
69. M. Palucki, P. Hanson and E.N. Jacobsen, *Tetrahedron Lett.*, **1992**, *33*, 7111.
70. K. Noda, N. Hosoya, K. Yanai, R. Irie and T. Katsuki, *Tetrahedron Lett.*, **1994**, *35*, 1887.
71. K. Noda, N. Hosoya, R. Irie, Y. Yamashita and T. Katsuki, *Tetrahedron*, **1994**, *50*, 9609.
72. D.A. Cogan, G. Liu, K. Kim, B.J. Backes and J.A. Ellman, *J. Am. Chem. Soc.*, **1998**, *120*, 8011.
73. N. Komatsu, M. Hashizume, T. Sugita and S. Uemura, *J. Org. Chem.*, **1993**, *58*, 7624.

6 Nucleophilic addition to carbonyl compounds

The addition of nucleophiles to carbonyl groups is a fundamental process in organic synthesis. This chapter reviews the rapid advances in the asymmetric catalytic versions of this process.

6.1 Addition of dialkylzincs to carbonyl compounds

In general, rather forcing conditions are required for the alkylation of aldehydes by dialkylzinc reagents. However, the addition of small amounts of amino alcohol (or other additives) catalyses the reaction. As long ago as 1984, the use of catalytic enantiomerically pure amino alcohols as additives was shown to give an asymmetric reaction.[1] Since then, several hundred ligands, mostly amino alcohols, have been examined for their efficiency in catalysing the enantioselective addition of dialkylzinc reagents to aldehydes, and much of this work has been reviewed.[2] Even now, new ligands are frequently reported, often with high selectivities, especially for the test-bed reaction of benzaldehyde (**6.01**) reacting with diethylzinc to give 1-phenylpropanol (**6.02**).[3]

A few of these ligands are illustrated, including the camphor-derived ligand (**6.03**), which was the first one to give high enantioselectivies,[4] the norephedrine-derived amino alcohol (**6.04**),[5] and the proline-derived ligand (**6.05**).[6]

Whilst much of the work that has been published in this field focuses on the addition of diethylzinc to simple arylaldehydes, many other possibilities have been reported, as demonstrated by the examples given in the schemes. Aldehydes (**6.06–6.09**) are all converted, with good yields, into the enantiomerically-enriched alcohols (**6.10–6.13**) by catalysed addition of simple dialkylzinc reagents. The asymmetric addition to ketones, whilst less well documented, gives rise to tertiary alcohols, with good enantioselectivity. The addition of methanol was found to assist in achieving good yields.[7] The ketone (6.10) is converted into the tertiary alcohol (6.15) by reaction with diphenylzinc and an amino alcohol catalyst.

The alkylation of imines can be achieved with high selectivity; however, low reactivity is seen when the amino alcohols are used in catalytic amounts.[8]

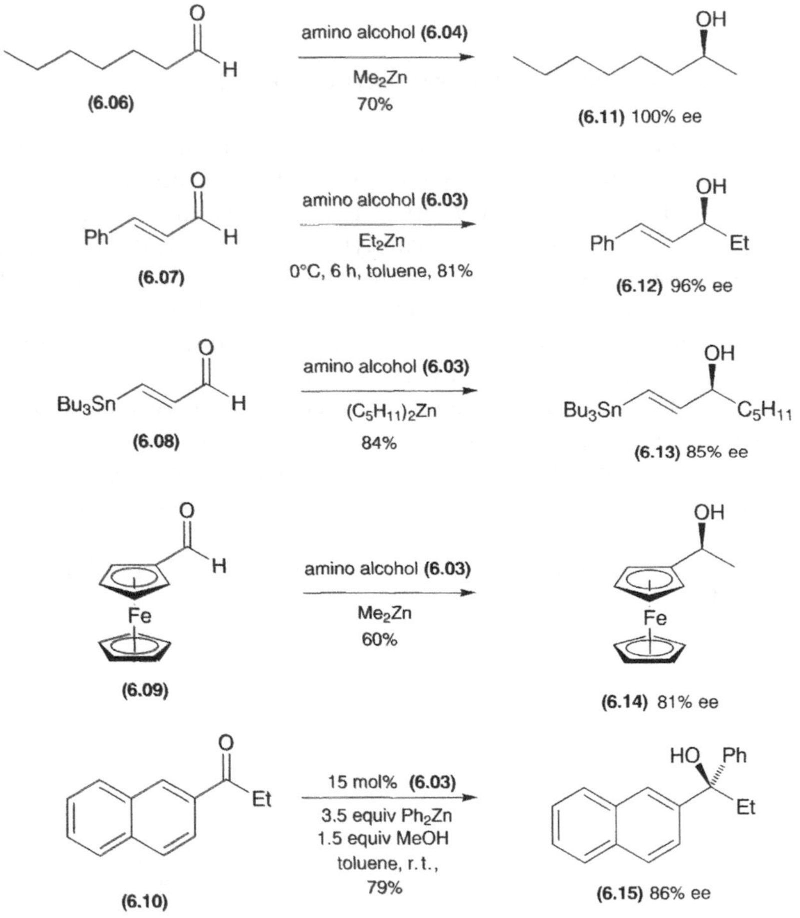

(6.06) → amino alcohol (6.04) / Me_2Zn / 70% → (6.11) 100% ee

(6.07) → amino alcohol (6.03) / Et_2Zn / 0°C, 6 h, toluene, 81% → (6.12) 96% ee

(6.08) Bu_3Sn → amino alcohol (6.03) / $(C_5H_{11})_2Zn$ / 84% → (6.13) 85% ee Bu_3Sn ... C_5H_{11}

(6.09) Fe → amino alcohol (6.03) / Me_2Zn / 60% → (6.14) 81% ee Fe

(6.10) Et → 15 mol% (6.03) / 3.5 equiv Ph_2Zn / 1.5 equiv MeOH / toluene, r.t., 79% → (6.15) 86% ee HO Ph Et

The mechanism of these reactions is generally considered to involve a key intermediate, where two zinc atoms are incorporated. In the case of the DAIB ligand (6.03), the favoured alkyl transfer in arrangement (6.16) (shown in Figure 6.1) correctly rationalises the stereochemical outcome observed.[9]

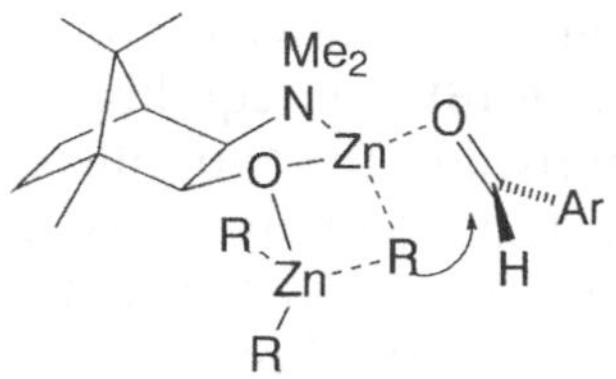

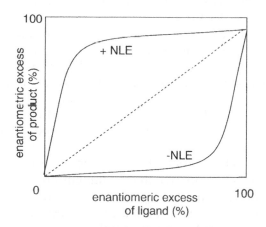

(6.16)

Figure 6.1 Orientation of alkyl addition to an aldehyde.

One of the most remarkable features of the amino alcohol-catalysed alkylation reaction is the chiral amplification observed. Thus, the alkylation of benzaldehyde with diethylzinc occurs with 95% ee, even when the ligand employed has a low enantiomeric excess (just 15% ee).[10] On first consideration, it would be reasonable to assume that the enantiomeric excess of the product would bear a direct linear correlation with the enantiomeric excess of the ligand. However, the remarkable positive linear effect can lead to high selectivities being observed in the product, with low enantiomeric excess. This is illustrated in Figure 6.2, where deviations from the expected linear relationship between enantiomeric excess of the ligand and the product are shown by positive non-linear effects (+NLE) and negative non-linear effects (-NLE).

Figure 6.2 Variation of enantiomeric excess of product and ligand. Abbreviations: +NLE, positive non-linear excess; −NLE, negative non-linear excess.

The topic of chiral amplification has been reviewed,[11] but is most well understood in the alkylation of aldehydes with dialkylzinc reagents. The origin of the remarkable positive non-linear effect is due to dimerisation of alkylzinc intermediates. In order for the reaction to occur, the aldehyde must be able to coordinate to the zinc atom of the monomer (6.18). However, the monomer (6.18) is in equilibrium with the dimer (6.17). When the dimer is comprised of the two different enantiomers of amino alcohol (heterochiral), it is particularly stable and the reaction does not proceed through it. However, when the dimer is comprised of amino alcohols of the same chirality (homochiral), it is more prone to dissociation and, therefore, the reaction pathway will preferentially use this material. For example, in the situation where the enantiomeric excess is 20%, i.e. 60% (R) and 40% (S), the more stable heterodimer will form selectively, using up 40% (R) and 40% (S). This will leave the (R)-enantiomer behind and, although it can form the homodimer, it is much more prone to dissociation to the monomer, and is therefore catalytically active.

(6.17)   (6.18)

In fact, it is not necessary for two zinc atoms to be present in the transition state in order for a selective reaction to occur. An alternative metal can be employed, including lithium salts and titanium complexes. Knochel and co-workers employed catalytic quantities of ligand (6.19) with titanium tetraisopropoxide to give highly selective alkylation reactions. In particular, they extended the range of reagents to include functionalised organozinc reagents and aldehydes.[12, 13, 14] The functionalised aldehydes (6.20–6.22) undergo reaction with the functionalised zinc reagents (6.23) and (6.24). The functionalised products obtained from these reactions are formed with good enantioselectivity. Clearly, these types of structure offer considerable synthetic opportunities.

Other ligands can be employed in the titanium-promoted version of the aldehyde addition reaction; the ligands bind to the titanium and not to zinc in the active complexes.[15] This has recently been demonstrated by an X-ray structure of related titanium complexes.[16]

(6.19)

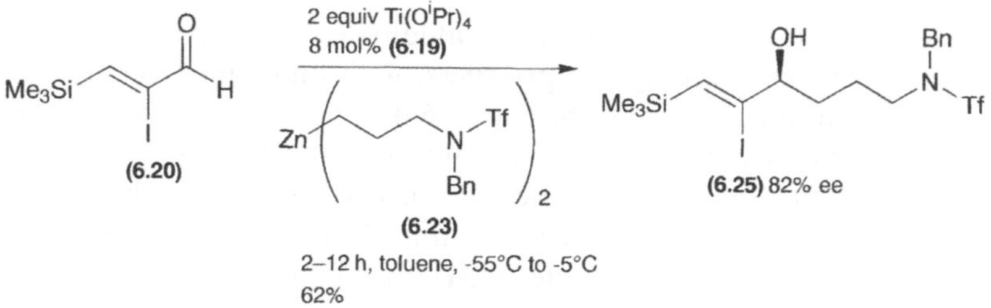

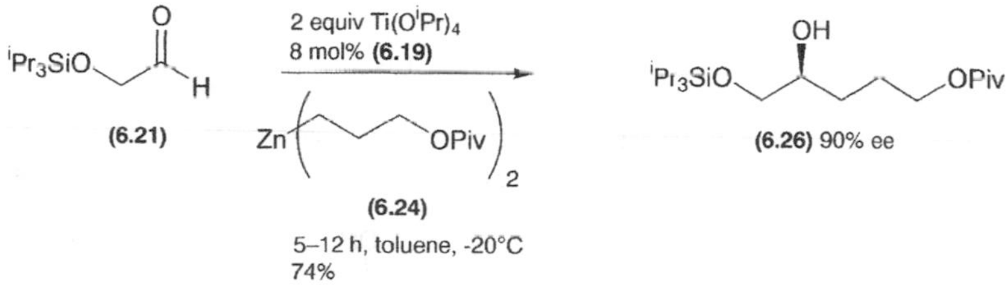

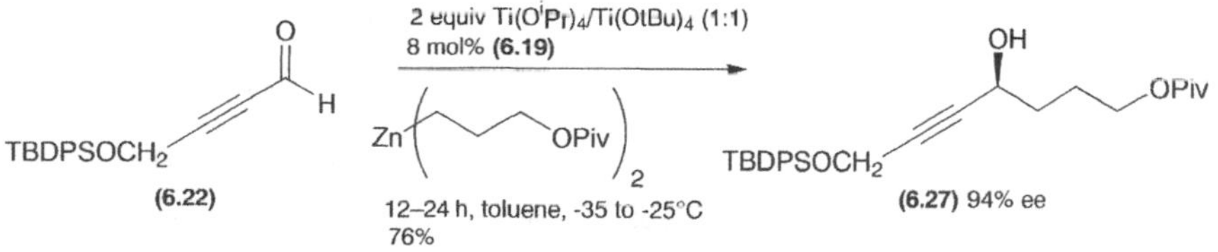

6.2 Addition of cyanide to aldehydes

The conversion of aldehydes (**6.28**) into the corresponding cyanohydrins (**6.29**) is an appealing synthetic transformation, since cyanohydrins are readily converted into α-hydroxy acids (**6.30**) and β-amino alcohols (**6.31**), and there are also other synthetic possibilities.

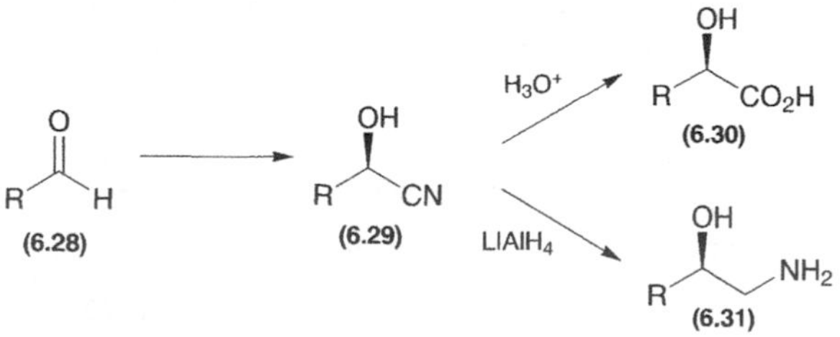

The most selective catalysts for the asymmetric transformation of aldehydes into cyanohydrins are oxynitrilase enzymes.[17] D-Oxynitrilase from almonds and the oxynitrilase isolated from sorghum give complementary enantioselectivity, as shown in the reaction of benzaldehyde (**6.01**).[18, 19] The oxynitrilase from almonds has been particularly well studied and can be used on aliphatic aldehydes[20] and even on pentan-2-one[21] to yield the cyanohydrin products, with high enantiomeric excess.

The bronchodilator, (R)-terbutaline (6.33), has recently been prepared using an enzymatic approach.[22] The stereochemistry is established in the first step in the formation of the cyanohydrin (6.35), with subsequent functional group manipulation providing (R)-terbutaline (6.33) in good overall yield, 44% from aldehyde (6.34), and with > 98% ee.

(S)-(6.32)   sorghum oxynitrilase   (6.01)   D-oxynitrilase   (R)-(6.32)
                    organic solvent   immobilised

(6.34)   D-oxynitrilase   (6.35) >98% ee   (6.33) (R)-terbutaline
                24 h, r.t., HCN/^{i}Pr$_2$O
                92%

Cyclic dipeptides containing a histidine residue are also able to catalyse the addition of hydrogen cyanide to aldehydes, with high enantioselectivity.[23, 24] In particular, cyclo-[(S)-Phe-(S)-His] (6.36) is an effective catalyst, especially for arylaldehydes (6.28) without an electron withdrawing group, although some aliphatic aldehydes also undergo reaction, with reasonable selectivity.[25] An assembly of reagents where the carbonyl group of the substrate is hydrogen-bonded to the catalyst has been suggested, and is represented by structure (6.37). The imidazole group is protonated to generate cyanide ion, which then attacks the carbonyl as indicated.

RCHO   +   HCN   2 mol% (6.36)
(6.28)   2 equiv   5–10 h, -20°C, toluene   R   CN
                                                           (6.29)

R = Ph	97% conversion	97% ee
R = 2-Naphth	61% conversion	91% ee
R = tBu	60% conversion	58% ee

(6.36)

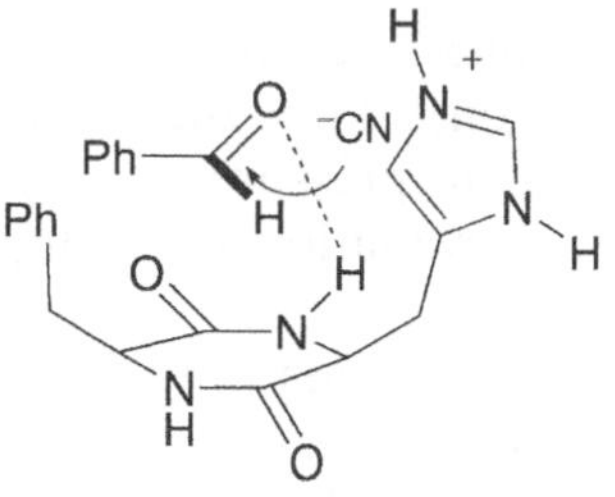

(6.37)

The Lewis acid-catalysed addition of trimethylsilyl cyanide (TMSCN) to aldehydes has been reported using several different catalysts. Titanium-based Lewis acids have proved to be particularly popular.[26, 27, 28, 29, 30] In a typical reaction, benzaldehyde (6.01) is converted into the cyanohydrin (6.02), usually after removal of the trimethylsilyl group by hydrolysis. Some of the complexes which have been used for this reaction are illustrated. These ligands, including those in complexes (6.38–6.42), typically contain oxygen and nitrogen donor atoms.

Catalyst (6.40) has been shown to be effective for a range of aldehydes, whilst catalyst (6.42) is notable for its high catalytic efficiency, since

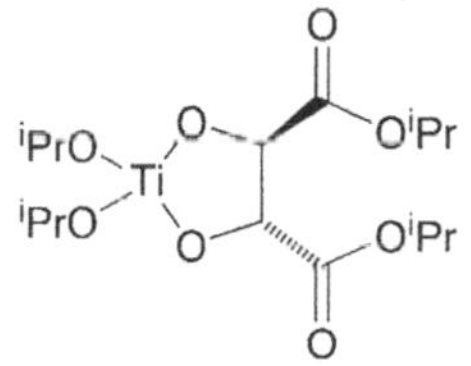

(6.38)

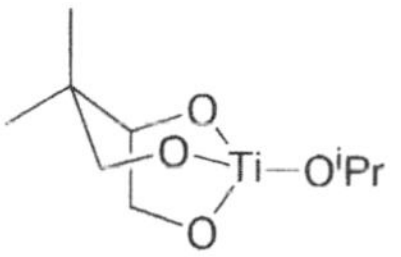

(6.39)

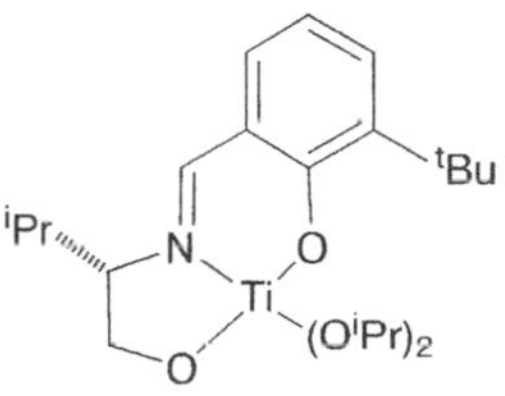

(6.40)

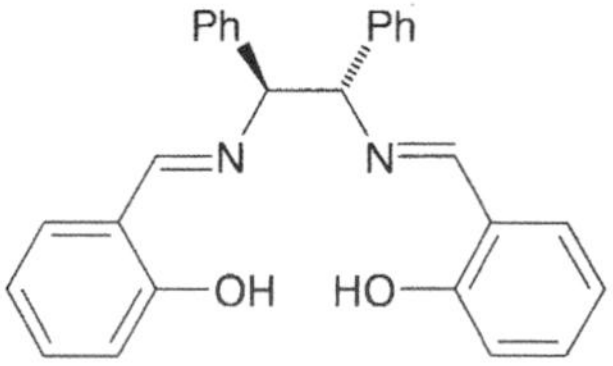

(6.41) + Ti(OiPr)$_4$

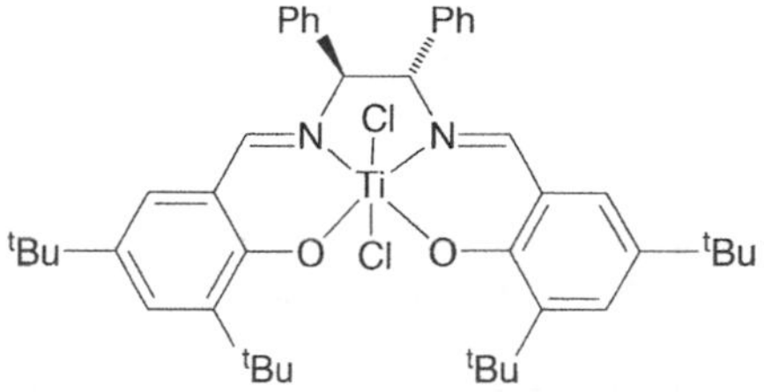

(6.42)

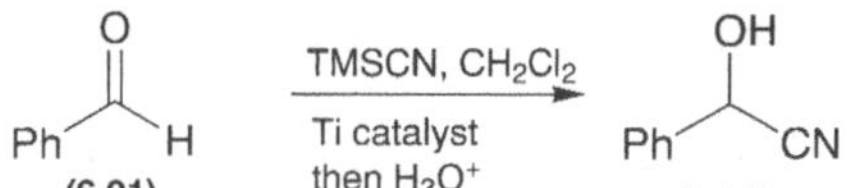

ligand	loading	yield	ee	conditions
(6.38)	20 mol%	84%	91%	0°C, 18 h
(6.39)	10 mol%	60%	38%	0°C, 5 h
(6.40)	20 mol%	67%	85%	-80°C, 36 h
(6.41)	10 mol%	72%	87%	-78°C, 24 h
(6.42)	0.1 mol%	100%	86%	r.t., 24 h

only 0.1 mol% is required. Ketones are not usually good substrates for catalysed cyanohydrin formation, since they are less reactive than aldehydes. However, under high pressure, catalyst (**6.39**) has been used to effect the trimethylsilyl cyanation of acetophenone with up to 60% ee.[31]

Corey and Wang reported an enantioselective magnesium bis-oxazoline-catalysed trimethylsilyl cyanation.[32] The reaction uses the magnesium complex (**6.43**) in addition to the bis-oxazoline (**6.44**), which is believed to provide a chiral environment for the HCN that is formed from TMSCN *in situ*. This represents an interesting concept, where the nucleophile and the electrophile are both in an asymmetric environment, and the two effects combine to afford high selectivity. The favoured catalytic pathway may proceed as shown in Figure 6.3.

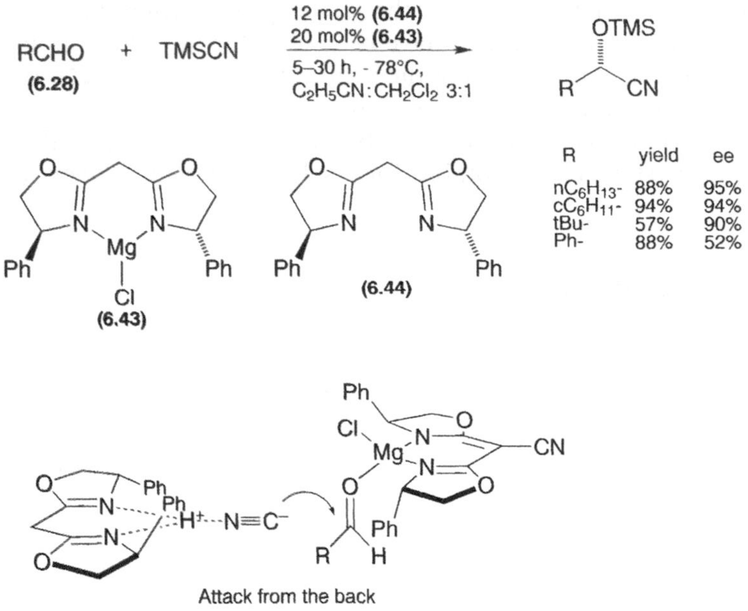

R	yield	ee
nC_6H_{13}-	88%	95%
cC_6H_{11}-	94%	94%
tBu-	57%	90%
Ph-	88%	52%

Figure 6.3 The use of two bis-oxazolines in HCN addition to aldehydes. Abbreviation: TMSCN=trimethylsilyl cyaride.

In the benzaldehyde 'test-bed' reaction, the unusual catalysts (**6.45**)[33] and (**6.46**)[34] have provided 84% ee (*R*) and 90% ee (*S*), respectively, in the cyanohydrin formation reaction, and were shown to work well over a range of electron-rich arylaldehydes. Both of these catalytic systems use fairly low catalyst loadings.

The addition of cyanide to imines forms the basis of the Strecker reaction, and can be used in the synthesis of amino acid derivatives by

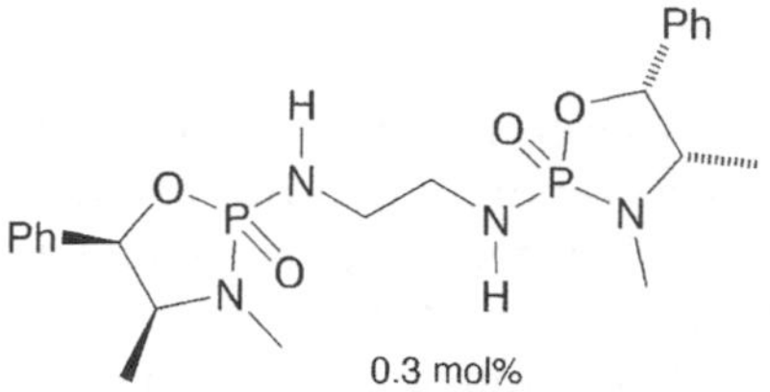

0.3 mol%
0.1 mol% SmCl$_3$
provides 84% ee

(6.45)

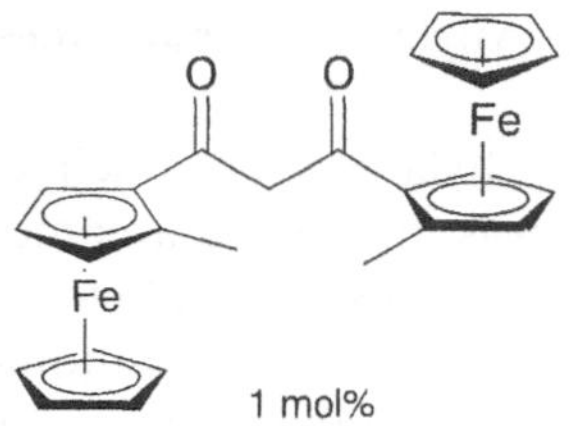

1 mol%
0.2 mol% Y$_5$(O)(O^{i}Pr)$_{13}$
provides 90% ee

(6.46)

hydrolysis of the nitrile to acid. Sigman and Jacobsen reported two approaches to this reaction.[35, 36] Firstly, a parallel library of 132 structures of type (6.47) was screened and compound (6.48) was identified as the best equivalent solution-phase catalyst (there is no metal involved in this reaction). Alternatively, an aluminium (salen) complex (6.49) has been shown to provide high enantioselectivities in the addition of HCN to imines, including the conversion of imine (6.50) into an aminonitrile, which is isolated as the trifluoroacetamide (6.51).

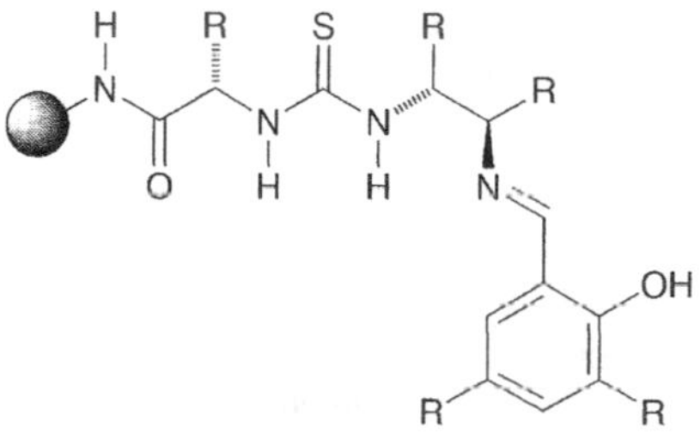

(6.47)

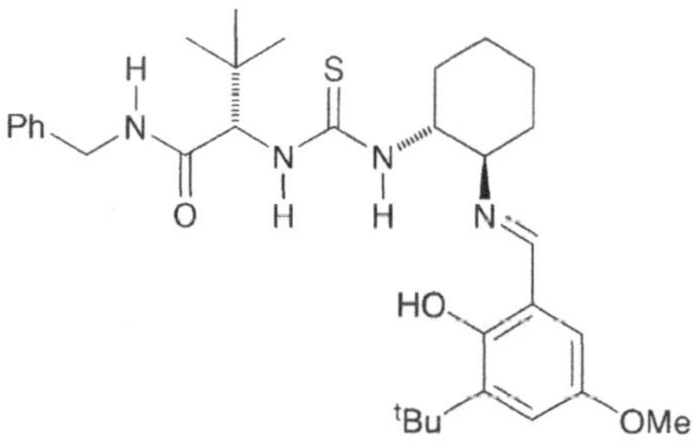

(6.48)

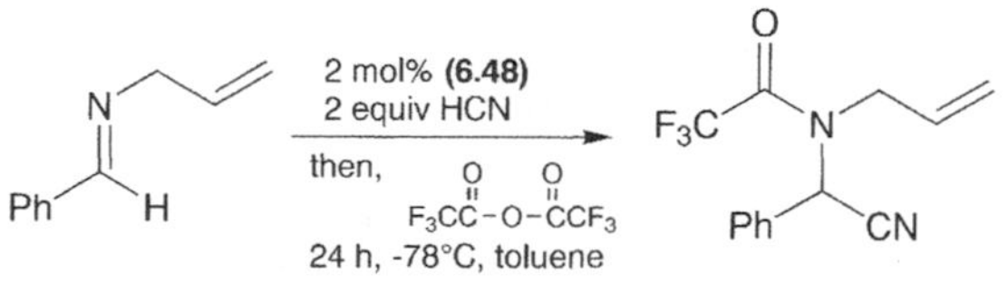

(6.50)

2 mol% (6.48)
2 equiv HCN
then, F$_3$CC-O-CCF$_3$
24 h, -78°C, toluene

(6.51) 91% ee

(6.50)

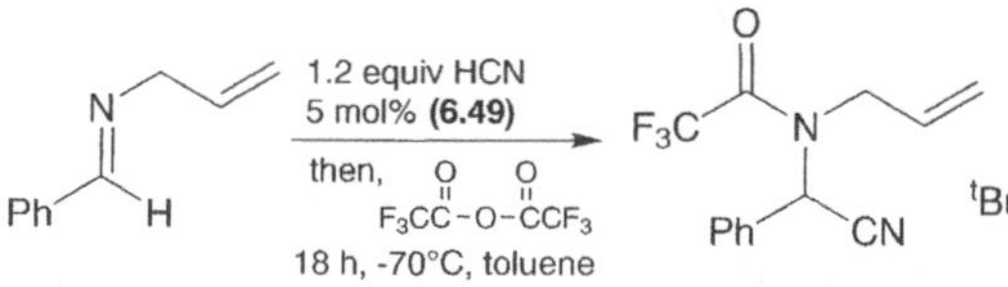

1.2 equiv HCN
5 mol% (6.49)
then, F$_3$CC-O-CCF$_3$
18 h, -70°C, toluene
91%

(6.51) 95% ee

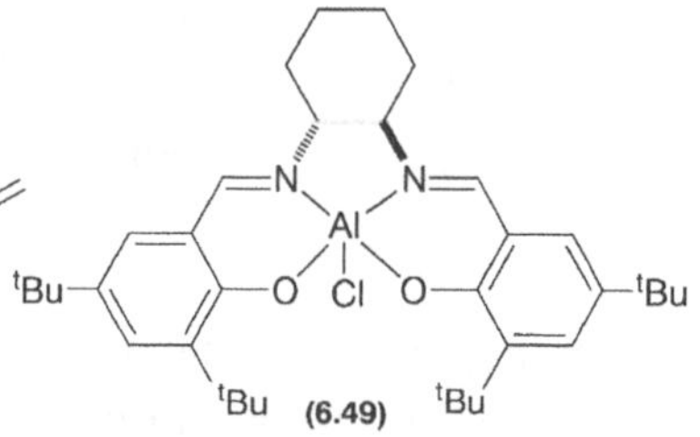

(6.49)

6.3   Allylation of aldehydes

The allylation of aldehydes provides a useful route to homoallylic alcohols. In 1991, Yamamoto and co-workers reported that the allylation of aldehydes with allylsilanes (the Sakurai-Hosomi reaction) could be achieved with the enantiomerically pure boron catalyst (6.52).[37, 38] Whilst the parent allylsilane (6.53) was not reactive enough to give the product (6.54) efficiently, alkyl substitution of the allylsilane (6.55) increased the reactivity of the silane towards benzaldehyde (6.01) and yielded high levels of enantioselectivity and diastereoselectivity in the product (6.56). The corresponding reactions with aliphatic aldehydes were less satisfactory.

Marshall and Tang extended the scope of the reaction to the use of allylstannane reagents, which are more reactive.[39] The addition of triflic anhydride as a promoter overcame a problem with catalyst deactivation and allowed the reaction of allylstannane (6.57) to take place efficiently and with fairly good stereocontrol in the product (6.58).

(6.52)

PhCHO   +   (6.53) SiMe$_3$ $\xrightarrow[\text{-78°C, EtCN, 46%}]{\text{20 mol% 6.52}}$ Ph (6.54) 55% ee

(6.01)

PhCHO   +   (6.55) SiMe$_3$ $\xrightarrow[\text{-78°C, EtCN, 46%}]{\text{20 mol% (6.52)}}$ Ph (6.56) 96% ee   97:3 syn:anti

(6.01)

PhCHO +   (6.57) SnBu$_3$ $\xrightarrow[\text{-78°C, EtCN, 88%}]{\begin{array}{c}\text{20 mol% (6.52)}\\\text{40 mol% (CF}_3\text{CO)}_2\text{O}\end{array}}$ Ph (6.58) 74% ee   85:15 syn:anti

(6.01)

Tributylallyltin (6.59) has subsequently been used in the allylation of aldehydes using various catalysts. In 1993, Umani-Ronchi and co-workers[40] and Keck and co-workers[41, 42] published related results using BINOL/titanium-derived catalysts. In the Umani-Ronchi system, BINOL-TiCl$_2$ catalyst (6.60) was employed and was shown to work well with aliphatic aldehydes, including octanal (6.61). The Keck system uses BINOL (6.62) with titanium tetraisopropoxide in a ratio of either 2:1 or 1:1. This combination is effective both for aliphatic and aromatic aldehydes, including cyclohexane carboxaldehyde (6.62). The BINOL:Ti-catalysed allylation of aldehydes has been shown to be subject to chiral amplification[43] and chiral poisoning.[44]

In an effort to improve catalyst efficiency, Yu and co-workers have shown that it is possible to use 'molecular accelerators' in these reactions.[45, 46] They showed that the addition of the alkylthioborane allows the use of lower catalyst loadings (typically 1 mol%) in the addition of tributylallyltin to aldehydes.

(6.60)   (6.62)

C$_7$H$_{15}$CHO + allyl—SnBu$_3$ → C$_7$H$_{15}$ (with OH)

(6.61)   2 equiv
                (6.59)

20 mol% (6.60)
4Å MS
24 h, -20°C 83%

(6.64) 97.4% ee

cyclohexyl—CHO + allyl—SnBu$_3$ →

(6.63)   3 equiv
                (6.59)

20 mol% (6.62)
10 mol% Ti(O^{i}Pr)$_4$
4Å MS
95%

(6.65) 92% ee

Silver catalysts have not been widely investigated for asymmetric synthesis. However, Yamamoto and co-workers have shown that the combination of silver triflate with BINAP (6.66) is effective for the allylation of aldehydes.[47] The reaction is represented by the conversion of furfural (6.67) into the addition product (6.69), with high yield and enantioselectivity. The use of crotylstannanes (6.68) was found to lead to the selectivity of anti-adducts (6.70), as shown in the reaction with benzaldehyde (6.01).[48] Enantiomerically pure zirconium BINOL complexes[49] and zinc bis-oxazoline complexes[50] have also been used to catalyse the asymmetric allylation of aldehydes.

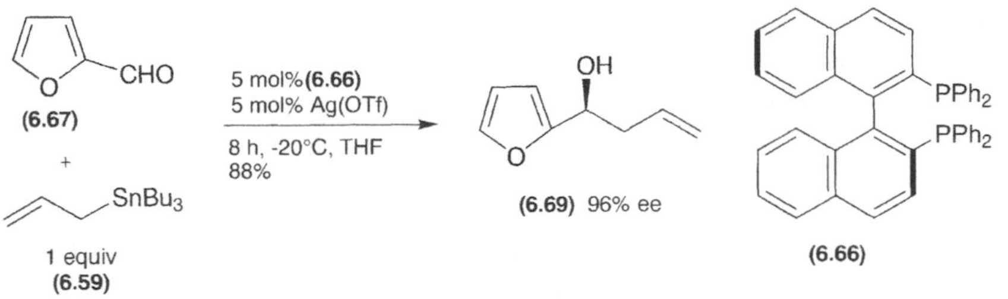
CHO
(6.67)
+
SnBu3
1 equiv
(6.59)
5 mol%(6.66)
5 mol% Ag(OTf)
8 h, -20°C, THF
88%
OH
(6.69) 96% ee
PPh2
PPh2
(6.66)

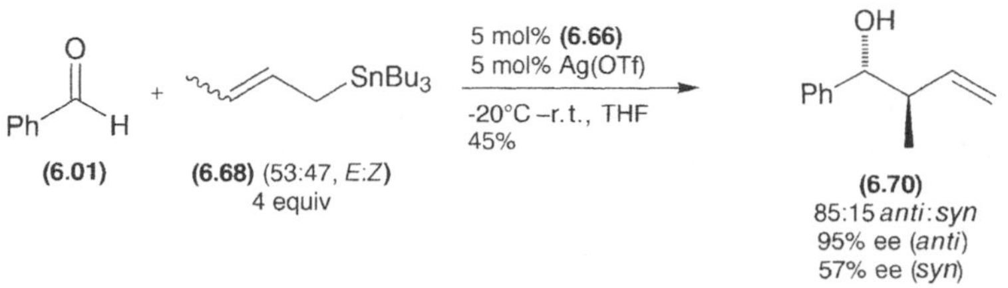
O
Ph
H
(6.01)
+
SnBu3
(6.68) (53:47, E:Z)
4 equiv
5 mol% (6.66)
5 mol% Ag(OTf)
-20°C –r.t., THF
45%
OH
Ph
(6.70)
85:15 anti:syn
95% ee (anti)
57% ee (syn)

Although allyltributylstannane has been the most widely-used allylating agent, allyltrimethylsilane is cheaper and less toxic. Gauthier and Carreira have shown that allyltrimethylsilane is an effective allylating agent in the presence of a catalyst derived from titanium tetrafluoride and BINOL.[51] Pivaldehyde (**6.71**) undergoes allylation to yield the homoallylic alcohol (**6.72**), with good enantioselectivity and yield (after hydrolysis of the trimethylsilyl ether).

In addition to allylation reactions catalysed by Lewis acids, it has been shown that the reaction can be catalysed by chiral Lewis bases. In 1994, Denmark and co-workers showed that allylation of benzaldehyde (**6.01**) with allyltrichlorosilane (**6.73**) could be achieved using phosphoramide (**6.74**) as a catalyst.[52]

Using a related phosphoramide (**6.75**), Iseki and co-workers have also developed a Lewis base-catalysed allylation reaction.[53] The geometry of the crotylsilane substrate, (**6.76**) or (**6.77**), was reflected in the major diastereomer of adduct formed, (**6.78**) or (**6.79**), indicating a cyclic chair transition state. For the (*E*)-crotylsilane (**6.77**), a transition state formulated as shown in Figure 6.4 has been described.

Enantiomerically pure formamides have also been used to catalyse asymmetric allylation with allyltrichlorosilanes,[54] whilst enantiomerically pure diamines have been employed in a similar reaction in conjunction with diallyltindibromide.[55]

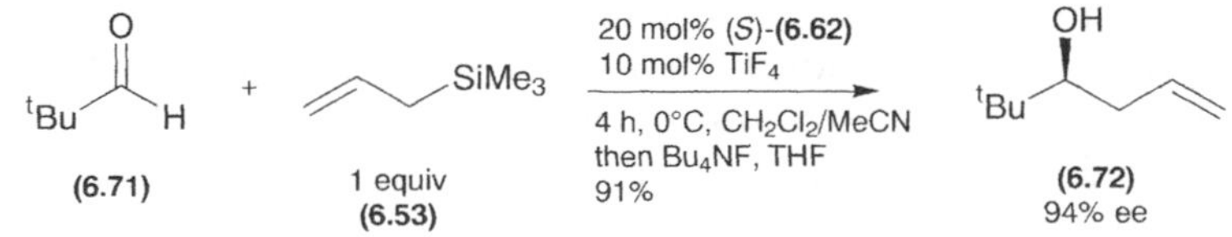
O
tBu
H
(6.71)
+
SiMe3
1 equiv
(6.53)
20 mol% (S)-(6.62)
10 mol% TiF4
4 h, 0°C, CH2Cl2/MeCN
then Bu4NF, THF
91%
OH
tBu
(6.72)
94% ee

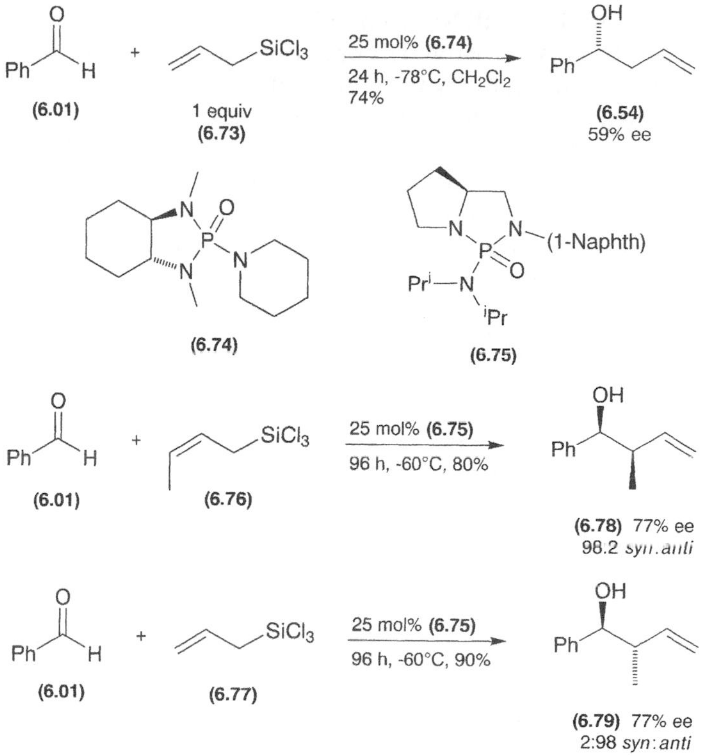

6.4   Hydrophosphonylation of aldehydes and imines

The catalytic enantioselective hydrophosphonylation of aldehydes was first reported in 1993 by Shibuya and co-workers.[56] These early results showed that the addition of diethylphosphonate (**6.80**) to benzaldehyde (**6.01**) could be catalysed by titanium complex (**6.38**), with moderate enantioselectivity in the α-hydroxyphosphonate product (**6.81**). The titanium catalyst serves to facilitate tautomerisation of the phosphonate into diethylphosphite, HOP(OEt)$_2$, and activates the aldehyde towards addition.

LaLi$_3$tris(binaphthoxide) (LLB) catalysts (see Section 7.1) have also been used for this reaction,[57, 58] although the method of preparation of the catalyst is important for very high selectivities.[59] The LLB catalyst (**6.84**) was prepared by mixing LaCl$_3$.7H$_2$O (1 equiv) BINOL dilithium salt (2.7 equiv) and sodium tert-butoxide (0.3 equiv) in THF at 50°C. When prepared in this way, the catalyst was effective for the hydrophosphonylation of aldehydes (**6.28**), using dimethylphosphite (**6.82**) to give α-hydroxyphosphonates (**6.83**).

In addition to La-Li-BINOL complexes, Shibasaki and co-workers have shown that Al-Li-BINOL complexes also catalyse hydrophos-phonylation of aldehydes, with good yields and enantioselectivities.[60] Shibasaki's heterobimetallic BINOL complexes also work well for the catalytic asymmetric hydrophosphonylation of imines.[61] In this case, lanthanum-potassium-BINOL complexes (6.85) have been found to provide the highest enantioselectivities for the hydrophosphonylation of acyclic imines (6.86). The hydrophosphonylation of cyclic imines using heterobimetallic lanthanoid complexes has recently been reported. Ytterbium and samarium complexes have shown the best results in the cases investigated so far.[62]

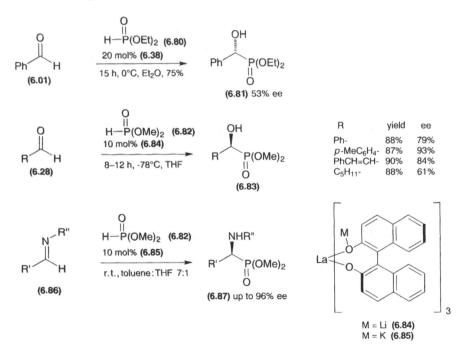

R	yield	ee
Ph-	88%	79%
p-MeC$_6$H$_4$-	87%	93%
PhCH=CH-	90%	84%
C$_5$H$_{11}$-	88%	61%

M = Li (6.84)
M = K (6.85)

References

1. N. Oguni and T. Omi, *Tetrahedron Lett.*, **1984**, *25*, 2823.
2. (a) R. Noyori and M. Kitamura, *Angew. Chem., Int. Ed. Engl.*, **1991**, *30*, 49. (b) K. Soai and S. Niwa, *Chem. Rev.*, **1992**, *92*, 833.
3. For a few recent examples, see: (a) C. Bolm, K.M. Fernández, A. Seger and G. Raabe, *Synlett*, **1997**, 1051. (b) M. Kossenjans and J. Martens, *Tetrahedron: Asymmetry*, **1998**, *9*, 1409. (c) G.B. Jones, M. Guzel and B.J. Chapman, *Tetrahedron: Asymmetry*, **1998**, *9*, 901. (d) G. Bringman and M. Breuning, *Tetrahedron: Asymmetry*, **1998**, *9*, 667. (e) J.C. Anderson and M. Harding, *J. Chem. Soc., Chem. Commun.*, **1998**, 393. (f) W.-S. Huang,

Q.-S. Hu and L. Pu, *J. Org. Chem.*, **1998**, *63*, 1364. (g) J.-M. Brunel, T. Constantieux, O. Legrand and G. Buono, *Tetrahedron Lett.*, **1998**, *39*, 2961. (h) P.I. Dosa, J.C. Ruble and G.C. Fu, *J. Org. Chem.*, **1997**, *62*, 444.

4. (a) M. Kitamura, S. Suga, K. Kawai and R. Noyori, *J. Am. Chem. Soc.*, **1986**, *108*, 6071. (b) R. Noyori, S. Suga, K. Kawai, M. Kitamura, N. Oguni, M. Hayashi, T. Kaneko and Y. Matsuda, *J. Organomet. Chem.*, **1990**, *382*, 19.

5. K. Soai, S. Yokoyama, K. Ebihara, and T. Hayasaka, *J. Chem. Soc., Chem. Commun.*, **1987**, 2405.

6. K. Soai, A. Ookawa, T. Kaba and K. Ogawa, *J. Am. Chem. Soc.*, **1987**, *109*, 7111.

7. (a) P.I. Dosa and G.C. Fu, *J. Am. Chem. Soc.*, **1998**, *120*, 445. (b) D.J. Ramón and M. Yus, *Tetrahedron Lett.*, **1998**, *139*, 1239.

8. (a) D. Guijarro, P. Pinho and P.G. Andersson, *J. Org. Chem.*, **1998**, *63*, 2530. (b) P.G. Andersson, D. Guijarro and D. Tanner, *J. Org. Chem.*, **1997**, *62*, 7364.

9. For a detailed discussion of the various possibilities of bicyclic transition states, see: R. Noyori, *Asymmetric Catalysis in Organic Synthesis*, John Wiley and Sons, **1994**, Chapter 5.

10. M. Kitamura, S. Okada, S. Suga and R. Noyori, *J. Am. Chem. Soc.*, **1989**, *111*, 4028.

11. (a) S. Mason, *Chem. Soc., Rev.*, **1988**, *17*, 347. (b) M. Avalos, R. Babiano, P. Cintas, J.L. Jiménez and J.C. Palacios, *Tetrahedron: Asymmetry*, **1997**, *8*, 2997.

12. R. Ostwald, P.-V. Chavant, H. Stadtmüller and P. Knochel, *J. Org. Chem.*, **1994**, *59*, 4143.

13. C. Eisenberg and P. Knochel, *J. Org. Chem.*, **1994**, *59*, 3760.

14. H. Lütjens, S. Nowotny and P. Knochel, *Tetrahedron: Asymmetry*, **1995**, *6*, 2675.

15. For example, see: X. Zhang and C. Guo, *Tetrahedron Lett.*, **1995**, *36*, 4947.

16. S. Pritchett, D.H. Woodmansee, P. Gantzel and P.J. Walsh, *J. Am. Chem. Soc.*, **1998**, *120*, 6423.

17. F. Effenberger, *Angew. Chem., Int. Ed. Engl.*, **1994**, *33*, 1555.

18. F. Effenberger, T. Ziegler and S. Forster, *Angew. Chem., Int. Ed. Engl.*, **1987**, *26*, 458.

19. F. Effenberger, B. Horsch, S. Forster and T. Ziegler, *Tetrahedron Lett.*, **1990**, *31*, 1249.

20. T.T. Huuhtanen and L.T. Kanerva, *Tetrahedron: Asymmetry*, **1992**, *3*, 1223.

21. F. Effenberger, B. Horsch, F. Weingart, T. Ziegler and S. Kuhner, *Tetrahedron Lett.*, **1991**, *32*, 2605.

22. F. Effenberger and J. Jager, *J. Org. Chem.*, **1997**, *62*, 3867.

23. J. Oku and S. Inoue, *J. Chem. Soc., Chem. Commun.*, **1981**, 229.

24. For a review on catalytic asymmetric cyanohydrin synthesis with particular attention to cyclic dipeptides, see: M. North, *Synlett*, **1993**, 807.

25. K. Tanaka, A. Mori and S. Inoue, *J. Org. Chem.*, **1990**, *55*, 181.

26. M. Hayashi, T. Matsuda and N. Oguni, *J. Chem. Soc., Perkin Trans.*, *1*, **1992**, 3135.

27. D. Callant, D. Stanssens and J.G. de Vries, *Tetrahedron: Asymmetry*, **1993**, *4*, 185.

28. M. Hayashi, Y. Miyamoto, T. Inoue and N. Oguni, *J. Org. Chem.*, **1993**, *58*, 1515.

29. W. Pan, X. Feng, L. Gong, W. Hu, Z. Li, A. Mi and Y. Jiang, *Synlett*, **1996**, 337.

30. Y. Belokon, M. Flego, N. Ikonnikov, M. Moscalenko, M. North, C. Orizu, V. Tararov and M. Tasinazzo, *J. Chem. Soc., Perkin Trans. 1*, **1997**, 1293.

31. M.C.K. Choi, S.S. Chan and K. Matsumoto, *Tetrahedron Lett.*, **1997**, *38*, 6669.

32. E.J. Corey and Z. Wang, *Tetrahedron Lett.*, **1993**, *34*, 4001.

33. W.-B. Yang and J.-M. Fang, *J. Org. Chem.*, **1998**, *63*, 1356.

34. A. Abiko and G.-Q. Wang, *J. Org. Chem.*, **1996**, *61*, 2264.

35. M.S. Sigman and E.N. Jacobsen, *J. Am. Chem. Soc.*, **1998**, *120*, 4901.

36. M.S. Sigman and E.N. Jacobsen, *J. Am. Chem. Soc.*, **1998**, *120*, 5315.

37. K. Furuta, M. Mouri and H. Yamamoto, *Synlett*, **1991**, 561.

38. K. Ishihara, M. Mouri, Q. Gao, T. Maruyama, K. Furuta and H. Yamamoto, *J. Am. Chem. Soc.*, **1993**, *115*, 11490.

39. J.A. Marshall and Y. Tang, *Synlett*, **1992**, 653.

40. A.L. Costa, M.G. Piazza, E. Tagliavini, C. Trombini and A. Umani-Ronchi, *J. Am. Chem. Soc.*, **1993**, *115*, 7001.
41. G.E. Keck, K.H. Tarbet and L.S. Geraci, *J. Am. Chem. Soc.*, **1993**, *115*, 8467.
42. G.E. Keck and L.S. Geraci, *Tetrahedron Lett.*, **1993**, *34*, 7827.
43. G.E. Keck, D. Krishnamurthy and M.C. Grier, *J. Org. Chem.*, **1993**, *58*, 6543.
44. J.W. Faller, D.W.I. Sams and X. Liu, *J. Am. Chem. Soc.*, **1996**, *118*, 1217.
45. C.-M. Yu, H.-S. Choi, W.-H. Jung and S.-S. Lee, *Tetrahedron Lett.*, **1996**, *37*, 7095.
46. C.-M. Yu, H.-S. Choi, W.-H. Jung, H.-J. Kim and J. Shin, *J. Chem. Soc., Chem. Commun.*, **1997**, 761.
47. A. Yanagisawa, H. Nakashima, A. Ishiba and H. Yamamoto, *J. Am. Chem. Soc.*, **1996**, *118*, 4723.
48. A. Yanagisawa, A. Ishiba, H. Nakashima and H. Yamamoto, *Synlett*, **1996**, 88.
49. P. Bedeschi, S. Casolari, A.L. Costa, E. Tagliavini and A. Umani-Ronchi, *Tetrahedron Lett.*, **1995**, *36*, 7897.
50. P.G. Cozzi, P. Orioli, E. Tagliavini and A. Umani-Ronchi, *Tetrahedron Lett.*, **1997**, *38*, 145.
51. D.R. Gauthier Jr. and E.M. Carreira, *Angew. Chem., Int. Ed. Engl.*, **1996**, *35*, 2363.
52. S.E. Denmark, D.M. Coe, N.E. Pratt and B.D. Griedel, *J. Org. Chem.*, **1994**, *59*, 6161.
53. K. Iseki, Y. Kuroki, M. Takahashi, S. Kishimoto and Y. Kobayashi, *Tetrahedron*, **1997**, *53*, 3513.
54. K. Iseki, S. Mizuno, Y. Kuroki and Y. Kobayashi, *Tetrahedron Lett.*, **1998**, *39*, 2767.
55. S. Kobayashi and K. Nishio, *Tetrahedron Lett.*, **1995**, *36*, 6729.
56. T. Yokomatsu, T. Yamagishi and S. Shibuya, *Tetrahedron: Asymmetry*, **1993**, *4*, 1779.
57. T. Yokomatsu, T. Yamagishi and S. Shibuya, *Tetrahedron: Asymmetry*, **1993**, *4*, 1783.
58. N.P. Rath and C.D. Spilling, *Tetrahedron Lett.*, **1994**, *35*, 227.
59. H. Sasai, M. Bougauchi, T. Arai and M. Shibasaki, *Tetrahedron Lett.*, **1997**, *38*, 2717.
60. T. Arai, M. Bougauchi, H. Sasai and M. Shibasaki, *J. Org. Chem.*, **1996**, *61*, 2926.
61. H. Sasai, S. Arai, Y. Tahara and M. Shibasaki, *J. Org. Chem.*, **1995**, *60*, 6656.
62. H. Gröger, Y. Saida, H. Sasai, K. Yamaguchi, J. Martens and M. Shibasaki, *J. Am. Chem. Soc.*, **1998**, *120*, 3089.

7   The aldol and related reactions

The standard aldol reaction involves the addition of an enolate to a ketone or an aldehyde. However, there are related reactions involving addition to C=N groups. As well as these mechanistically-related processes, the carbonyl-ene reaction is also discussed here. Whilst the mechanism of the carbonyl-ene reaction is different from the aldol reaction, the synthetic result is rather similar and, perhaps, fits most comfortably into the present chapter.

7.1   The aldol reaction

The aldol reaction, and related processes have been of considerable importance in organic synthesis. The control of *syn/anti* diastereoselectivity, enantioselectivity and chemoselectivity has now reached impressive levels. The use of catalysts is a relatively recent addition to the story of the aldol reaction. Such catalysts usually behave as Lewis acids, and a typical catalytic cycle is presented in Figure 7.1, where aldehyde (**7.01**) coordinates to the catalytic Lewis acid, which encourages addition of the silyl enol ether (**7.02**). Release of the Lewis acid yields the aldol product, often as the silyl ether (**7.03**). The asymmetric catalytic aldol reaction was reviewed in 1998.[1]

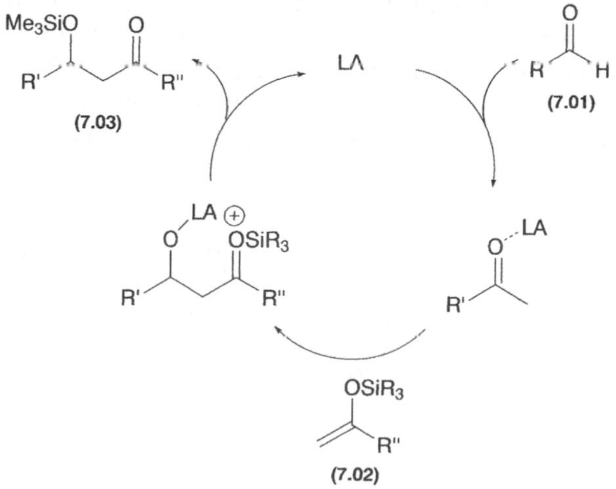

Figure 7.1 General catalytic cycle for aldol reactions of silyl enol ethers. Abbreviation: LA=Lewis acid.

Amongst the first catalytic aldol reactions were the tin triflate-catalysed reactions reported by Kobayashi and co-workers.[2,3] The tin complex of ligand (7.04) catalyses the addition of reactive silyl enol ethers, such as compound (7.05), to give very good enantioselectivity in some cases. The methodology has been used in the preparation of natural products, including sphingofungins. The initial asymmetric induction was achieved by coupling aldehyde (7.07) and ketene acetal (7.08).[4]

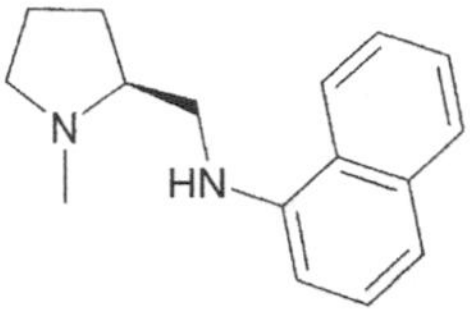

(7.04)

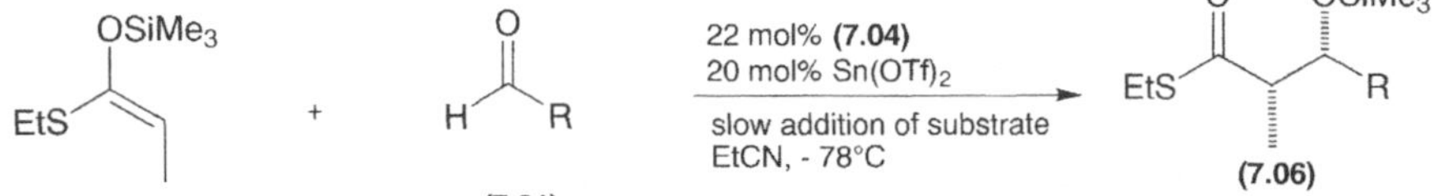

22 mol% (7.04)
20 mol% Sn(OTf)$_2$

slow addition of substrate
EtCN, - 78°C

(7.05)   (7.01)   (7.06)

R = Ph   77% 93:7 syn:anti   90% ee
R = nC$_7$H$_{15}$ 80% 100:0 syn:anti >98% ee

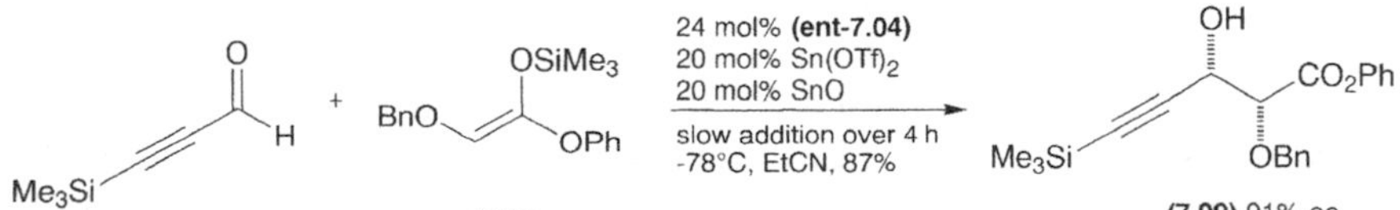

24 mol% (ent-7.04)
20 mol% Sn(OTf)$_2$
20 mol% SnO

slow addition over 4 h
-78°C, EtCN, 87%

(7.07)   (7.08)   (7.09) 91% ee
97:3 syn:anti

Enantiomerically pure boron-based Lewis acids have also been used successfully in catalytic aldol reactions. Corey's catalyst (7.10) provides good enantioselectivity with ketone-derived silyl enol ethers, including compound (7.11).[5] Other oxazaborolidine complexes (7.13) derived from α,α-disubstituted α-amino acids give particularly high enantioselectivity,[6] especially with the disubstituted ketene acetal (7.14). Yamamoto's CAB catalysts (see Section 8.1) have also been used in catalytic aldol reactions.

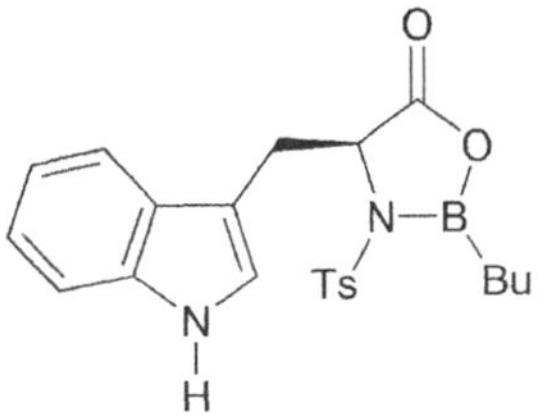

(7.10) 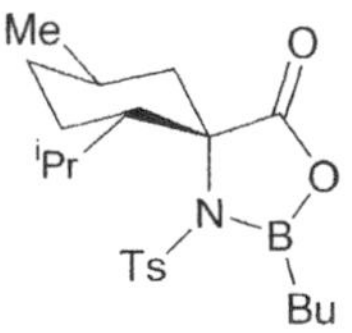   (7.13) 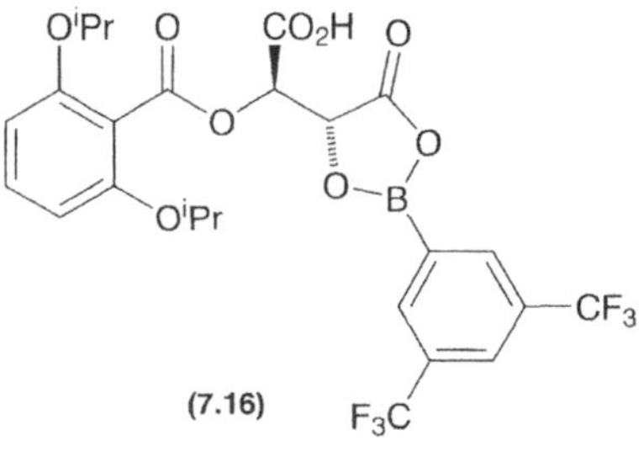   (7.16)

OSiMe$_3$

Ph (7.11)

+

H R O (7.01)

20 mol% (7.10)
EtCN
14 h, -78°C
then H$_2$O/H$_3$O$^+$

Ph O OH R (7.12)

R = Ph   82%, 89% ee
R = nC$_6$H$_{11}$ 67%, 93% ee

OSiMe$_3$

EtO (7.14)

+

H R O (7.01)

20 mol% (7.13)
EtCN
1 h, -78°C
68 – 89%

EtO O OH R (7.15) 91–99% ee

OSiMe$_3$

nBu (7.18)

+

H R O (7.17)

20 mol% (7.16)
EtCN, -78°C
97%

Bu O OSiMe$_3$ Ph (7.19) 94% ee
93:7 syn:anti

Either geometry of silyl enol ether (7.18) yields *syn* selectivity in the product, (7.19), indicating an open transition state.[7]

Titanium complexes are often encountered in Lewis acid-catalysed reactions. This is certainly true for catalysed aldol reactions. Mikami and Matsukawa demonstrated that titanium/BINOL complexes, e.g. complex (7.20), provided high yield and enantioselectivity in aldol reactions.[8] The thioester-derived enol ether (7.21) reacts with various aldehydes (7.01) to give the aldol products (7.22). The authors invoke a silatropic ene transition state, structure (7.23), substantiated by suitable cross-over experiments.

Further improvements to enantioselectivity were made by using improved silyl migrating groups (silacyclobutanes) and employing penta-fluorophenol as an achiral ligand associated with the titanium.[9] Keck and Krishnamurthy also used a titanium/BINOL catalyst in aldol reactions to give high enantioselectivty.[10] The aldol reaction and the 'normal' ene reaction are clearly related (the ene process is discussed in Section 7.6).

Carreira and co-workers reported a highly selective catalytic system for aldol reactions of silyl ketene acetals. Their catalyst (7.24) is derived from an amino-BINOL variant, along with an achiral salicylate ligand.[11,12] Very high enantioselectivities have been achieved using this catalyst for a range of aldehyde substrates (7.01) with enol ethers (7.26), often using relatively low catalyst loading. The reaction has also been applied to aldol reactions of the dienolate (7.27).[13] Aldehydes, including substrate (7.25), were used to give high enantioselectivities (80–94% ee) in the products (7.29).

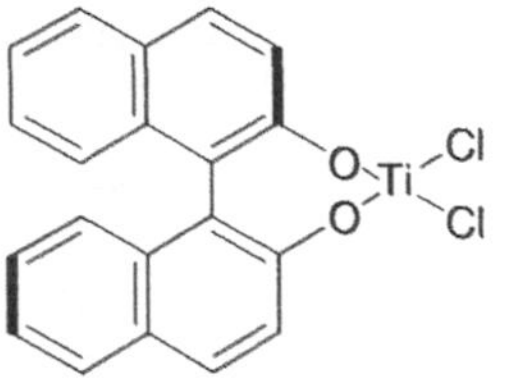

(7.20)

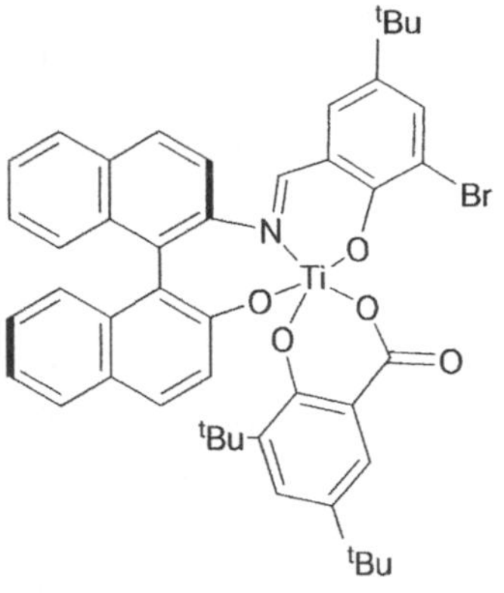

(7.24)

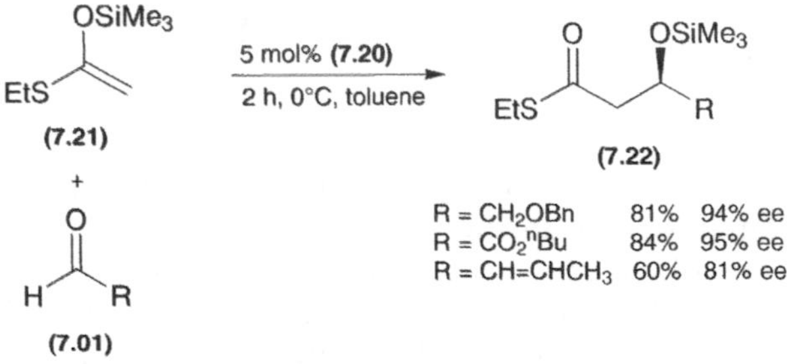
OSiMe3
EtS
(7.21)
+
H R
(7.01)
5 mol% (7.20)
2 h, 0°C, toluene
OSiMe3
EtS R
(7.22)
R = CH2OBn 81% 94% ee
R = CO2nBu 84% 95% ee
R = CH=CHCH3 60% 81% ee

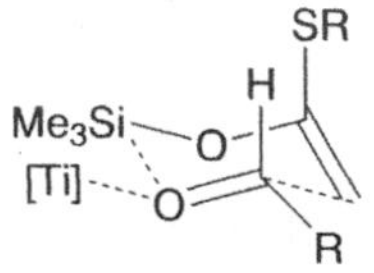
SR
H
Me3Si
[Ti]
R

(7.23)

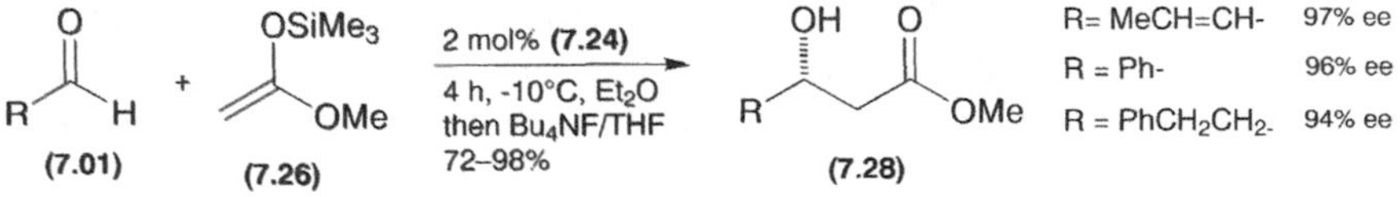
R H
(7.01)
OSiMe3
OMe
(7.26)
2 mol% (7.24)
4 h, -10°C, Et2O
then Bu4NF/THF
72–98%
OH
R OMe
(7.28)
R= MeCH=CH- 97% ee
R = Ph- 96% ee
R = PhCH2CH2- 94% ee

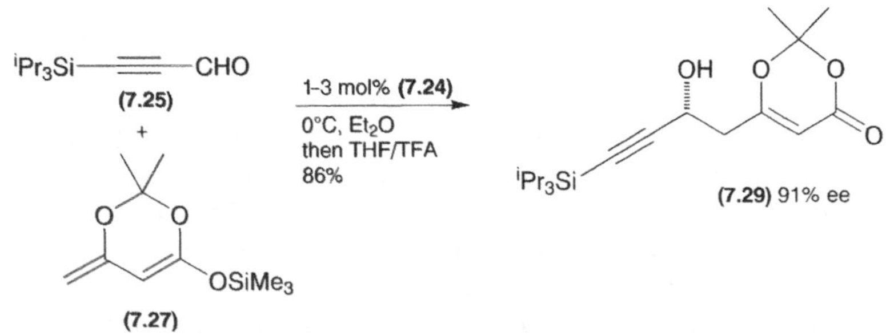
iPr3Si CHO
(7.25)
+
OSiMe3
(7.27)
1-3 mol% (7.24)
0°C, Et2O
then THF/TFA
86%
OH
iPr3Si
(7.29) 91% ee

Evans and co-workers reported that copper(II) complexes of ligand **(7.30)** are very efficient catalysts for aldol reactions, using benzyloxy-acetaldehyde **(7.31)** as the electrophilic partner. This aldehyde is able to

achieve two-point binding to the copper catalyst, and good enantioselectivities are obtained in the copper(II)-catalysed aldol reaction with silyl
enol ethers (7.32) and (7.33).[14] Pyruvate esters can also enjoy the
advantages of two-point binding and have proved to be good substrates
for the copper-catalysed reaction.[15] Evans and co-workers also reported
the use of tin complexes of ligand (7.30) for catalytic aldol reactions,
including the reaction of pyruvate ester (7.36) with silyl enol ether (7.37).[16]

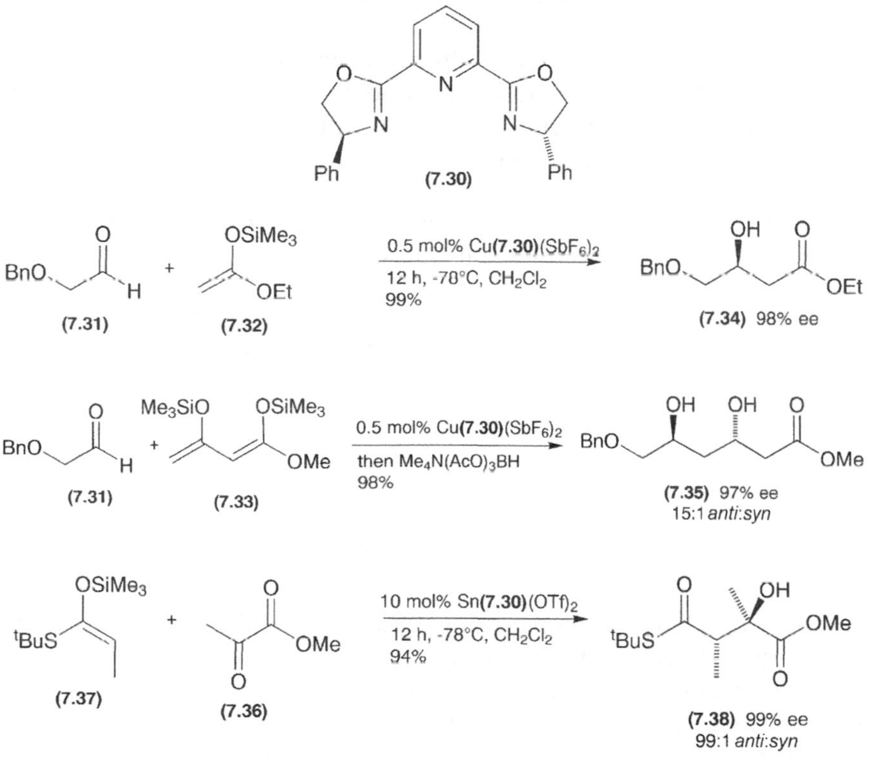

An alternative use of copper(II) catalysts was reported by Krüger and
Carreira.[17] Using a fluoride counterion, they reason that the copper
enolate is formed rather than acting (solely) as a Lewis acid. Using
TolBINAP as the ligand, the catalyst provides high enantioselectivities.
The reaction worked well with most of the aldehydes that were reported
(65–95% ee), including the reaction of benzaldehyde (7.17).

Palladium catalysts have also been reported to give good enantioselectivities in aldol reactions, again *via* catalytic formation of an
enantiomerically pure enolate.[18,19]

(7.17)   (7.27)   (7.39) 94%

Whilst most studies concerning catalytic aldol reactions employ silyl enol ethers as the nucleophilic component, an interesting aldol reaction between unmodified aldehydes and ketones has been reported.[20] Shibasaki's heterobimetallic complexes are able to act as a Lewis acid to activate the aldehyde towards attack, as well as promoting enolisation of the ketone component. Early results are encouraging, although reaction times tend to be long. For example, pivaldehyde (7.40) and acetophenone (7.41) undergo aldol reaction to provide the adduct (7.42), using the heterobimetallic complex (7.43).

Of course, enzyme-catalysed aldol reactions do not require silyl enol ethers! There are several aldolase enzymes, which provide high yields and enantioselectivities for certain substrates.[21]

(7.40)   (7.41)   (7.42)

(7.43)

Catalytic aldol reactions using enantiomerically pure bases and nucleophiles have also been reported, including an asymmetric dimerisation of methyl ketone,[22] and the use of phosphoramides with trichlorosilyl enol ethers.[23] In fact, one of the earliest examples of asymmetric catalysis used (S)-proline (7.44) as a catalyst for an aldol reaction (the Hajos-Wiechert reaction).[24] The achiral triketone (7.45) cyclises to give the aldol product (7.46), with good enantioselectivity. The reaction proceeds via temporary formation of an enamine from the remote ketone, which is sufficiently reactive to effect the cyclisation.

(7.45)   (7.46) 93.4% ee   (7.44)

7.2 Isocyanide aldol reactions

Isocyanoacetic esters, such as compound **(7.47)**, were amongst the first nucleophiles employed in aldol-type reactions to give high enantioselectivity.[25] Although gold-catalysed reactions have not been widely investigated in asymmetric catalysis, they do work well in this reaction. The isonitrile group coordinates to the gold, whilst the side-arm in ligand **(7.48)** assists in deprotonation and stabilisation of the nucleophilic enolate.

The reaction also works well with aliphatic aldehydes (85–97% ee, 85–100% *trans*). The reaction has been used with α-isocyano Weinreb amides, such as compound **(7.50)**. This allows subsequent manipulation of the carbonyl compound in product **(7.52)** into aldehydes and ketones.[26] The aldehydes used gave very high enantioselectivities in the products (93–99% ee, 90–96% de). An impressive range of substrates has been investigated in the reaction of related substrates. Progress was reviewed in 1993,[27] along with other work in catalytic aldol reactions achieved at that stage.

Cyanopropionates have also been employed in catalytic aldol reactions. The enolisation of the nucleophile **(7.53)** by the rhodium complex of TRAP ligand **(7.54)** is the basis for the catalysis.[28] The use of bulky esters affords high selectivity in the aldol reaction with formaldehyde **(7.55)**, although only moderate *anti:syn* selectivity was observed when alternative aldehydes were employed.

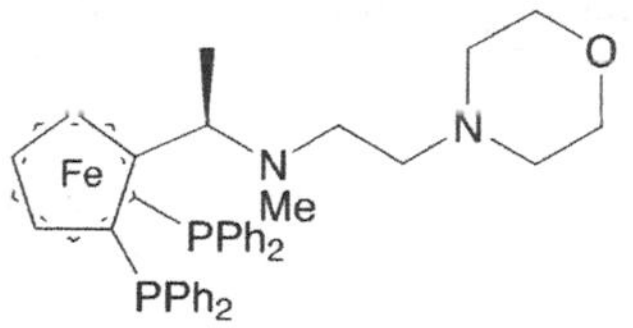

Fe
N
Me
PPh2
PPh2
N
O

(7.48)

Ph2P
Fe
Fe
PPh2

(7.54)

O
Ph
H
(7.17)
+
CN-CH2-CO2Me
(7.47)
1 mol% Au(C6H11NC)2]+BF4−
1 mol% (7.48)
CH2Cl2
Ph
CO2Me
O
N

(7.49) 95% ee, 90% de

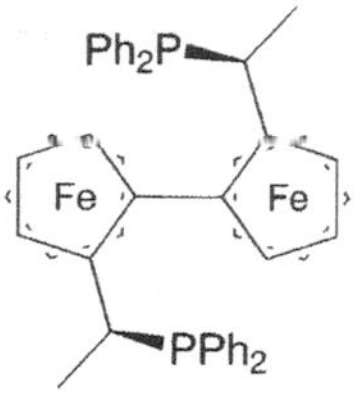

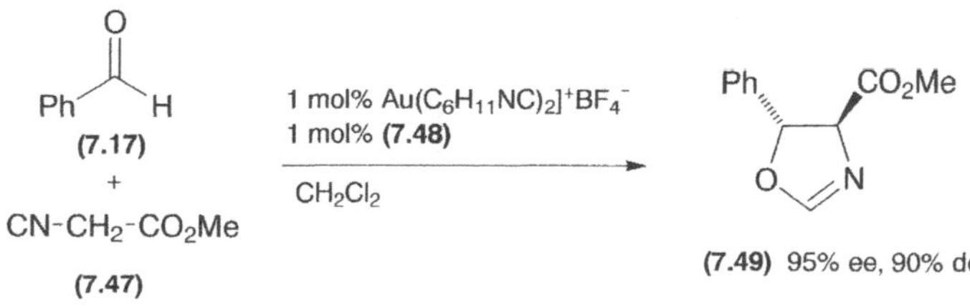

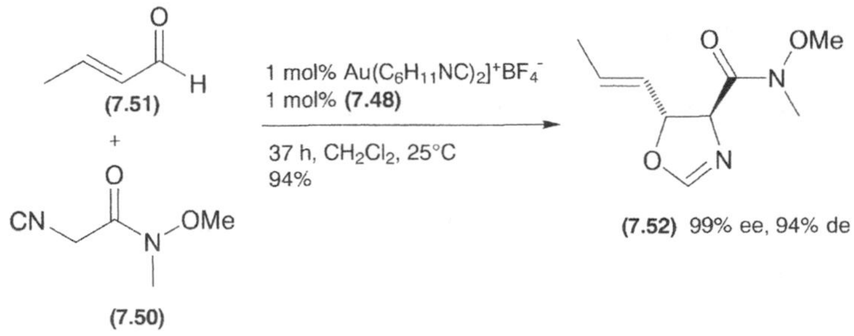

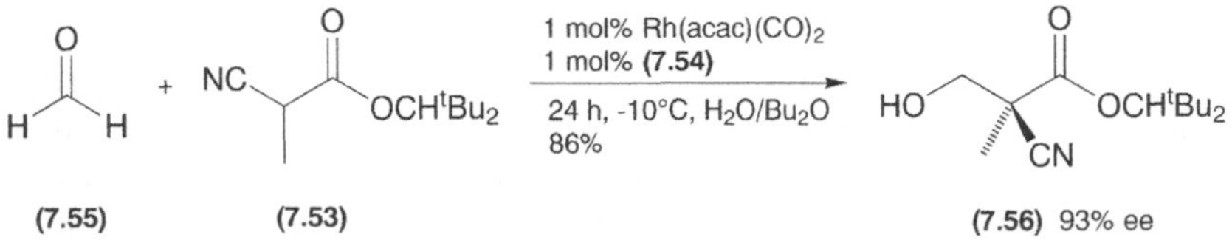

O
H
(7.51)
+
O
CN
N
OMe
(7.50)
1 mol% Au(C6H11NC)2]+BF4-
1 mol% (7.48)
37 h, CH2Cl2, 25°C
94%
O
OMe
N
O
N
(7.52) 99% ee, 94% de

O
H
H
+
NC
O
OCHtBu2
1 mol% Rh(acac)(CO)2
1 mol% (7.54)
24 h, -10°C, H2O/Bu2O
86%
HO
O
OCHtBu2
CN
(7.55)
(7.53)
(7.56) 93% ee

7.3 Addition of enolates to imines

The replacement of the aldehyde by an imine in the aldol reaction has received much less attention. However, in 1998, there were three independent reports of this process giving products in over 90% enantiomeric excess.

Kobayashi and co-workers used the zirconium complex (**7.57**) to catalyse the addition of α-silyloxy ketene acetals (**7.58**) to imines.[29] High *syn* selectivity and enantiomeric excess were seen with arylimines (**7.59**). Sodeoka and co-workers employed the palladium TolBINAP complexes used in normal aldol reactions, providing up to 90% ee in the formation of the β-aminoketone (**7.62**).[30] The use of various late transition metal BINAP complexes (**7.63–7.65**) provided high enantiomeric excesses in the imine/aldol reaction forming product (**7.67**), especially with copper(I) complexes.[31]

It remains to be seen what range of imines may be used in these Mannich-type reactions. In some cases, direct conversion into β-lactams has been achieved using ester enolates.[32]

Br
Br
Br
Br
L
O
O
Zr
O
O
L

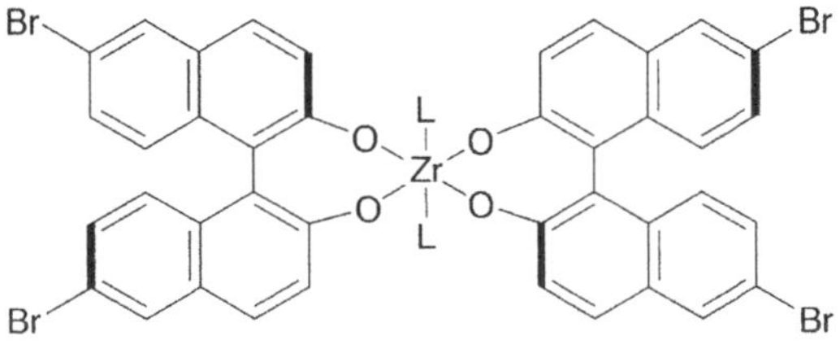

L = 1,2-dimethylimidazole

(7.57)

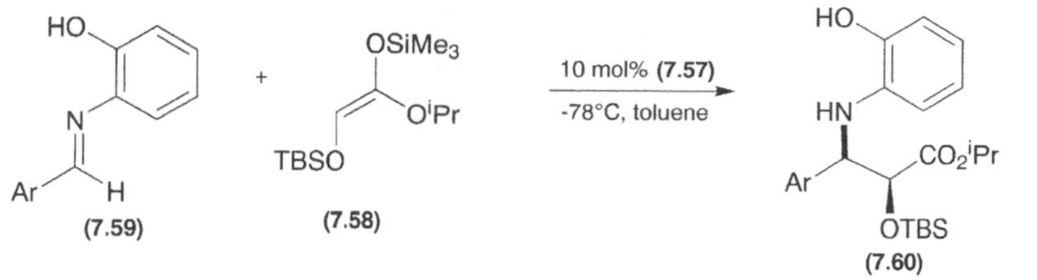

Ph   100%, 96:4 *syn:anti* 95% ee
1-Naphth 65% >99:1 *syn:anti* 91% ee

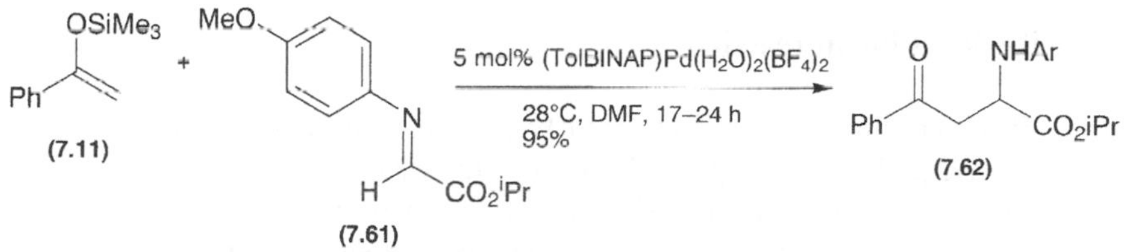

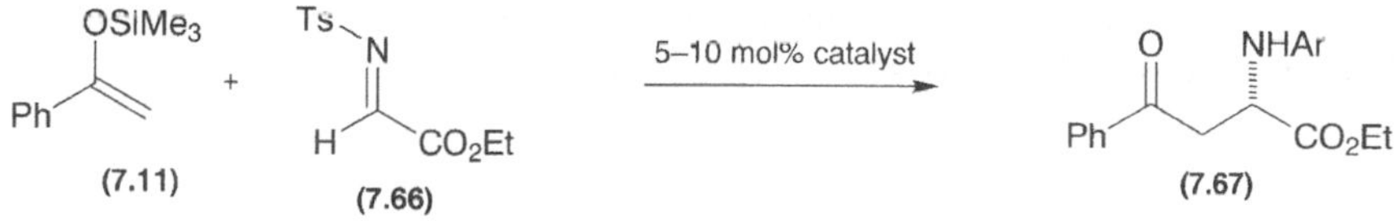

10 mol% [(*R*)-BINAP]AgSbF$_6$ (7.63) THF, −80°C   95%, 90% ee
5 mol% [(*R*)-BINAP]Pd(ClO$_4$)$_2$ (7.64) CH$_2$Cl$_2$, −80°C 91%, 80% ee
10 mol% [(*R*)-TolBINAP]CuClO$_4$ (7.65) THF, 0°C   91%, 98% ee

7.4   Darzens condensation

Darzens condensation is an aldol-like reaction, in which the aldolate product closes to give an epoxy ketone. The reaction has been achieved with moderate (up to 58%) enantiomeric excess using chiral crown ethers.[33,34] The phase-transfer catalyst (7.68) has been used to give higher enantiomeric excesses with various aldehydes (42–79% ee), including the reaction of aliphatic aldehyde (7.69) with the α-chloroketone (7.70).[35] The epoxide (7.71) is formed by displacement of the α-chloro substituent by the β-hydroxy group.

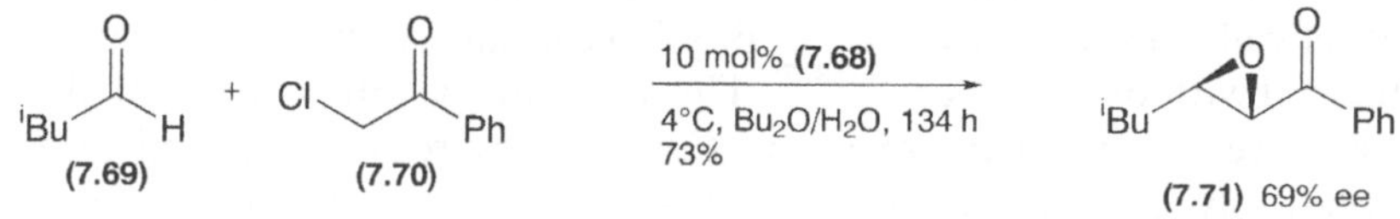

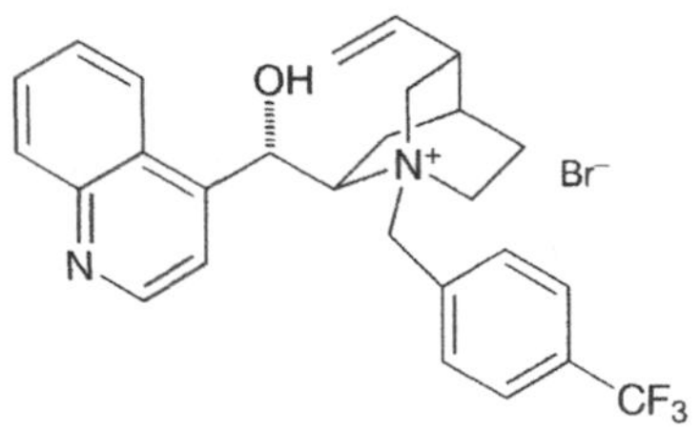

(7.68)

7.5   Baylis-Hillman reaction

The Baylis-Hillman reaction involves the conversion of an α,β-unsaturated carbonyl compound into an aldol-like adduct.[36] The reaction is catalysed by tertiary amines (7.72), which form an intermediate enolate (7.74) by conjugate addition (rather than by direct deprotonation of the α-proton). The enolate undergoes an aldol reaction with an aldehyde (7.01), followed by loss of the amine catalyst to provide the Baylis-Hillman adduct (7.75), as shown in Figure 7.3.

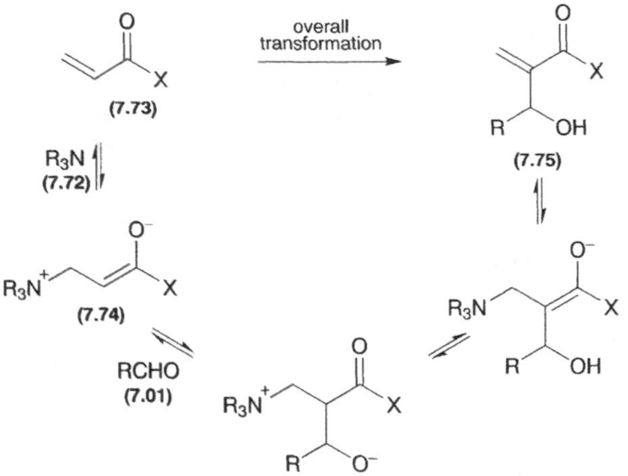

Figure 7.3   Catalytic cycle of the Baylis-Hillman reaction.

The use of an enantiomerically pure amine (or phosphine) to catalyse the reaction is an interesting prospect, since the products would be synthetically useful. However, a Baylis-Hillman reaction has yet to be catalysed with over 50% ee.[37, 38, 39] This may, at least in part, be attributed to the reversibility of the reaction. For entropy reasons, racemisation is energetically favourable and, therefore, any enantioselective catalytic reaction must not be under complete thermodynamic control, since products will tend to become racemised.

Barrett and Kamimura have developed a stepwise asymmetric catalytic variant of the Baylis-Hillman reaction.[40] A stoichiometric amount of nucleophile, Me_3Si-SPh or Me_3Si-SePh (7.76), was added but only a catalytic amount of Lewis acid (7.79) in the reaction between methyl vinyl ketone (7.77) and acetaldehyde (7.78). The aldol adduct (7.80) was subjected to oxidation and selenoxide elimination to give the product (7.81), which is the same as the product from a Baylis-Hillman reaction.

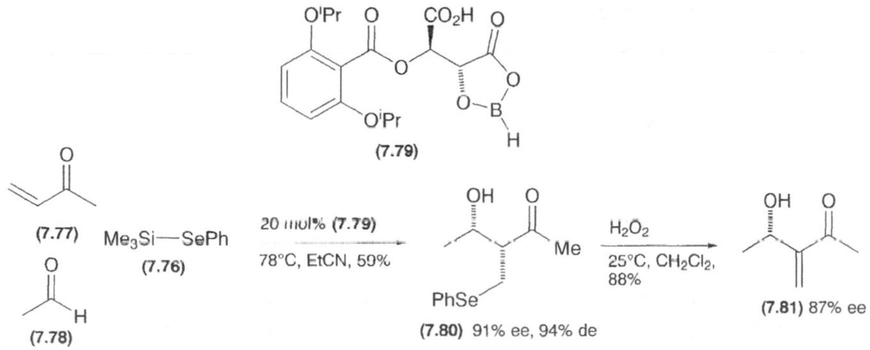

7.6 Carbonyl-ene reactions

The ene reaction can be accelerated by catalysis. In particular, the carbonyl-ene reaction, represented in Figure 7.4, can be accelerated by Lewis acids. The reaction can be synthetically equivalent to an aldol reaction (when the ene component is a vinyl ether), and is considered in this section at the end of aldol reactions and before concerted reactions. The catalysed carbonyl-ene reaction frequently employs reactive aldehydes, especially glyoxalate esters. Mikami and co-workers studied the titanium/BINOL catalysed carbonyl-ene reaction in considerable detail and published a detailed account of their research, in 1992.[41] Typically, the catalyst is prepared *in situ* from diisopropoxytitanium dihalide and BINOL in the presence of 4Å molecular sieves (MS). Thus, alkenes (7.83)

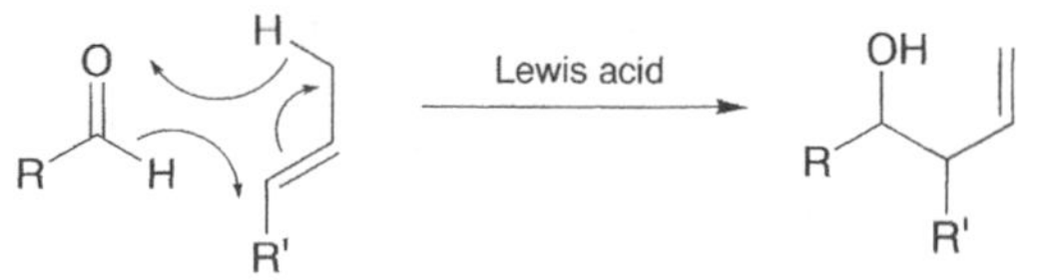

Figure 7.4 General mechanism of the carbonyl-ene reaction.

and (7.84) are converted into the homoallylic alcohols (7.85) and (7.86), with high enantioselectivity. Typical examples use up to 10 mol% of catalysts but variation in the catalyst preparation allows the use of only 0.2 mol%.[42]

The titanium/BINOL-catalysed ene reaction is subject to a strong positive nonlinear effect (see Section 6.1). Thus, the use of BINOL of only 33% ee still provides product (7.86) with 91% ee.

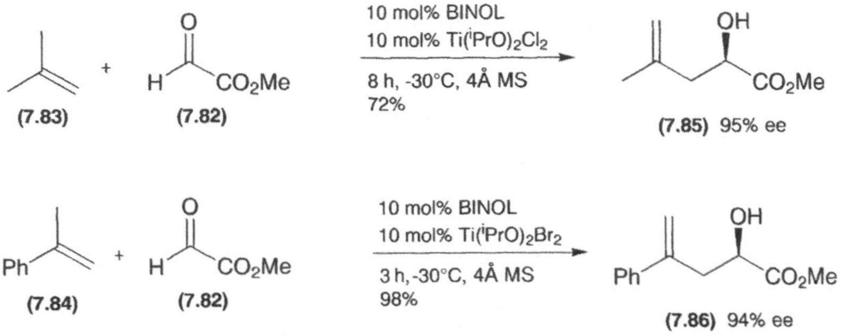

Normally, silyl enol ethers are considered to react with aldol substrates *via* an aldol mechanism, but Mikami and co-workers, in their examples, showed that the reaction involves an ene mechanism.[43] This is clear from the regiochemistry of the product (7.87) that is isolated from the reaction of silyl enol ether (7.85) and aldehyde (7.86) before hydrolysis. The same catalyst system has been used for the asymmetric desymmetrisation of the diene (7.88) with aldehyde (7.89). The product (7.90) is obtained with superb diastereo- and enantiocontrol.[44] A transition state model rationalising this remarkable selectivity has been proposed by Corey and co-workers.[45]

As well as glyoxylate substrates, Mikami and co-workers have employed fluoral (CF_3CHO) as a reactive aldehyde component of a carbonyl-ene reaction, still with very high selectivities.[46]

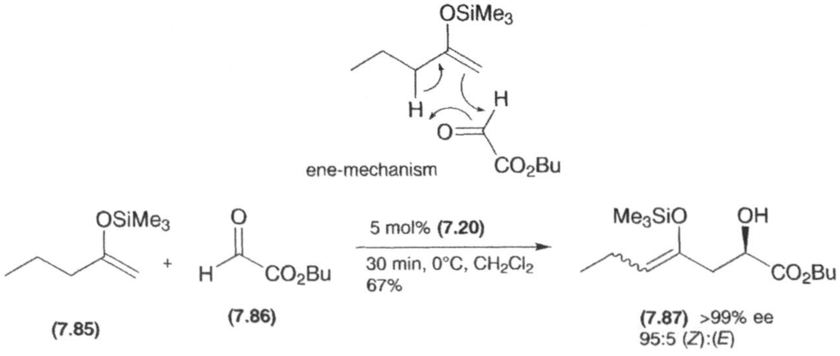

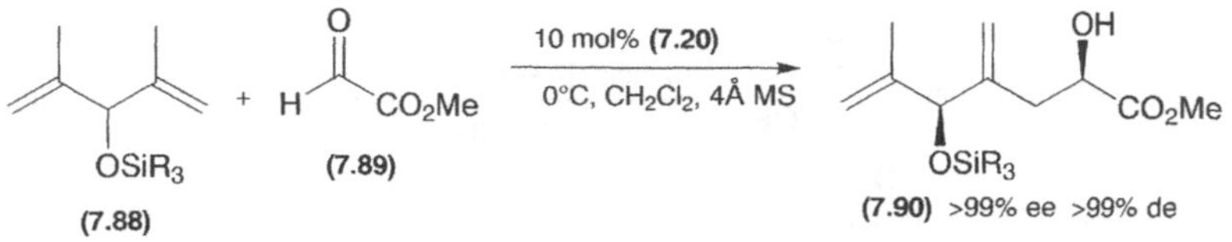

OSiR3
(7.88)
+ H CO2Me
O
(7.89)
10 mol% (7.20)
0°C, CH2Cl2, 4Å MS
OH
CO2Me
OSiR3
(7.90) >99% ee >99% de

Carriera and co-workers employed titanium catalysts (see Section 7.1) in an ene reaction of 2-methoxypropene (**7.91**) with various aldehydes (**7.01**) (not especially reactive ones).[47] The overall synthetic strategy of using 2-methoxypropene (**7.91**) has significant merit in comparison with aldol reactions, because it is a cheap starting material and means that a silyl enol ether does not have to be prepared prior to an aldol reaction.

The Evans copper(II) bis-oxazoline catalysts (**7.94**) and (**7.95**) (see Section 7.1) have also been used effectively for glyoxalate ene reactions.[48] Even monosubstituted alkenes can be used as the ene component, where the alkene (**7.96**) reacts with ethyl glyoxalate (**7.97**) to give the α-hydroxy ester (**7.98**), with very high enantioselectivity. Other alkenes were also effective, providing enantiomeric excesses of over 90%, including alkene (**7.99**), which is converted into the ene-product (**7.100**).

The sense of asymmetric induction observed is consistent with a square planar copper(II)-glyoxylate complex, with approach of the ene from the face opposite to the nearby tert-butyl substituent. Use of copper(II) bisoxazoline complexes as catalysts was independently reported by Vederas and co-workers.[49]

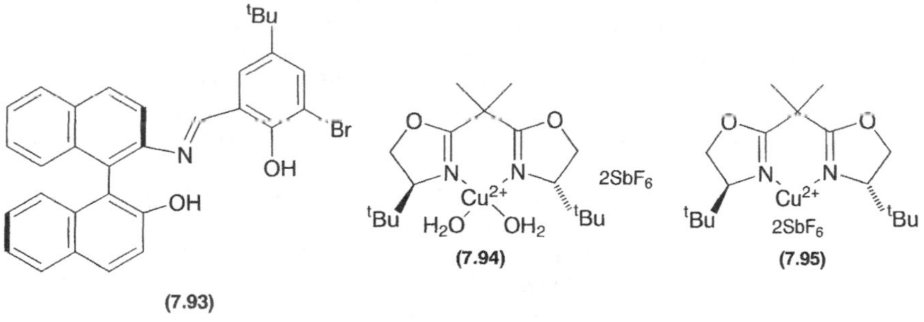

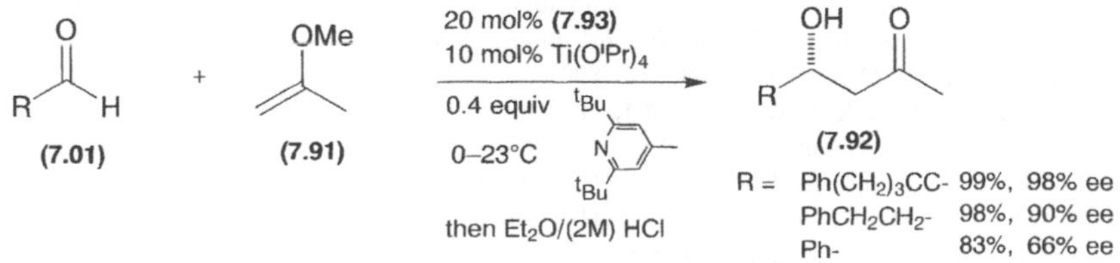

tBu
Br
N
OH
OH
(7.93)
O
O
N
N
Cu2+
tBu
H2O
OH2
tBu
2SbF6
(7.94)
O
O
N
N
Cu2+
tBu
2SbF6
tBu
(7.95)

O
R H
(7.01)
+
OMe
(7.91)
20 mol% (7.93)
10 mol% Ti(OiPr)4
0.4 equiv
tBu
N
tBu
0–23°C
then Et2O/(2M) HCl
OH O
R
(7.92)
R = Ph(CH2)3CC- 99%, 98% ee
PhCH2CH2- 98%, 90% ee
Ph- 83%, 66% ee

H₇C₃ ╱╲╱ (7.96) + H–C(=O)–C(=O)–OEt (7.97) →[10 mol% (7.94)][25°C, CH₂Cl₂, 96%] H₇C₃ ╱═╲╱–CH(OH)–C(=O)–OEt

(7.96) **(7.97)** **(7.98)** 97% ee 96:4 *E:Z*

(7.99) methylenecyclohexane + H–C(=O)–C(=O)–OEt (7.97) →[1 mol% (7.95)][0°C, CH₂Cl₂, 90%] cyclohexenyl–CH₂–CH(OH)–C(=O)–OEt

(7.99) **(7.97)** **(7.100)** 97% ee

References

1. S.G. Nelson, *Tetrahedron: Asymmetry*, **1998**, *9*, 357.
2. S. Kobayashi, Y. Fujishita and T. Mukaiyama, *Chem. Lett.*, **1990**, 1455.
3. S. Kobayashi, M. Furuya, A. Ohtsubo and T. Mukaiyama, *Tetrahedron: Asymmetry*, **1991**, *2*, 635.
4. S. Kobayashi, T. Furuta, T. Hayashi, M. Nishijima and K. Hanada, *J. Am. Chem. Soc.*, **1998**, *120*, 908.
5. E.J. Corey, C.L. Cywin and T.D. Roper, *Tetrahedron Lett.*, **1992**, *33*, 6907.
6. E.R. Parmee, O. Tempkin, S. Masamune and A. Abiko, *J. Am. Chem. Soc.*, **1991**, *113*, 9365.
7. K. Furuta, T. Maruyama and H. Yamamoto, *J. Am. Chem. Soc.*, **1991**, *113*, 1041.
8. K. Mikami and S. Matsukawa, *J. Am. Chem. Soc.*, **1994**, *116*, 4077.
9. S. Matsukawa and K. Mikami, *Tetrahedron: Asymmetry*, **1995**, *6*, 2571.
10. G.E. Keck and D. Krishnamurthy, *J. Am. Chem. Soc.*, **1995**, *117*, 2363.
11. E.M. Carreira, R.A. Singer and W. Lee, *J. Am. Chem. Soc.*, **1994**, *116*, 8837.
12. R.A. Singer, M.S. Shephard and E.M. Carreira, *Tetrahedron*, **1998**, *54*, 7025.
13. R.A. Singer and E.M. Carreira, *J. Am. Chem. Soc.*, **1995**, *117*, 12360.
14. D.A. Evans, J.A. Murry and M.C. Kozlowski, *J. Am. Chem. Soc.*, **1996**, *118*, 5814.
15. D.A. Evans, M.C. Kozlowski, C.S. Burgey and D.W.C. MacMillan, *J. Am. Chem. Soc.*, **1997**, *119*, 7893.
16. D.A. Evans, D.W.C. MacMillan and K.R. Campos, *J. Am. Chem. Soc.*, **1997**, *119*, 10859.
17. J. Krüger and E.M. Carreira, *J. Am. Chem. Soc.*, **1998**, *120*, 837.
18. M. Sodeoka, K. Ohrai and M. Shibasaki, *J. Org. Chem.*, **1995**, *60*, 2648.
19. M. Sodeoka, R. Tokunoh, F. Miyazaki, E. Hagiwara and M. Shibasaki, *Synlett*, **1997**, 463.
20. Y.M.A. Yamada, N. Yoshikawa, H. Sasai and M. Shibasaki, *Angew. Chem., Int. Ed. Engl.*, **1997**, *36*, 1871.
21. For example, see: W.-D. Fessner, A. Schneider, H. Held, G. Sinerius, C. Walter, M. Hixon and J.V. Schloss, *Angew. Chem., Int. Ed. Engl.*, **1996**, *35*, 2219.
22. M.A. Calter, *J. Org. Chem.*, **1996**, *67*, 8006.
23. S.E. Denmark, R.A. Stavenger and K.-T. Wong, *J. Org. Chem.*, **1998**, *63*, 918.
24. (a) U. Eder, G. Sauer and R. Wiechert, *Angew. Chem., Int. Ed. Engl.*, **1971**, *10*, 496.
    (b) Z. G. Hajos and D.R. Parrish, *J. Org. Chem.*, **1974**, *39*, 1615.
25. (a) Y. Ito, M. Sawamura and T. Hayashi, *J. Am. Chem. Soc.*, **1986**, *108*, 6405.
    (b) T. Hayashi, M. Sawamura and Y. Ito, *Tetrahedron*, **1992**, *48*, 1999.

26. M. Sawamura, Y. Nakayama, T. Kato and Y. Ito, *J. Org. Chem.*, **1995**, *60*, 1727.
27. M. Sawamura and Y. Ito, in *Catalytic Asymmetric Synthesis*, (I. Ojima, ed.) VCH, New York, 1993, Chapter 7.2, 367.
28. R. Kuwano, H. Miyazaki and Y. Ito, *J. Chem. Soc., Chem. Commun.*, **1998**, 71.
29. (a) S. Kobayashi, H. Ishitani and M. Ueno, *J. Am. Chem. Soc.*, **1998**, *120*, 431. (b) H. Ishitani, M. Ueno and S. Kobayashi, *J. Am. Chem. Soc.*, **1997**, *119*, 7153.
30. E. Hagiwar, A. Fujii and M. Sodeoka, *J. Am. Chem. Soc.*, **1998**, *120*, 2474.
31. D. Ferraris, B. Young, T. Dudding and T. Lectka, *J. Am. Chem. Soc.*, **1998**, *120*, 4548.
32. H. Fujieda, M. Kanai, T. Kambara, A. Iida and K. Tomioka, *J. Am. Chem. Soc.*, **1997**, *119*, 2060.
33. P. Bakó, Á. Szöllõsy, P. Bombicz and L. Tõke, *Synlett*, **1997**, 291.
34. P. Bakó, K. Vizvárdi, Z. Bajor and L. Tõke, *J. Chem. Soc., Chem. Commun.*, **1998**, 1193.
35. S. Arai and T. Shioiri, *Tetrahedron Lett.*, **1998**, *39*, 2145.
36. D. Basavaiah, P.D. Rao and R.S. Hyma, *Tetrahedron*, **1996**, *52*, 8001.
37. T. Oishi, H. Oguri and M. Hirama, *Tetrahedron: Asymmetry*, **1995**, *6*, 1241.
38. I.E. Markó, P.R. Giles and N.J. Hindley, *Tetrahedron*, **1997**, *53*, 1015.
39. T. Hayase, T. Shibata, K. Soai and Y. Wakatsuki, *J. Chem. Soc., Chem. Commun.*, **1998**, 1271.
40. A.G.M. Barrett and A. Kamimura, *J. Chem. Soc., Chem. Commun.*, **1995**, 1755.
41. (a) K. Mikami, M. Terada, S. Narisawa and T. Nakai, *Synlett*, **1991**, 255. (b) D.J. Berrisford and C. Bolm, *Angew. Chem., Int. Ed. Engl.*, **1995**, *34*, 1717.
42. M. Terada and K. Mikami, *J. Chem. Soc., Chem. Commun.*, **1994**, 833.
43. K. Mikami and S. Matsukawa, *J. Am. Chem. Soc.*, **1993**, *115*, 7039.
44. K. Mikami, S. Narisawa, M. Shimizu and M. Terada, *J. Am. Chem. Soc.*, **1992**, *114*, 6566.
45. E.J. Corey, D. Barnes-Seeman, T.W. Lee and S.N. Goodman, *Tetrahedron Lett.*, **1997**, *38*, 6513.
46. K. Mikami, T. Yajima, M. Terada, E. Kato and M. Maruta, *Tetrahedron: Asymmetry*, **1994**, *5*, 1087.
47. E.M. Carreira, W. Lee and R.A. Singer, *J. Am. Chem. Soc.*, **1995**, *117*, 3649.
48. D.A. Evans, C.S. Burgey, N.A. Paras, T. Vojkovsky and S.W. Tregay, *J. Am. Chem. Soc.*, **1998**, *120*, 5824.
49. Y. Gao, P. Lane-Bell and J.C. Vederas *J. Org. Chem.*, **1998**, *63*, 2133.

8   The Diels-Alder reaction and related transformations

This chapter describes some of the results obtained using Lewis acids as catalysts for Diels-Alder and hetero-Diels-Alder reactions. This is a field which has been well studied for some substrates, and several examples of near perfect levels of selectivity have been recorded.

8.1   Diels-Alder reactions

Whilst the whole area of Lewis acid-catalysed enantioselective reactions has received considerable attention, it is the Diels-Alder reaction which has been subjected to the most intense research. The result is that several effective asymmetric catalytic Diels-Alder reactions are now available.[1,2] This section considers, in turn, the different metals employed as Lewis acids, including catalysts based on boron, aluminium and titanium, as well as a host of other transition metal complexes.

8.1.1   Boron-based Lewis acids

Oxazaborolidines of the general type (**8.01**) were reported by Takasu and Yamamoto[3] and by Helmchen and co-workers[4] to give good asymmetric induction as catalysts in Diels-Alder reactions. However, Corey and Loh demonstrated that the N-tosyl tryptophan-derived catalyst (**8.02**) was particularly effective in providing high enantioselectivity.[5]

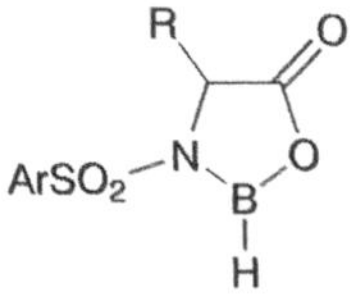

(8.01) 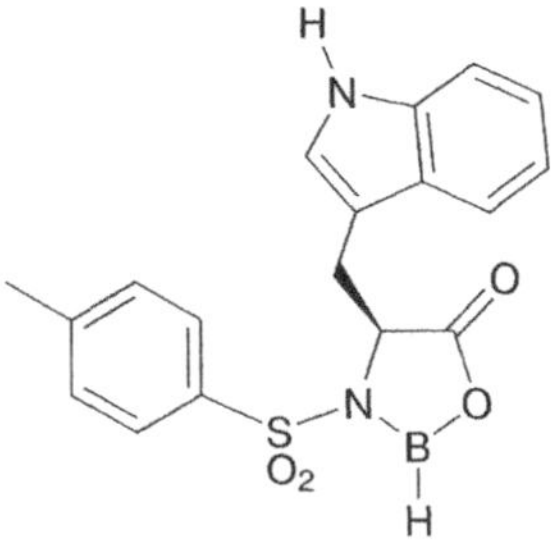   (8.02)

The indole unit, which is a π-base, is believed to have an attractive interaction with the dienophile, which is a π-acid. When the dienophile

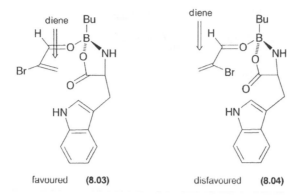

favoured   (8.03)   disfavoured   (8.04)

Figure 8.1   The favoured s-*cis* transition state with oxazaborolidine catalysts.

has associated to the Lewis acidic boron, the enal may adopt either the s-*cis* (**8.03**) or s-*trans* (**8.04**) conformation, as illustrated in Figure 8.1. Complex (**8.04**) (s-*trans*) undergoes an unfavourable interaction between the bromine atom and indole ring as the diene approaches, and the bromine atom is forced downwards. The reaction preferentially proceeds through the s-*cis* conformation (**8.03**) and the model accounts for the very high selectivity observed.[6]

Bromoacrolein (**8.05**) is a good substrate for the enantioselective Diels-Alder reaction, and reacts with cyclopentadiene (**8.06**) to give the Diels-Alder adduct (**8.07**), with good yield and excellent selectivity. Corey and co-workers subsequently used this and related catalysts, including oxazaborolidine (**8.08**), to provide precursors to various natural products.[7] For example, the reaction of bromoacrolein (**8.05**) with the elaborated cyclopentadiene (**8.09**) yields the product (**8.10**), with excellent diastereo- and enantiocontrol; the product was used in synthesis of a gibberellic

(8.06)   (8.05)

5 mol% (**8.02**)
―――――――――――
1 h, -78°C, CH₂Cl₂
95%

(8.07) 99% ee
96:4 *exo* CHO : *endo*

(8.09)   (8.05)

10 mol% (**8.02**)
―――――――――――
16 h, -78°C, CH₂Cl₂
81%

(8.10) 99% ee
99:1 *exo* CHO:*endo*

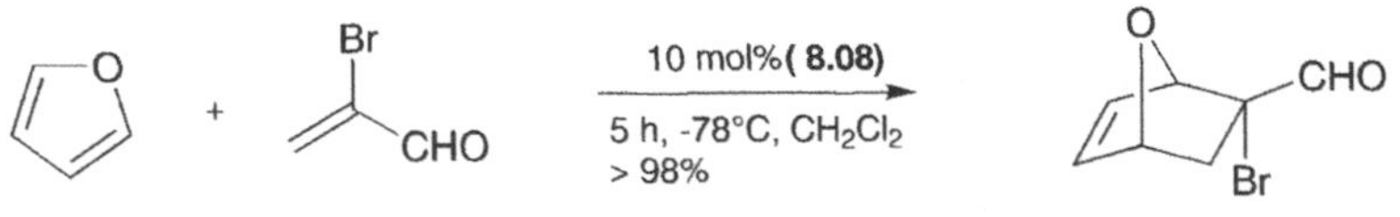

(8.11)   (8.05)

(8.12) 92% ee
99:1 *exo* CHO:*endo*

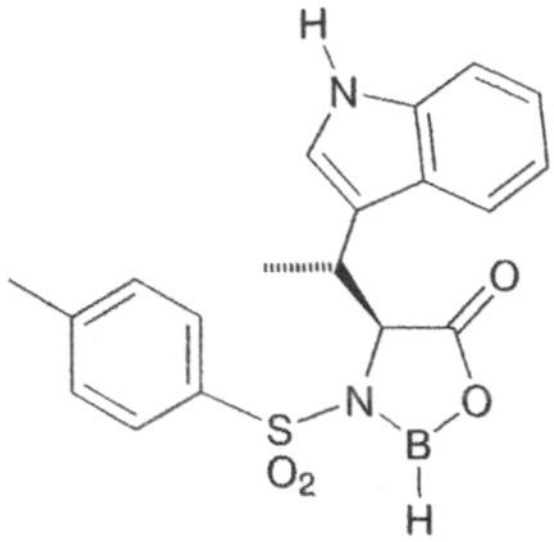

10 mol%(**8.08**)

5 h, -78°C, CH₂Cl₂
> 98%

(8.08)

acid. The catalyst also lends itself well to Diels-Alder reactions with furan
(**8.11**) as the dienophile;[8] the oxabicyclic product (**8.12**) is a useful
synthetic building block.

In 1988, Yamamoto and co-workers reported the use of chiral
(acyloxy)borane (CAB) catalysts (**8.13**) for the enantioselective Diels-
Alder reaction.[9] These catalysts are derived from tartaric acid and, again,
their reactions have been particularly selective with α-substituted enals as
substrate.[10] The dienophile (**8.14**) undergoes a highly selective Diels-
Alder reaction, yielding the adduct (**8.15**), which contains four new
stereocentres controlled by the catalysed reaction. The CAB catalysts
were successfully applied to a cyclic example, where the acyclic substrate
(**8.16**) gave rise to a bicyclic product (**8.17**).[11]

The solution conformations of CAB complexes with methacrolein and
crotonaldehyde were investigated by Nuclear Overhauser Effect (NOE)
spectroscopy.[12] The s-*trans* conformation was found to be favoured in
most cases, although this does not necessarily prove that it is the most
reactive conformation.

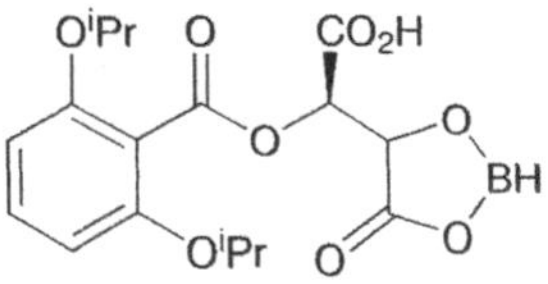

(8.13)

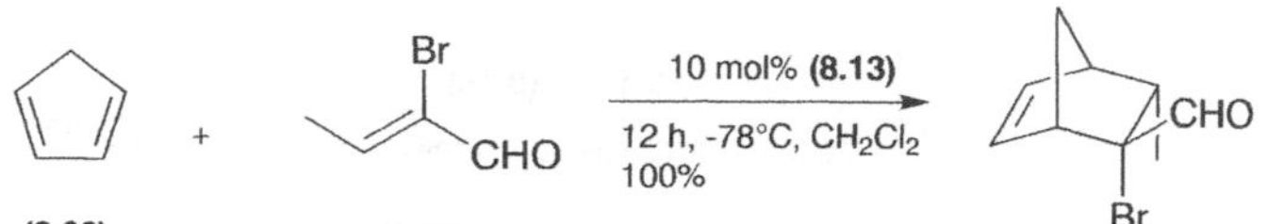

(8.06)   (8.14)

(8.15) 98% ee
99:1 *exo* CHO:*endo*

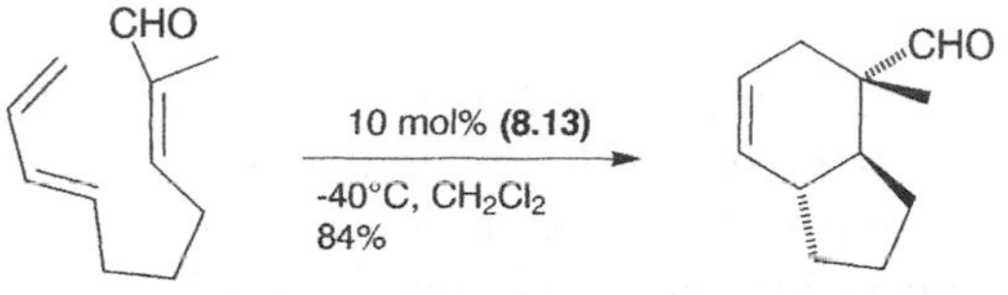

(8.16)

(8.17) 92% ee
99:1 *endo:exo*

Yamamoto and Ishihara used the unusual catalyst (8.18), which they describe as a 'Brønsted acid assisted chiral Lewis acid', for the Diels-Alder reaction. The catalyst provides superb stereocontrol with the α-substituted enals reported.[13] In the transition state assembly, one of the phenoxy groups is protonated, which allows the carbonyl group of the dienophile to coordinate to the Lewis acidic boron atom. Bromoacrolein (8.05), methacrolein (8.19) and dienophile (8.20) all react with essentially complete stereoselectivity to give the Diels-Alder adducts (8.22–8.24).

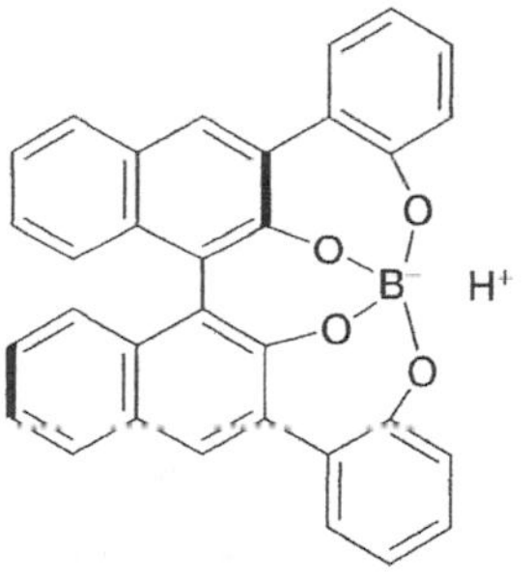

(8.18)

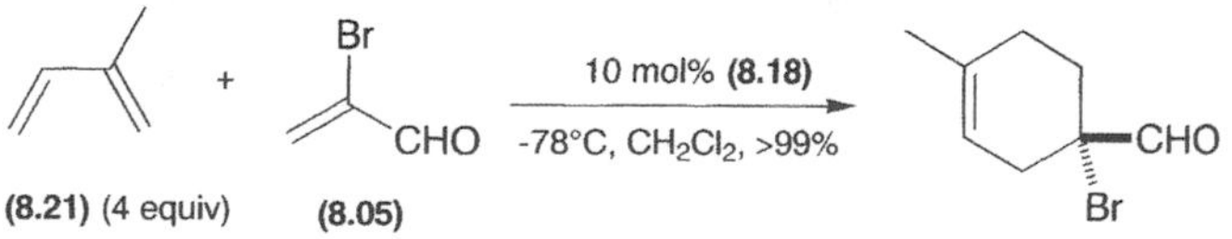

(8.21) (4 equiv)   (8.05)

(8.22) 98% ee

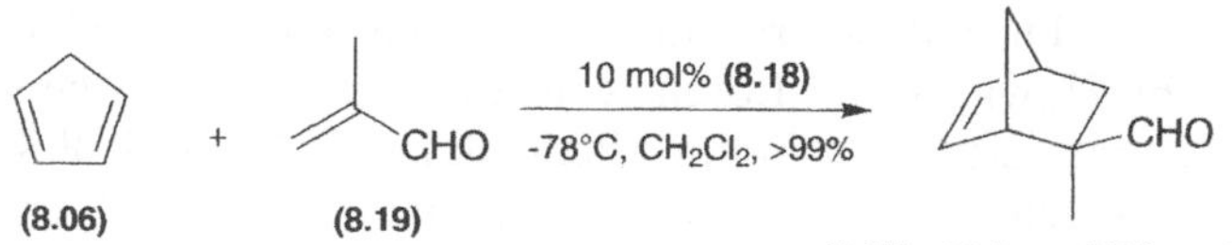

(8.06)   (8.19)

(8.23) >99:1 *exo* CHO:*endo*
98% ee

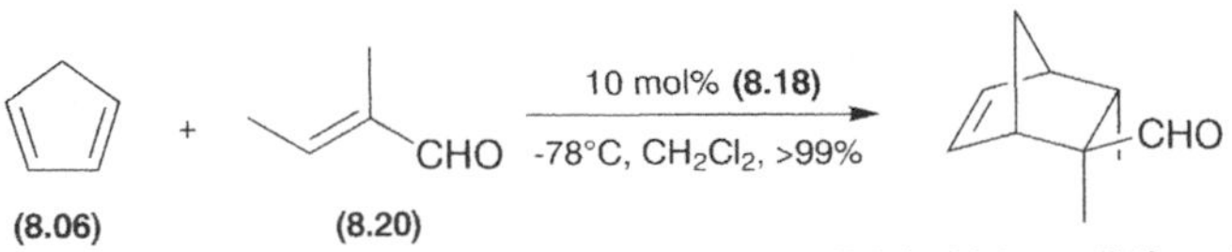

(8.06)   **(8.20)**

(8.24) >99:1 *exo* CHO:*endo*
98% ee

The cationic oxazaborinane **(8.25)** is remarkable because of its reactivity. Even at −94°C, relatively unreactive dienes, such as butadiene **(8.26)** and cyclohexadiene **(8.27)**, undergo reaction with bromoacrolein **(8.05)**, with excellent yields and good selectivities.[14]

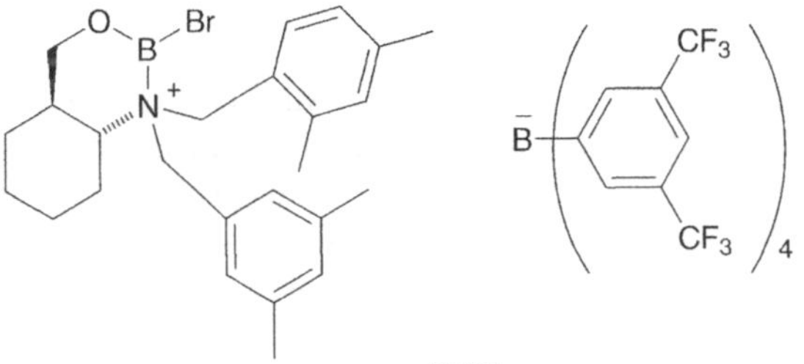

(8.25)

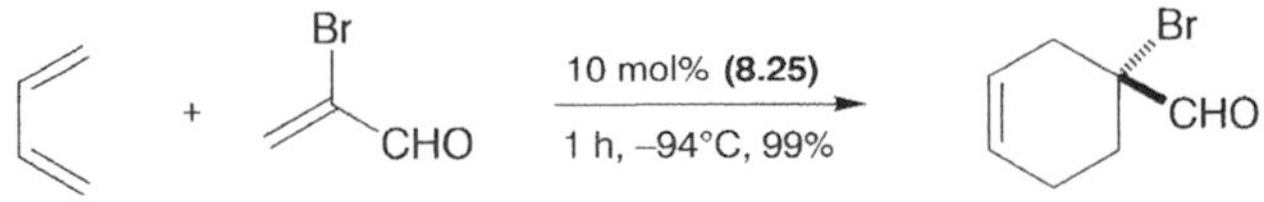

(8.26)   **(8.05)**

(8.28) 94% ee

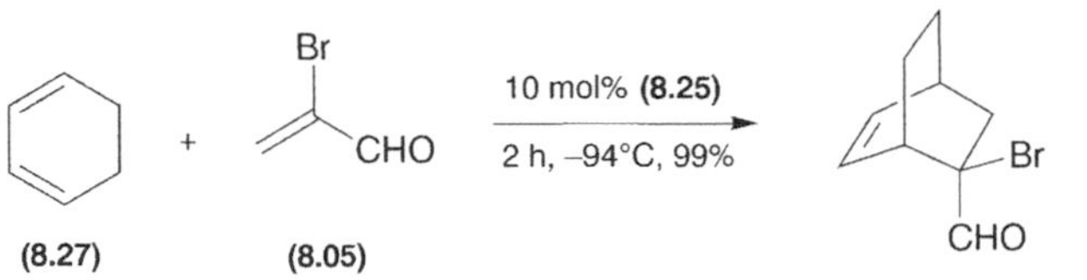

(8.27)   **(8.05)**

(8.29)
exo:*endo* CHO 4:96
93% ee

Many other enantiomerically pure boron-based Lewis acid catalysts are used in the Diels-Alder reaction.[15] Amongst these, dichloroborane **(8.30)** used by Hawkins and Loren is noteworthy because the substrates used were enoate esters, including dienophiles **(8.31)** and **(8.32)**, rather than the enals generally tested in Diels-Alder reactions.[16] The bicyclic adducts were formed, with good to excellent enantioselectivities.

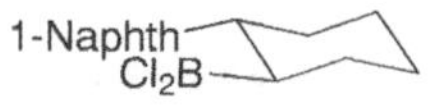

1-Naphth
Cl$_2$B

(8.30)

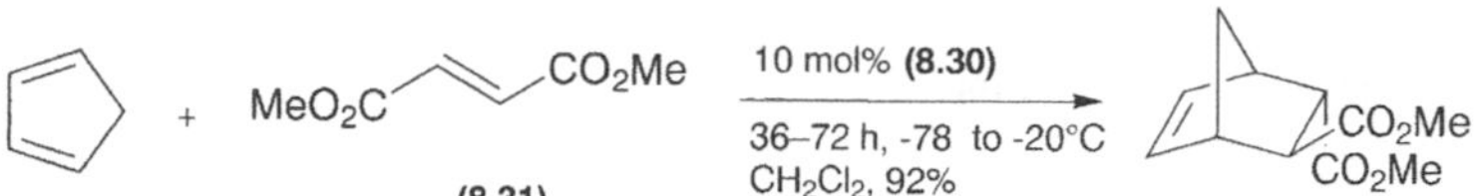

(8.06) + MeO$_2$C ⟶ CO$_2$Me (8.31)

10 mol% **(8.30)**
⟶
36–72 h, -78 to -20°C
CH$_2$Cl$_2$, 92%

CO$_2$Me
CO$_2$Me

(8.33) 90% ee

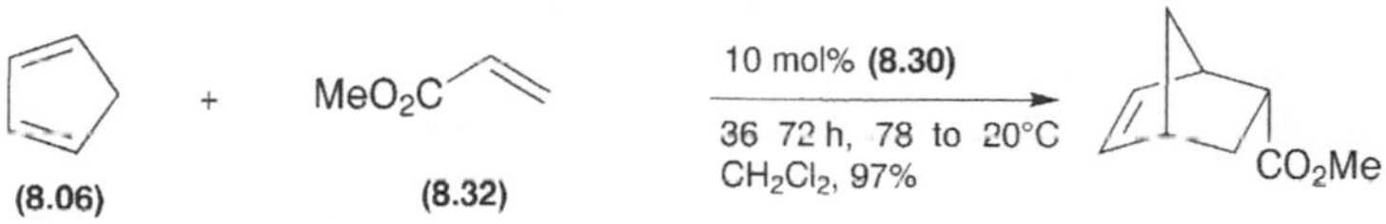

(8.06) + MeO$_2$C (8.32)

10 mol% **(8.30)**
⟶
36 72 h, 78 to 20°C
CH$_2$Cl$_2$, 97%

CO$_2$Me

(8.34) 97% ee

8.1.2 *Aluminium-based Lewis acids*

Some of the earliest work with catalytic asymmetric Diels-Alder reaction used aluminium catalysts.[17] Corey and co-workers used an aluminium catalyst (**8.35**) in the Diels-Alder reaction between a substituted cyclopentadiene (**8.36**) and the acryloyl oxazolidinone (**8.37**).[18] The cycloadduct (**8.38**) was obtained, with good selectivity. A particularly impressive example of the catalytic asymmetric Diels-Alder reaction was provided by Wulff and co-workers, who used the vaulted BINOL-aluminium complex (**8.39**) as a catalyst.[19] Not only is the reaction highly

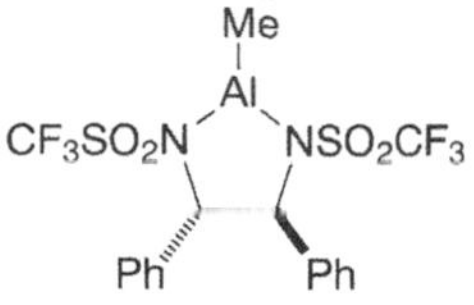

Me
Al
CF$_3$SO$_2$N   NSO$_2$CF$_3$
Ph   Ph

(8.35)

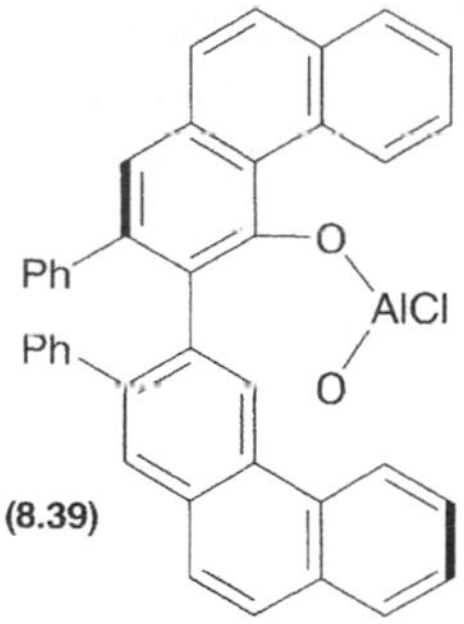

Ph
Ph
O
AlCl
O

(8.39)

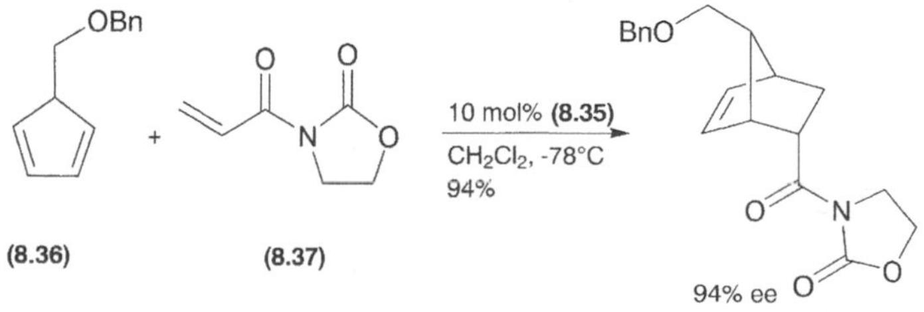

OBn

(8.36) + acryloyl oxazolidinone (8.37)

10 mol% **(8.35)**
⟶
CH$_2$Cl$_2$, -78°C
94%

BnO

94% ee

(8.38)

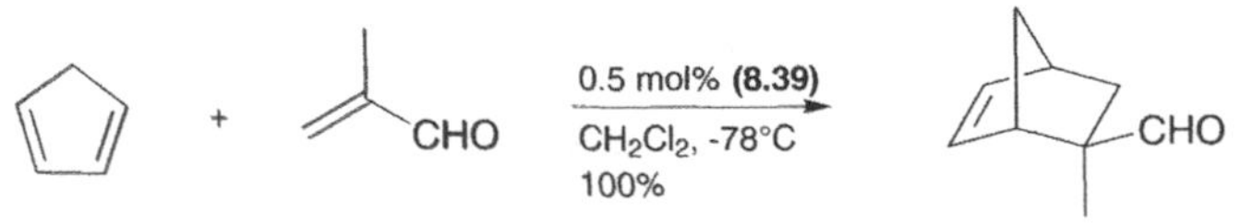

(8.06)   **(8.19)**

(8.23)
97:3 *exo*:CHO *endo*
97.7% ee

selective, it uses an unusually low catalyst loading of just 0.5 mol%. Amongst Lewis acid-catalysed reactions in general and Diels-Alder reactions in particular, this is a remarkably small amount of catalyst.

8.1.3   Titanium-based Lewis acids

There are several common Lewis acids based around titanium, including titanium tetrachloride and titanium tetraiopropoxide. Enantiomerically pure variants of these Lewis acids provide the basis for catalytic asymmetric reactions.

Narasaka and co-workers used the titanium complex of the TADDOL ($\alpha,\alpha,\alpha',\alpha'$-tetraaryl-1,3-dioxolane-4,5-dimethanol) ligand **(8.40)** to catalyse Diels-Alder reactions of acyloxazolidinones.[20] Thus, the crotonyl derivative **(8.41)** was converted into the Diels-Alder adduct **(8.42)** upon reaction with cyclopentadiene. The use of oxazolidinone substrates with titanium catalysts allows two-point binding, as indicated by structure **(8.43)**, and the bidentate nature of the interaction offers rigidity and, therefore, good selectivity. In 1995, three papers were published, independently, discussing the mechanism of Ti-TADDOLate-catalysed Diels-Alder reactions.[21]

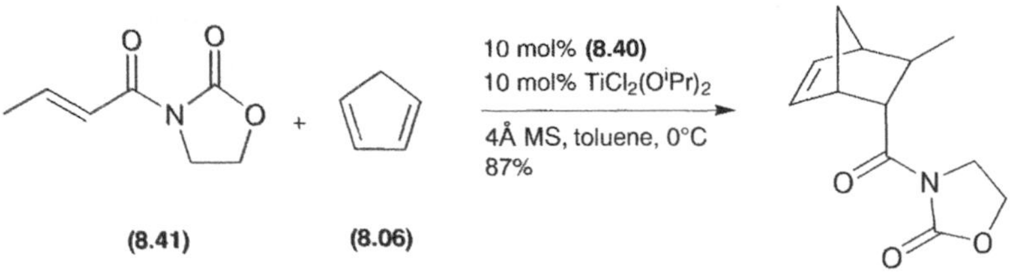

(8.41)   **(8.06)**

10 mol% **(8.40)**
10 mol% TiCl$_2$(O^{i}Pr)$_2$

4Å MS, toluene, 0°C
87%

(8.42)
92:8 *endo:exo*
91% ee

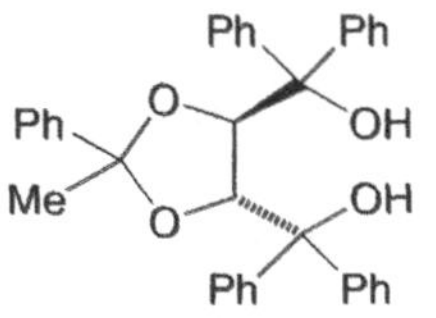

(8.40) 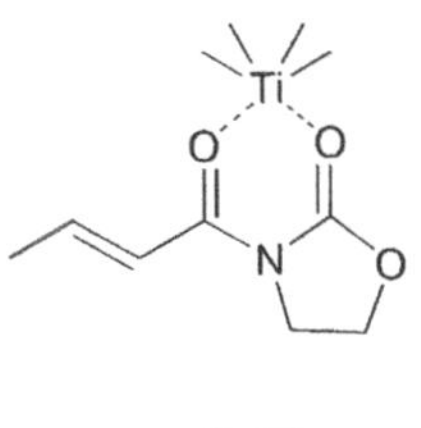   **(8.43)**

BINOL/Titanium complexes have also provided high enantioselectivities in Diels-Alder reactions.[22] The 6,6'-dibromoBINOL/titanium complex (8.44) gives slightly improved selectivities in some cases.[23] For example, the Diels-Alder reaction between diene (8.45) and methacrolein (8.19) is catalysed by this titanium complex, providing the product (8.46) with very good selectivity.

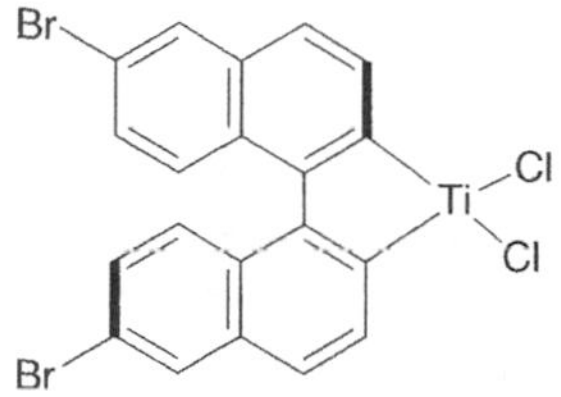

(8.44)

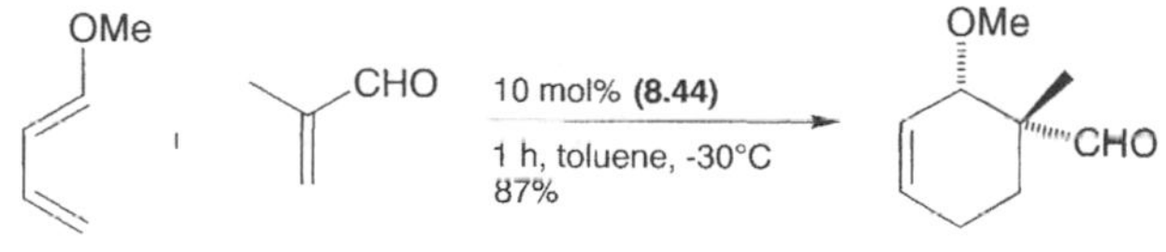

(8.45)   (8.19)   10 mol% (8.44)

1 h, toluene, -30°C
87%

(8.46)
99:1 endo:exo
94% ee

Other titanium complexes have been used to give good enantioselectivities in the catalysed Diels-Alder reaction, including the titanocene complex (8.47),[24] although the zirconocene analogue was slightly superior. Also, the elaborated BINOL ligand complex (8.48),[25] and the sulfonamide complex (8.49) have given good selectivities in the Diels-Alder reactions of cyclopentadiende with dienophiles (8.37), (8.50) and (8.05).[26]

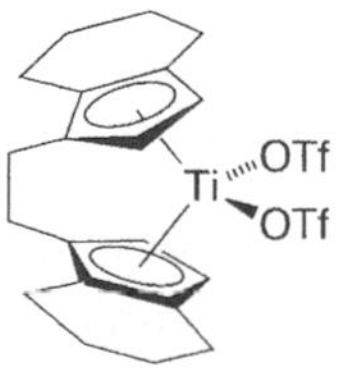

(8.47)

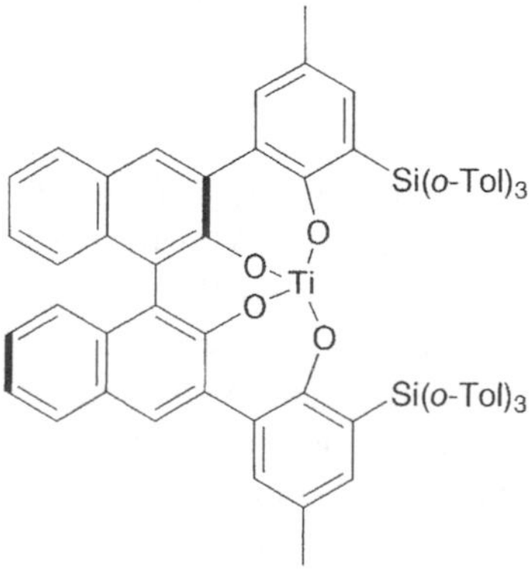

(8.48)

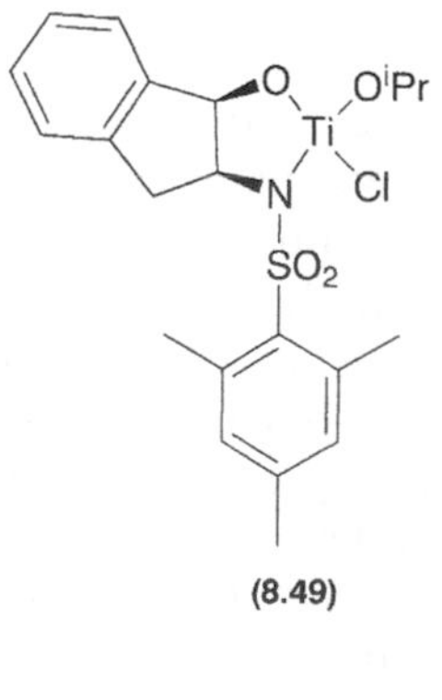

(8.49)

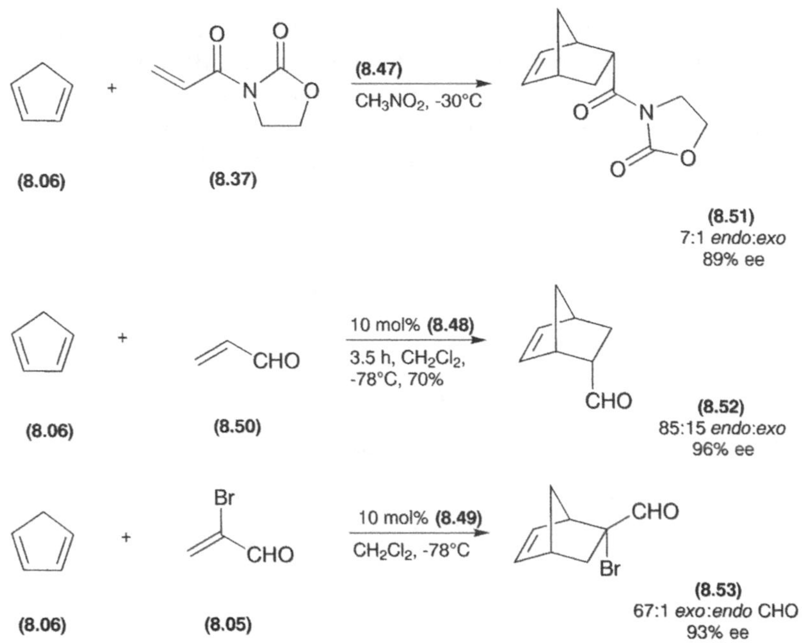

8.1.4   Metal/oxazoline catalysts

Ligands based around the oxazoline unit have been used successfully in several metal-catalysed enantioselective processes. They have certainly shown their value as ligands in the Diels-Alder reaction. The first report of a bis-oxazoline being used as a ligand in a Diels-Alder reaction was published by Corey and co-workers in 1991.[27] They showed an enantio-selectivity of up to 86% ee with 99:1 *endo*:*exo* selectivity for the reaction betwen cyclopentadiene (8.06) and acryloyloxazolidinone (8.37) with a bis-oxazoline/iron (III) catalyst. The use of a similar magnesium complex of ligand (8.54) afforded slightly higher enantioselectivity.[28] The coordination of the substrate to the magnesium complex offers one face of the alkene selectivity for reaction with cyclopentadiene, which preferentially approaches from the top face of the transition state assembly, as shown in Figure 8.2. Alternative magnesium/oxazoline complexes have also been reported by other researchers.[29, 30]

Evans and co-workers examined the use of copper(II) complexes of bis-oxazoline (8.55) as catalysts for Diels-Alder reactions. These complexes can provide very high asymmetric induction, as shown in the Diels-Alder reaction of the acryloyloxazolidinone (8.37) with cyclopentadiene (8.06).[31] In this case, a square planar complex explains the stereochemical outcome, as shown in Figure 8.2.

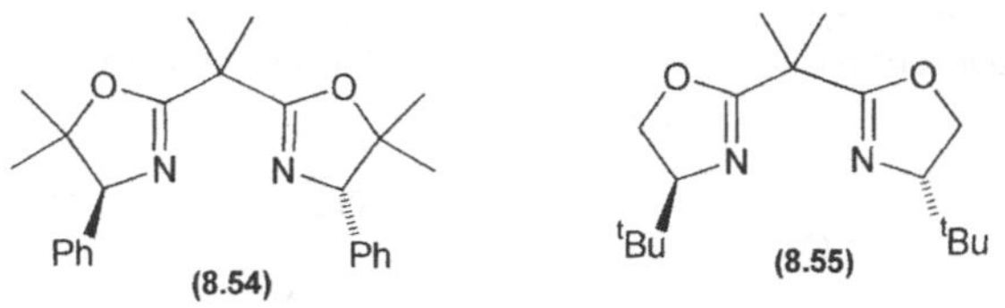
Ph
Ph
(8.54)
tBu
tBu
(8.55)

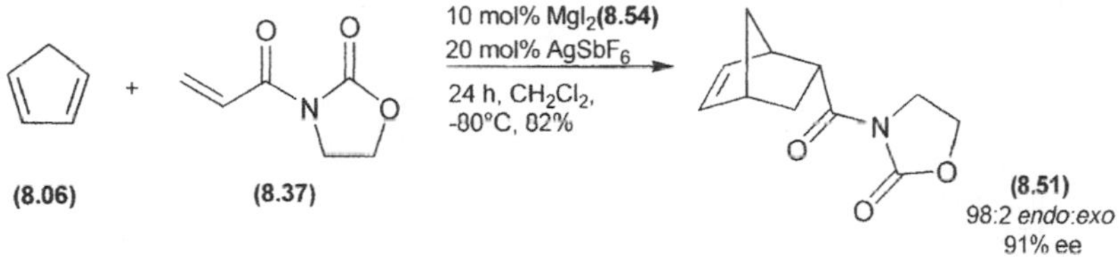
10 mol% MgI2(8.54)
20 mol% AgSbF6
24 h, CH2Cl2,
-80°C, 82%
(8.06)
(8.37)
(8.51)
98:2 endo:exo
91% ee

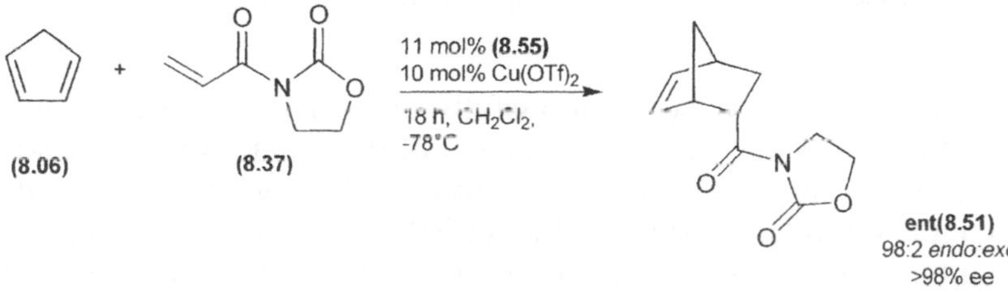
11 mol% (8.55)
10 mol% Cu(OTf)2
18 h, CH2Cl2,
-78°C
(8.06)
(8.37)
ent(8.51)
98:2 endo:exo
>98% ee

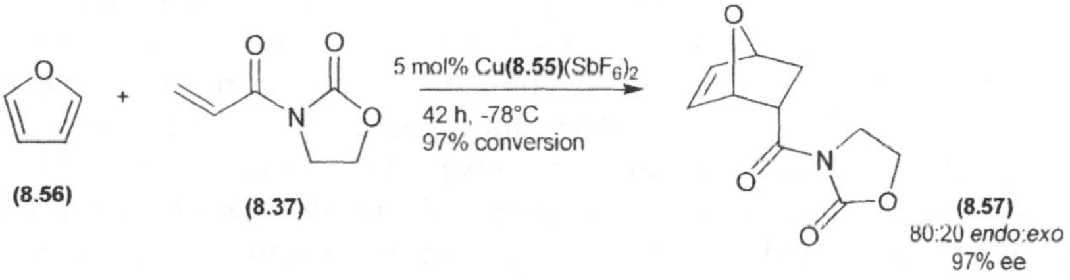
5 mol% Cu(8.55)(SbF6)2
42 h, -78°C
97% conversion
(8.56)
(8.37)
(8.57)
80:20 endo:exo
97% ee

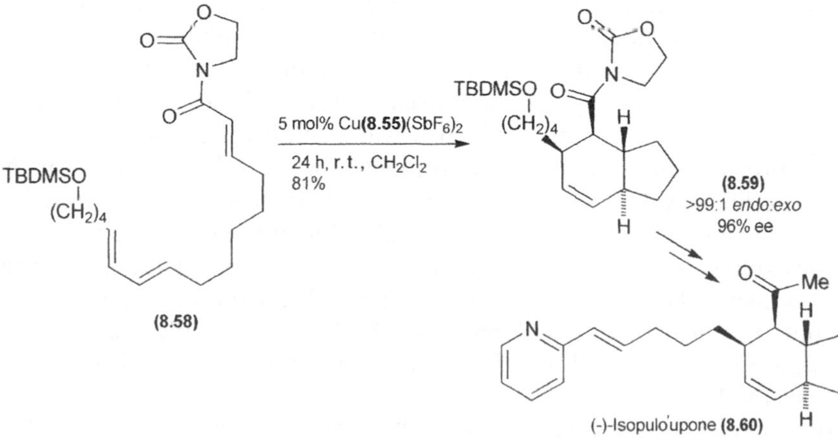
TBDMSO
(CH2)4
5 mol% Cu(8.55)(SbF6)2
24 h, r.t., CH2Cl2
81%
(8.58)
TBDMSO
(CH2)4
(8.59)
>99:1 endo:exo
96% ee
Me
(-)-Isopulo'upone (8.60)

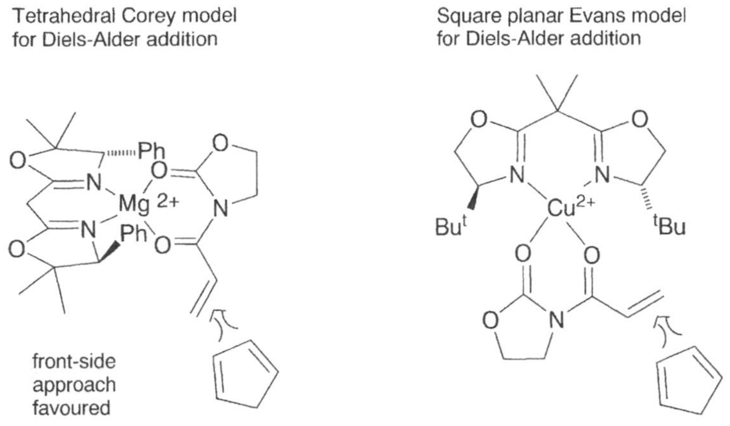

Figure 8.2 Corey and Evans models for asymmetric induction.

The reaction has also been successfully applied to the use of furan (**8.56**) as the dienophile, where replacement of the triflate counterion with SbF_6^- provides a more reactive catalyst.[32] At higher temperatures ($-20°C$), the furan Diels-Alder adduct (**8.57**) was isolated after 24 h with no enantiomeric excess, although after 2–5 h a 59% ee was seen, indicating that racemisation of the product was occurring under these conditions. The lower temperature ($-78°C$) circumvented this problem.

The copper(II) catalyst has been applied to the intramolecular Diels-Alder reaction.[33] The precursor (**8.58**) undergoes an intramolecular Diels-Alder reaction (IMDA reaction) to give the product (**8.59**), which proceeds with remarkable selectivity to give the product (**8.59**), which was subsequently converted into the marine toxin (−)-isopulo'upone (**8.60**).

Other oxazoline ligands have also been used successfully in copper-catalysed Diels-Alder reactions[34], including phosphino-oxazoline ligands.[35] Rhodium[36] and ruthenium[37] oxazoline complexes have been reported to give fairly good enantioselectives. In 1998, a truly remarkable paper was published, in which it was demonstrated that the ligand (**8.61**) provides an asymmetric environment for several metal salts for the catalytic Diels-Alder reaction.[38] In each case, the catalyst is prepared *in situ* by mixing of the ligand and metal salt. One of the key features of this bis-oxazoline is its ability to act as a *trans*-chelator in an octahedral environment. The change in coordination chemistry means that even though the bis-oxazolines (**8.55**) and (**8.61**) possess differing absolute configurations, they afford the same enantiomer of the Diels-Alder adduct (**8.51**).

In the nickel-catalysed reaction, the use of ligand with only 20% enantiomeric excess still produced a Diels-Alder adduct with up to

96% ee, which is a highly effective example of chiral amplification (see Section 6.1).

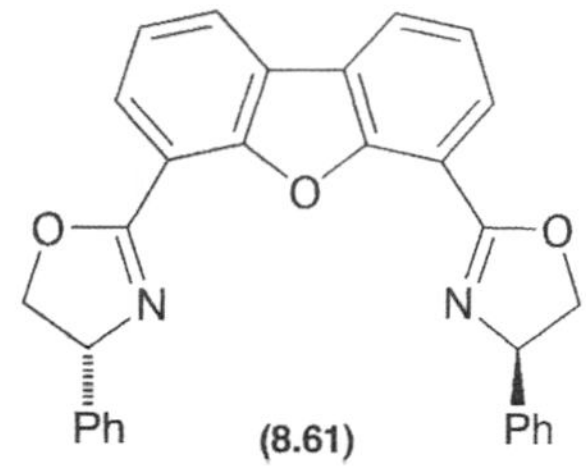

(8.61)

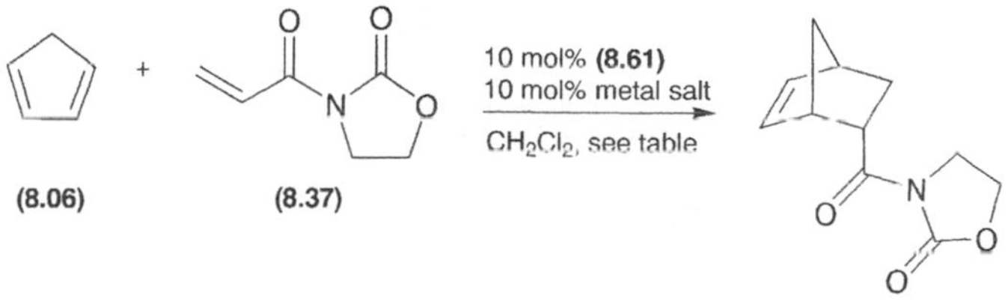

(8.06)   +   (8.37)

10 mol% (8.61)
10 mol% metal salt
CH$_2$Cl$_2$, see table

(ent-8.51)

metal salt	conditions	yield (%)	endo:exo	ee (%)
Mg(ClO$_4$)$_2$	-40°C, 10 h	100	97:3	91
Ni(ClO$_4$)$_2$.6H$_2$O	-40°C, 14 h	96	97:3	>99
Mn(ClO$_4$)$_2$.6H$_2$O	-40°C, 96 h	96	97:3	83
Fe(ClO$_4$)$_2$	-40°C, 48 h	90	99:1	98
Co(ClO$_4$)$_2$.6H$_2$O	-40°C, 48 h	97	97:3	99
Cu(ClO$_4$)$_2$ + 3H$_2$O	-40°C, 15 h	99	97.3	96
Zn(ClO$_4$)$_2$	-40°C, 48 h	99	98:2	97

8.1.5   Other catalysts

There have been many other reports of catalytic asymmetric Diels-Alder reactions, from the Yamashita and Katsuki Mn(salen) complexes[39] to the heterobimetallic complexes of Shibasaki and co-workers.[40] Amongst this array of catalysts, many provide high selectivity. Two further examples are given, which show particularly impressive selectivity. Kündig and co-workers used the cationic iron catalyst (8.62), for example in the Diels-Alder reaction to give cycloadduct (8.64).[41] Kobayashi and co-workers have shown that lanthanide and scandium triflate complexes of BINOL with 1,2,6-trimethylpiperidine, formulated as complex (8.65), are effective with oxazolidinone-based substrates.[42, 43, 44] Thus, the ytterbium complex, formulated as structure (8.65), gives good selectivity in the formation of the Diels-Alder product (8.42).

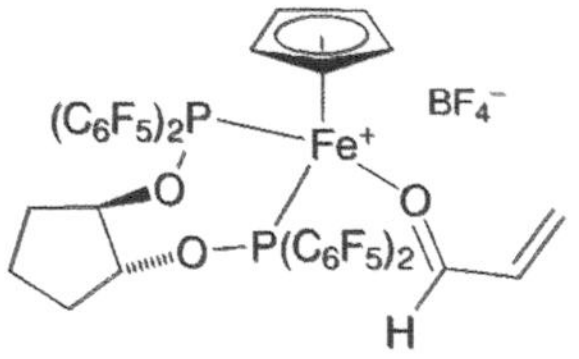

(8.62)

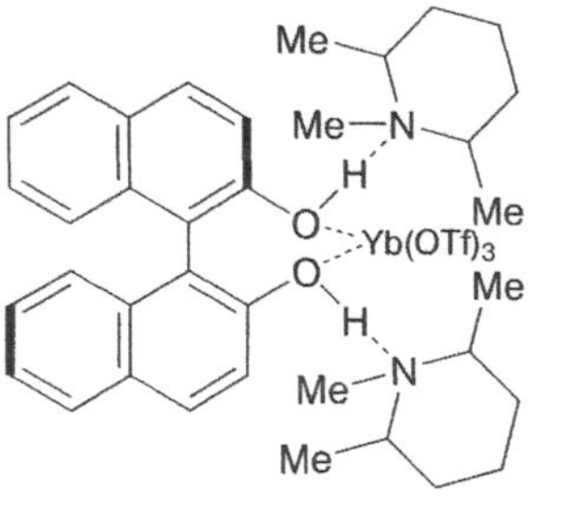

(8.65)

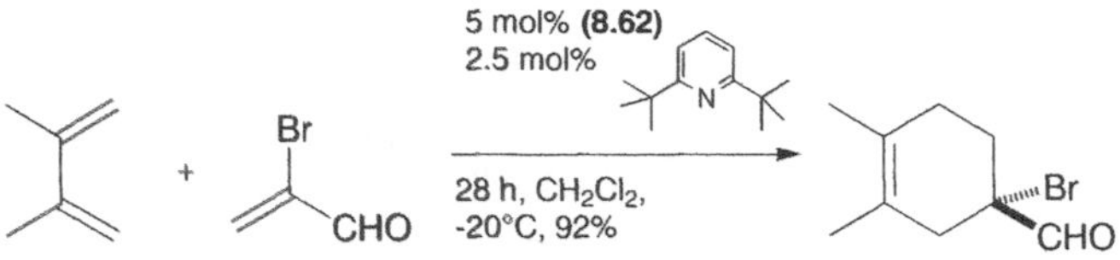

5 mol% (8.62)
2.5 mol%

28 h, CH₂Cl₂,
-20°C, 92%

(8.63)   (8.05)

(8.64)
97–98% ee

20 mol% (8.65)
4Å MS

30 min, CH₂Cl₂,
0°C

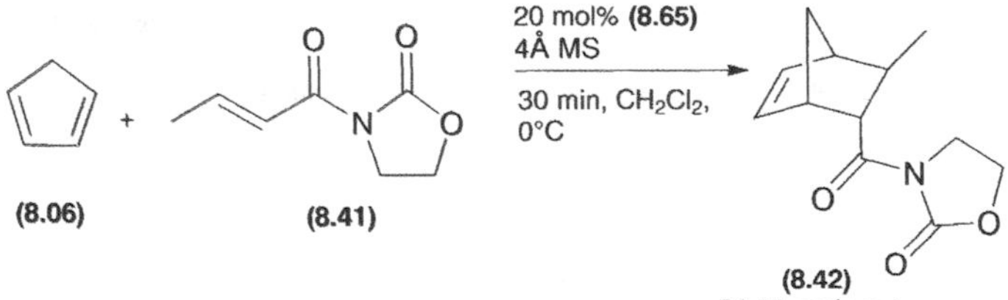

(8.06)   (8.41)

(8.42)
89:11 endo:exo
95% ee

Homo-Diels-Alder reactions have also been achieved, with very high selectivity in some cases. In 1990, two groups independently reported achieving the homo-Diels-Alder reaction between norbornadiene (8.66) and phenylacetylene (8.67), yielding the deltacyclene product (8.68).[45] When the phosphine used in this reaction was Norphos, the product was formed with remarkable selectivity.

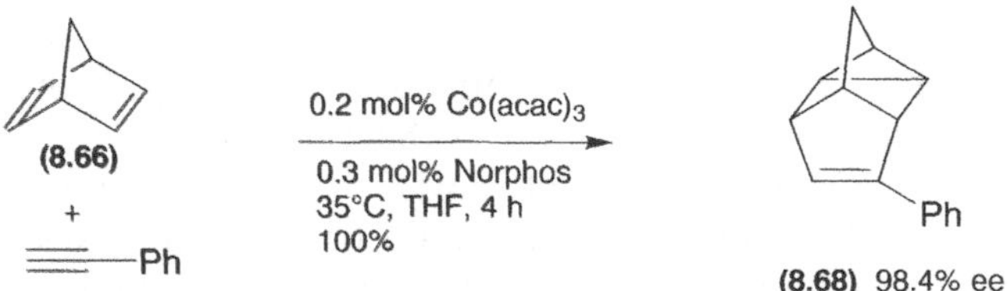

(8.66)

0.2 mol% Co(acac)₃

0.3 mol% Norphos
35°C, THF, 4 h
100%

+

≡—Ph
(8.67)

(8.68) 98.4% ee

The challenges remaining for catalytic asymmetric Diels-Alder reactions include the preparation of catalysts with higher turnover numbers (lower catalyst loading) and the wider use of dienophiles other than α-substituted enals and oxazolidinones.

8.2   Inverse electron demand Diels-Alder

Inverse electron demand Diels-Alder reactions involve a cycloaddition between an electron-rich dienophile and a electron-poor diene, the opposite electronic requirements from a normal Diels-Alder.

Markó and Evans used ytterbium triflate complexes of BINOL (8.69) and found that vinyl sulfide (8.70) provided the highest enantioselectivity in the inverse electron demand Diels-Alder reaction with the diene (8.71).[46]

Posner and co-workers used titanium complexes of BINOL (8.69) to promote inverse electron demand Diels-Alder reactions between the same diene (8.71) and vinyl ethers.[47] Most of this work involved the use of stoichiometric Lewis acid to give 95–98% ee. Curiously, catalytic conditions (10 mol%) afforded the opposite enantiomer in 50–60% ee.

(8.71)   (8.70)   10 mol% Yb(OTf)₃
                                10 mol% (R)-BINOL (8.69)
                                2 mol% ⁱPr₂NEt
                                CH₂Cl₂, 91%   (8.72) >95% ee

8.3   Hetero-Diels-Alder

The hetero-Diels-Alder reaction provides the opportunity to incorporate a heteroatom into the Diels-Alder product. Most commonly, the catalytic asymmetric version of this reaction involves the reaction between an aldehyde (8.73) and a reactive diene (8.74) (typically with one or two oxygen substituents attached). Normally, the products isolated, after acidic work-up, are the enones (8.75). The products can be formed either by a direct cycloaddition or via a two-step aldol-Michael sequence, according to Figure 8.3.

The earliest work on an enantioselective variant of this reaction was performed by Danishefsky and co-workers.[48] Enantiomerically pure europium complexes provided good enantioselectivities, after optimisation for the hetero-Diels-Alder reaction between diene (8.76) and benzaldehyde (8.77), which after hydrolysis gives the enone (8.78).

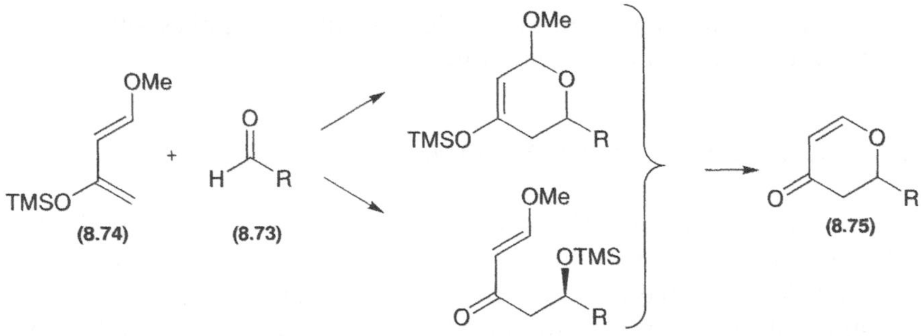

Figure 8.3   Alternative routes for the hetero-Diels-Alder reaction.

Examples of other catalysts used include the hindered BINOL ligand **(8.79)**, as its aluminium complex.[49] The same reaction has been achieved using an enantiomerically pure vanadium complex.[50] Yamamoto's CAB catalysts (see Section 8.1) have been used to good effect; it was found that the group attached to the boron was important for high enantioselectivities.[51] With an appropriate substituent, as shown in catalyst **(8.81)**, the formation of the hetero-Diels-Alder product **(8.84)** could be formed with excellent stereocontrol.

Cu(II)-bis-oxazoline complexes have been reported to be useful catalysts for hetero-Diels-Alder reactions.[52] The choice of solvent was shown by Johannsen and Jørgensen to be important, with nitroalkanes providing the best results. The authors proposed that the use of a polar solvent provides a higher degree of dissociation of the counterion from the metal. Under these conditions, the reaction of diene **(8.27)** with the activated aldehyde **(8.85)** affords the bicyclic adduct **(8.86)**, with high selectivity.[53] The reaction has also been applied to the hetero-Diels-Alder reacton of ketones, albeit activated ketones, such as ethyl pyruvate **(8.87)**.[54]

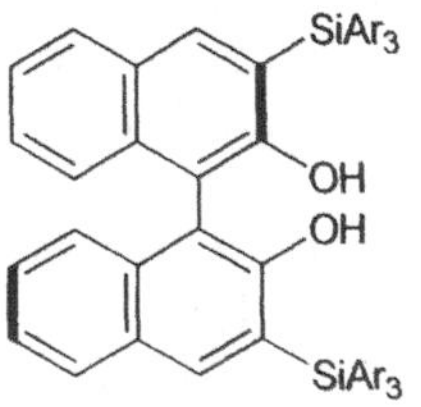

Ar = 3,5-(CH$_3$)$_2$C$_6$H$_3$-

(8.79)   **(8.81)**

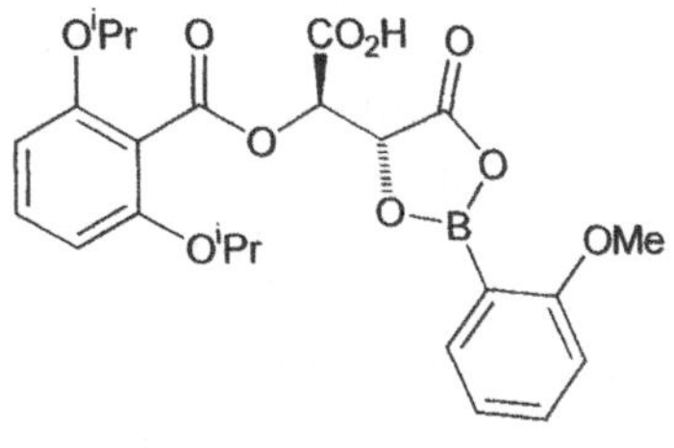

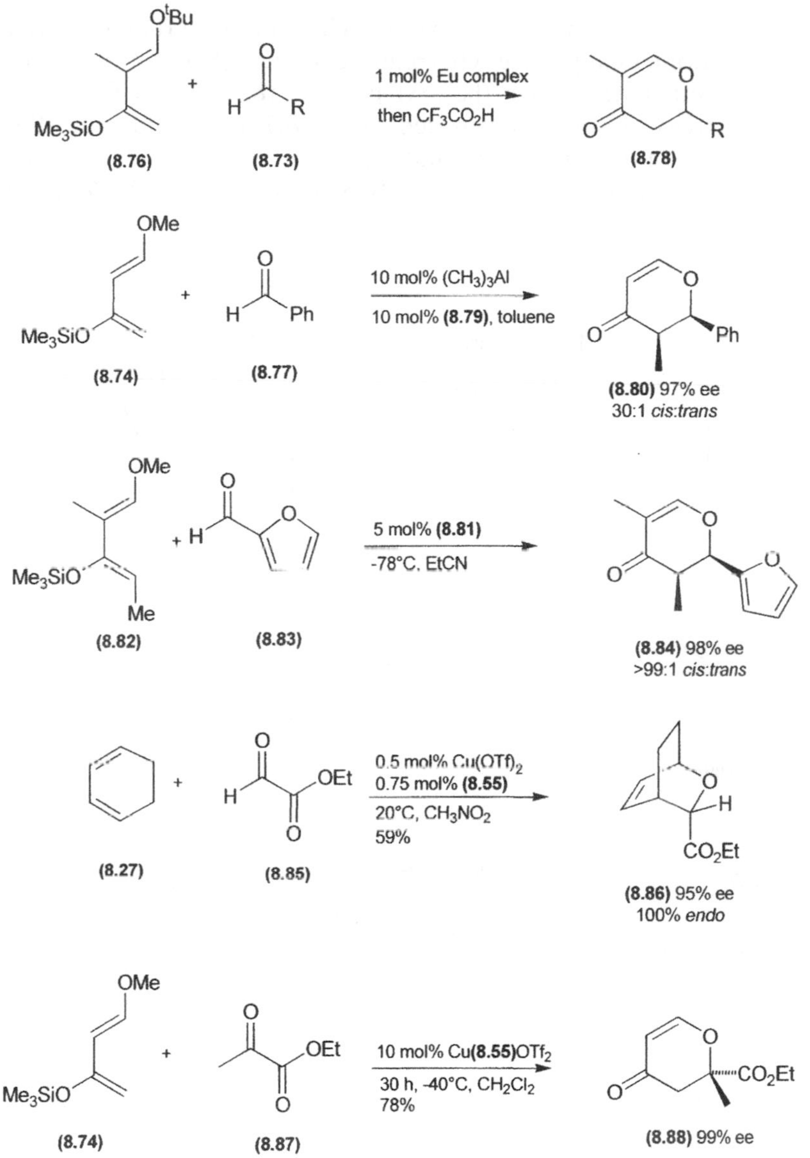

(8.76) + (8.73) → 1 mol% Eu complex, then CF₃CO₂H → (8.78)

(8.74) + (8.77) → 10 mol% (CH₃)₃Al, 10 mol% (8.79), toluene → (8.80) 97% ee, 30:1 *cis:trans*

(8.82) + (8.83) → 5 mol% (8.81), -78°C, EtCN → (8.84) 98% ee, >99:1 *cis:trans*

(8.27) + (8.85) → 0.5 mol% Cu(OTf)₂, 0.75 mol% (8.55), 20°C, CH₃NO₂, 59% → (8.86) 95% ee, 100% *endo*

(8.74) + (8.87) → 10 mol% Cu(8.55)OTf₂, 30 h, -40°C, CH₂Cl₂, 78% → (8.88) 99% ee

In addition to the hetero-Diels-Alder reaction, where the heteroatom is contained in the dienophile,[55] heterodienes can also be employed.[56] Evans demonstrated this to great effect using the phosphonate (8.89) with ethyl vinyl ether (8.90), as well as variations on these substrates.[57]

Aza-Diels-Alder reactions have received relatively little attention using asymmetric catalysts. Perhaps surprisingly, the imine (8.92) actually behaves as a diene in the reaction with cyclopentadiene (**8.06**) catalysed by an enantiomerically pure ytterbium complex (**8.93**).[58] The initial product rearomatises to give the aryl product (**8.94**), with fairly good enantioselectivity. Using electron-rich diene (**8.74**), the aldimine (**8.95**) now behaves as a dienophile and undergoes an enantioselective reaction using a zirconium catalyst with the BINOL derivative (**8.96**) as the ligand.[59]

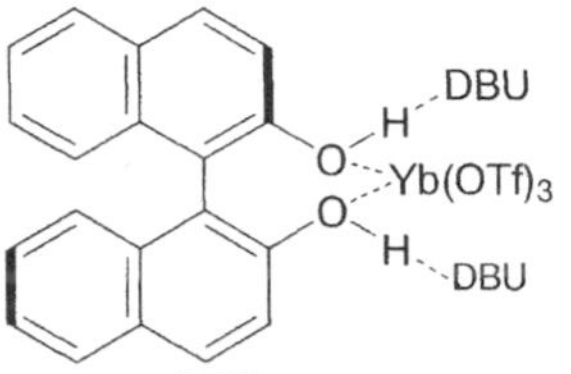

(8.93)
DBU = 1,8-diazabicyclo[5.4.0]undec-7-ene

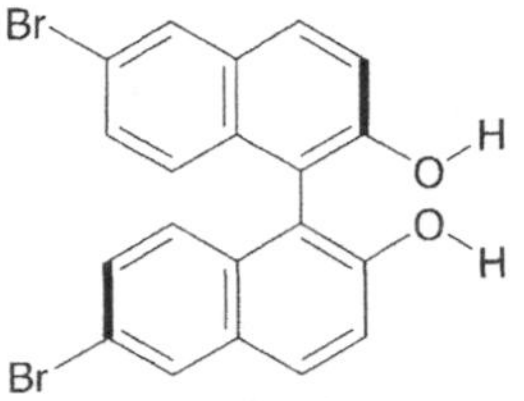

(8.96)

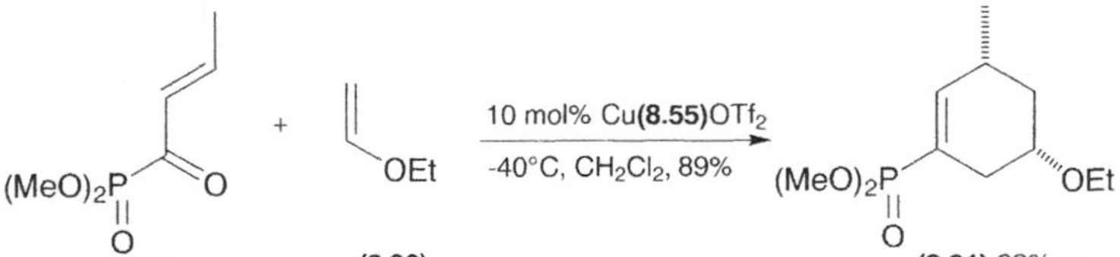

(MeO)$_2$P — (**8.89**) + OEt (**8.90**) → 10 mol% Cu(**8.55**)OTf$_2$ -40°C, CH$_2$Cl$_2$, 89% → (MeO)$_2$P — OEt **(8.91)** 99% ee 99:1 *endo:exo*

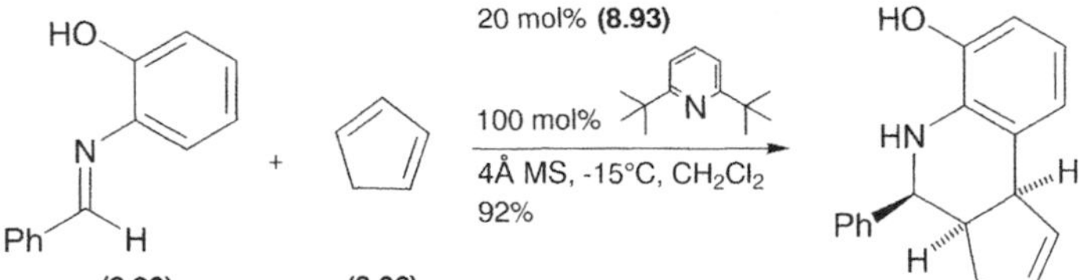

(**8.92**) + (**8.06**) → 20 mol% (**8.93**) 100 mol% 4Å MS, -15°C, CH$_2$Cl$_2$ 92% → **(8.94)** 71% ee 99:1 *cis:trans*

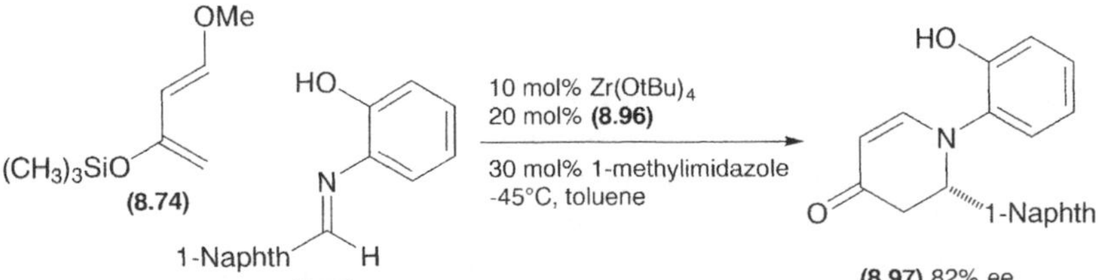

(CH$_3$)$_3$SiO (**8.74**) + 1-Naphth (**8.95**) → 10 mol% Zr(OtBu)$_4$ 20 mol% (**8.96**) 30 mol% 1-methylimidazole -45°C, toluene → **(8.97)** 82% ee

8.4   1,3-dipolar cycloaddition reactions

Whilst there are many cycloaddition reactions which could be subjected to asymetric catalysis, the majority of work has been involved with Diels-Alder and related reactions. Nevertheless, 1,3-dipolar cycloadditions have provided fairly good results as well. Asymmetric catalytic 1,3-dipolar cycloadditions have focused on the reactions of nitrones with alkenes.[60] In 1994, the first examples of such catalytic cycloaddition reactions were reported, independently, by Scheeren and co-workers[61] and Gothelf and Jørgensen.[62]

Scheeren and co-workers used the boron catalyst (8.97), which provided reasonable enantioselectivities in the reaction between nitrone (8.98) and the electron-rich alkene (8.99). The initial cycloadduct (8.100) readily undergoes N-O reduction to give the β-aminoester (8.101).[61] Gothelf and Jørgensen used titanium TADDOL complexes (8.102) to catalyse 1,3-dipolar cycloaddition of the nitrone (8.103) and alkene (8.41) to give similar levels of enantioselectivity in the cycloadduct (8.104).[62] Subsequent improvements and variations of the boron[63] and titanium[64] catalysts have been reported.

Furukawa and co-workers used palladium/BINAP complexes to catalyse 1,3-dipolar cycloaddition of nitrones to alkenes and, whilst enantioselectivity is high (up to 91% ee), control of diastereoselectivity (endo *vs* exo) is poor.[65]

The most selective results, so far, were published in 1998 by Kobayashi and Kawamura using the ytterbium catalyst (8.105).[66] The BINOL controls the sense of asymmetric induction with achiral amines (up to 78% ee). However, the correct enantiomer of the amine (8.107) enhanced the enantioselectivity still further, such that the cycloadduct (8.108) was formed with up to 96% ee.

Asymmetric catalysis of 1,3-dipolar cycloaddition reactions has now reached high levels of enantioselectivity. Since the cycloadducts are synthetically useful (after N-O reduction), it will be interesting to see how future research in this area develops.

(8.97)   (8.102)

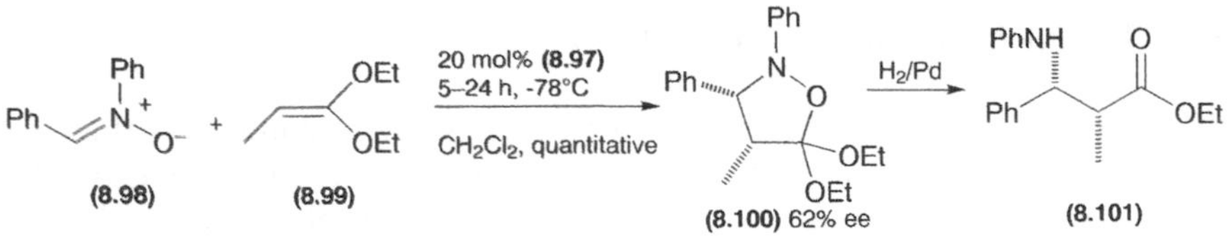

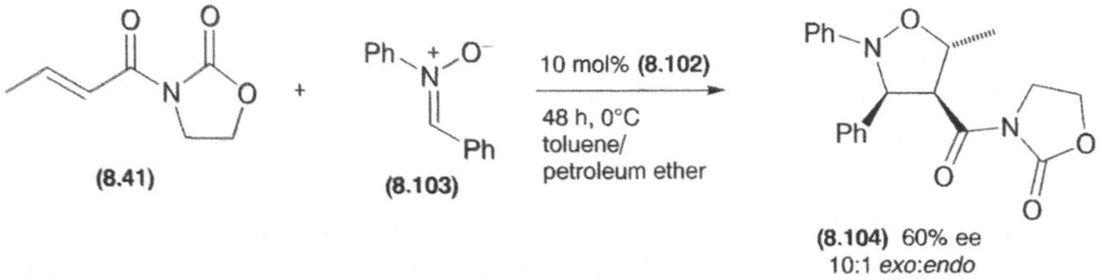

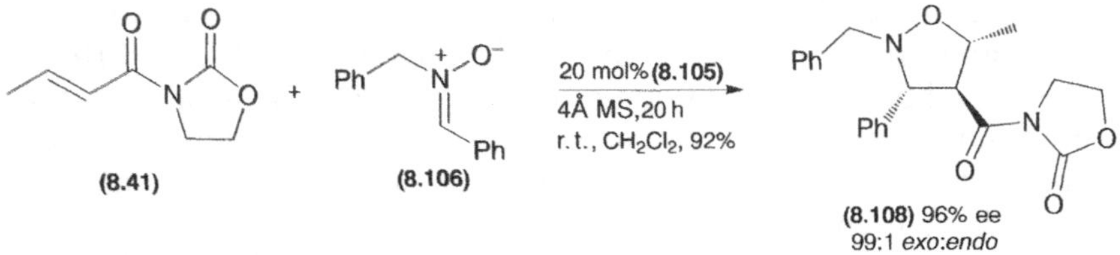

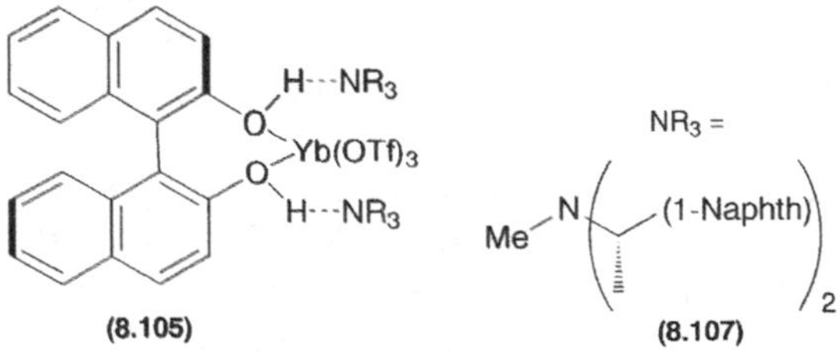

(8.98) (8.99)

20 mol% (8.97)
5–24 h, -78°C
CH₂Cl₂, quantitative

(8.100) 62% ee

H₂/Pd

(8.101)

(8.41) (8.103)

10 mol% (8.102)

48 h, 0°C
toluene/
petroleum ether

(8.104) 60% ee
10:1 exo:endo

(8.41) (8.106)

20 mol% (8.105)
4Å MS, 20 h
r. t., CH₂Cl₂, 92%

(8.108) 96% ee
99:1 exo:endo

NR₃ =

(8.105) (8.107)

References

1. H.B. Kagan and O. Riant, *Chem. Rev.*, **1992**, *92*, 1007.
2. L.C. Dias, *J. Braz. Chem. Soc.*, **1997**, *8*, 289.
3. M. Takasu and H. Yamamoto, *Synlett*, **1990**, 194.
4. D. Sartor, J. Saffrich and G. Helmchen, *Synlett*, **1990**, 197.
5. E.J. Corey and T.-P. Loh, *J. Am. Chem. Soc.*, **1991**, *113*, 8966.
6. E.J. Corey, T.-P. Loh, T.D. Roper, M.D. Azimioara and M.C. Noe, *J. Am. Chem. Soc.*, **1992**, *114*, 8290.
7. E.J. Corey, A. Guzman-Perez and T.-P. Loh, *J. Am. Chem. Soc.*, **1994**, *116*, 3611.
8. E.J. Corey and T.-P. Loh, *Tetrahedron Lett.*, **1993**, *34*, 3979.
9. K. Furuta, Y. Miwa, K. Iwanaga and H. Yamamoto, *J. Am. Chem. Soc.*, **1988**, *110*, 6254.
10. K. Ishihara, Q. Gao and H. Yamamoto, *J. Org. Chem.*, **1993**, *58*, 6917.
11. K. Furuta, A. Kanematsu, H. Yamamoto and S. Takaoka, *Tetrahedron Lett.*, **1989**, *30*, 7231.
12. K. Ishihara, Q. Gao and H. Yamamoto, *J. Am. Chem. Soc.*, **1993**, *115*, 10412.

13. K. Ishihara and H. Yamamoto, *J. Am. Chem. Soc.*, **1994**, *116*, 1561.
14. Y. Hayashi, J.J. Rohde and E.J. Corey, *J. Am. Chem. Soc.*, **1996**, *118*, 5502.
15. For example, see: (a) S. Kobayashi, M. Murakami, T. Harada and T. Mukaiyama, *Chem. Lett.*, **1991**, 1341. (b) D. Kaufmann and R. Boese, *Angew. Chem., Int. Ed. Engl.*, **1990**, *29*, 545. (c) K. Ishihara, H. Kurihara and H. Yamamoto, *J. Am. Chem. Soc.*, **1996**, *118*, 3049.
16. J.M. Hawkins and S. Loren, *J. Am. Chem. Soc.*, **1991**, *113*, 7794.
17. S. Hashimoto, N. Komeshima and K. Koga, *J. Chem. Soc., Chem. Commun.*, **1979**, 437.
18. E.J. Corey, R. Imwinkelried, S. Pikul and Y.B. Xiang, *J. Am. Chem. Soc.*, **1989**, *111*, 5493.
19. J. Bao, W.D. Wulff and A.L. Rheingold, *J. Am. Chem. Soc.*, **1993**, *115*, 3814.
20. K. Narasaka, N. Iwasawa, M. Inoue, T. Yamada, M. Nakashima and J. Sugimori, *J. Am. Chem. Soc.*, **1989**, *111*, 5340.
21. (a) C. Haase, C.R. Sarko and M. DiMare, *J. Org. Chem.*, **1995**, *60*, 1777. (b) D. Seebach, R. Dahinden, R.E. Marti, A.K. Beck, D.A. Plattner and F.N.M. Kühnle, *J. Org. Chem.*, **1995**, *60*, 1788. (c) K.V. Gothelf and K.A. Jørgensen, *J. Org. Chem.*, **1995**, *60*, 6847.
22. K. Mikami, Y. Motoyama and M. Terada, *J. Am. Chem. Soc.*, **1994**, *116*, 2812.
23. Y. Motoyama, M. Terada and K. Mikami, *Synlett*, **1995**, 967.
24. J.B. Jaquith, J. Guan, S. Wang and S. Collins, *Organometallics*, **1995**, *14*, 1079.
25. K. Maruoka, N. Murase and H. Yamamoto, *J. Org. Chem.*, **1993**, *58*, 2938.
26. E.J. Corey, T.D. Roper, K. Ishihara and G. Sarakinos, *Tetrahedron Lett.*, **1993**, *34*, 8399.
27. E.J. Corey, N. Imai and H.-Y. Zhang, *J. Am. Chem. Soc.*, **1991**, *113*, 728.
28. E.J. Corey and K. Ishihara, *Tetrahedron Lett.*, **1992**, *33*, 6807.
29. T. Fujisawa, T. Ichiyanagi and M. Shimizu, *Tetrahedron Lett.*, **1995**, *36*, 5031.
30. G. Desimoni, G. Faita and P.P. Righetti, *Tetrahedron Lett.*, **1996**, *37*, 3027.
31. D.A. Evans, S.J. Miller and T. Lectka, *J. Am. Chem. Soc.*, **1993**, *115*, 6460.
32. D.A. Evans and D.M. Barnes, *Tetrahedron Lett.*, **1997**, *38*, 57.
33. D.A. Evans and J.S. Johnson, *J. Org. Chem.*, **1997**, *62*, 786.
34. (a) D.A. Evans, J.A. Murry, P. von Matt, R.D. Norcross and S.J. Miller, *Angew. Chem., Int. Ed. Engl.*, **1995**, *34*, 798. (b) A.K. Ghosh, P. Mathivanan and J. Cappiello, *Tetrahedron Lett.*, **1996**, *37*, 3815. (c) I.W. Davies, L. Gerena, D. Cai, R.D. Larsen, T.R. Verhoeven and P.J. Reider, *Tetrahedron Lett.*, **1997**, *38*, 1145.
35. I. Sagasser and G. Helmchen, *Tetrahedron Lett.*, **1998**, *39*, 261.
36. A.J. Davenport, D.L. Davies, J. Fawcett, S.A. Garrett, L. Lad and D.R. Russell, *J. Chem. Soc., Chem. Commun.*, **1997**, 2347.
37. D.L. Davies, J. Fawcett, S.A. Garratt and D.R. Russell, *J. Chem. Soc., Chem. Commun.*, **1997**, 1351.
38. S. Kanemasa, Y. Oderaotoshi, S.-I. Sakaguchi, H. Yamamoto, J. Tanaka, E. Wada and D.P. Curran, *J. Am. Chem. Soc.*, **1998**, *120*, 3074.
39. Y. Yamashita and T. Katsuki, *Synlett*, **1995**, 829.
40. T. Morita, T. Arai, H. Sasai and M. Shibasaki, *Tetrahedron: Asymmetry*, **1998**, *9*, 1445.
41. E.P. Kündig, B. Bourdin and G. Bernardinelli, *Angew. Chem., Int. Ed. Engl.*, **1994**, *33*, 1856.
42. S. Kobayashi, H. Ishitani, M. Araki and I. Hachiya, *Tetrahedron Lett.*, **1994**, *35*, 6325.
43. S. Kobayashi, M. Araki and I. Hachiya, *J. Org. Chem.*, **1994**, *59*, 3758.
44. S. Kobayashi, H. Ishitani, I. Hachiya and M. Araki, *Tetrahedron*, **1994**, *50*, 11623.
45. (a) H. Brunner, M. Muschiol and F. Prester, *Angew. Chem., Int. Ed. Engl.*, **1990**, *29*, 652.
    (b) M. Lautens, J.C. Lautens and A.C. Smith, *J. Am. Chem. Soc.*, **1990**, *112*, 5627.
46. I.E. Markó and G.R. Evans, *Tetrahedron Lett.*, **1994**, *35*, 2771.
47. G.H. Posner, F. Eydoux, J.K. Lee and D.S. Bull, *Tetrahedron Lett.*, **1994**, *35*, 7541.
48. M. Bednarski, C. Maring and S. Danishefsky, *Tetrahedron Lett.*, **1983**, *24*, 3451.
49. K. Maruoka, T. Itoh, T. Shirasaka and H. Yamamoto, *J. Am. Chem. Soc.*, **1988**, *110*, 310.
50. A. Togni, *Organometallics*, **1990**, *9*, 3106.

51. (a) Q. Gao, K. Ishihara, T. Maruyama, M. Mouri and H. Yamamoto, *Tetrahedron*, **1994**, *50*, 979. (b) Q. Gao, T. Maruyama, M. Mouri and H. Yamamoto, *J. Org. Chem.*, **1992**, *57*, 1951.

52. M. Johannsen and K.A. Jørgensen, *J. Org. Chem.*, **1995**, *60*, 5757.

53. M. Johannsen and K.A. Jørgensen, *Tetrahedron*, **1996**, *52*, 7321.

54. (a) M. Johannsen, S. Yao and K.A. Jørgensen, *J. Chem. Soc., Chem. Commun.*, **1997**, 2169. (b) S. Yao, M. Johannsen, H. Audrain, R.G. Hazell and K.A. Jørgensen, *J. Am. Chem. Soc.*, **1998**, *120*, 8599.

55. For further examples of the hetero-Diels-Alder, see: (a) S.E. Schaus, J. Brånalt and E.N. Jacobsen, *J. Org. Chem.*, **1998**, *63*, 403. (b) T. Hanamoto, H. Furuno, Y. Sugimoto and J. Inanaga, *Synlett*, **1997**, 79. (c) M. Terada, K. Mikami and T. Nakai, *Tetrahedron Lett.*, **1991**, *32*, 935. (d) G.E. Keck, X.-Y. Li and D. Krishnamurthy, *J. Org. Chem.*, **1995**, *60*, 5998.

56. E. Wada, H. Yasuoka and S. Kanemasa, *Chem. Lett.*, **1994**, 1637.

57. D.A. Evans and J.S. Johnson, *J. Am. Chem. Soc.*, **1998**, *120*, 4895.

58. H. Ishitani and S. Kobayashi, *Tetrahedron Lett.*, **1996**, *37*, 7357.

59. S. Kobayashi, S. Komiyama and H. Ishitani, *Angew. Chem., Int. Ed. Engl.*, **1998**, *37*, 979.

60. For a review on asymmetric cycloaddition reactions of nitrones with alkenes, but not specifically catalytic applications, see: M. Frederickson, *Tetrahedron*, **1997**, *53*, 403.

61. J.-P.G. Seerden, A.W.A. Scholte op Reimer and H.W. Scheeren, *Tetrahedron Lett.*, **1994**, *35*, 4419.

62. K.V. Gothelf and K.A. Jørgensen, *J. Org. Chem.*, **1994**, *59*, 5687.

63. J.-P.G. Seerden, M.M.M. Kuypers and H.W. Scheeren, *Tetrahedron Lett.*, **1995**, *6*, 1441.

64. (a) K.B. Jensen, K.V. Gothelf, R.G. Hazell and K.A. Jørgensen, *J. Org. Chem.*, **1997**, *62*, 2471. (b) Using Mg and Cu bis-oxazolines as catalysts: K.V. Gothelf, R.G. Hazell and Jørgensen, *J. Org. Chem.*, **1996**, *61*, 346.

65. K. Hori, H. Kodama, T. Ohta and I. Furukawa, *Tetrahedron Lett.*, **1996**, *37*, 5947.

66. S. Kobayashi and M. Kawamura, *J. Am. Chem. Soc.*, **1998**, *120*, 5840.

9 Catalytic reactions involving metal carbenoids

The decomposition of diazo compounds with enantiomerically pure metal catalysts has led to catalytic asymmetric routes to cyclopropanes, various C-H insertion reactions as well as ylide reactions and rearrangements. The whole area was reviewed by Doyle in 1998.[1]

9.1 Copper-catalysed cyclopropanation

In 1966, Nozaki and co-workers reported a cyclopropanation reaction catalysed by an enantiomerically pure copper complex, albeit with low enantioselectivity (6% ee) by today's standards.[2] This was the first example of a homogeneous synthetic catalyst providing enantioselectivity in an organic reaction. The ligand design was elaborated by Aratani to provide complex (9.01), which is highly selective for certain cyclopropanation reactions.[3] The Aratani catalyst set the standard for catalytic cyclopropanation reactions. The next significant advance was made by Pfaltz and co-workers, who reported the use of semicorrin ligands (9.02)[4] and azasemicorrins (9.03)[5] for cyclopropanation reactions. Styrene (9.04) undergoes cyclopropanation with these catalysts, with good enantioselectivity, and the diastereoselectivity is reasonably good using bulky diazoesters (9.06).

(9.01)   (9.02)   (9.03)

(9.04)

1 mol% catalyst
N₂CHCO₂ᵗBu (9.06)

(9.05)

catalyst (9.02) 93% ee 81:19 trans:cis
catalyst (9.03) 96% ee 86:14 trans:cis

Bis-oxazolines provide a similar stereochemical environment around the copper but are generally easier to prepare. At about the same time, Masamune and co-workers,[6] Evans and co-workers[7] and Pfaltz and co-workers[8] all reported the use of bis-oxazolines for copper catalysed cyclopropanation. These ligands, including structures (9.07) and (9.08), provide excellent stereocontrol for many intermolecular examples of cyclopropanation. The use of the very hindered diazoester (9.09) enhances the *trans* selectivity of the cyclopropanation reaction of styrene (9.04). Mono-cyclopropanation of diene (9.11) was also achieved, with very good stereocontrol.[9]

Alternative bis-oxazoline ligands have been reported, including (9.13)[10,11] and (9.14),[12] but with no benefits over the more straightforward bis-oxazolines already described. Other di-nitrogen ligands, including binaphthamine-derived iminophosphoranes,[13] diamines[14] and bipyridines,[15] have all been shown to provide very good enantioselectivity in copper-catalysed cyclopropanation, with the best results often

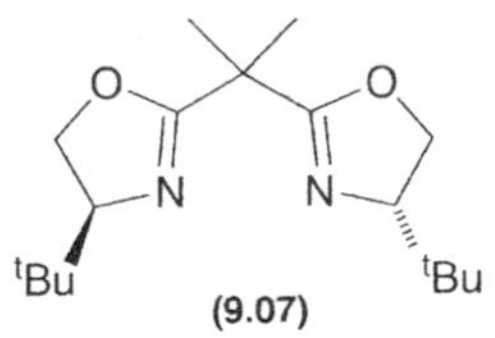

(9.07)

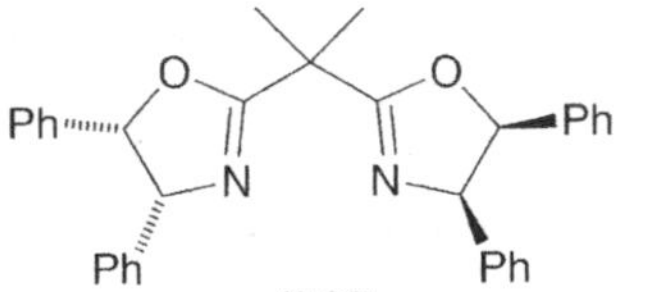

(9.08)

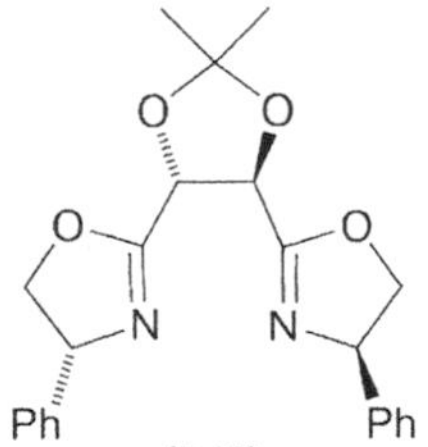

(9.13)

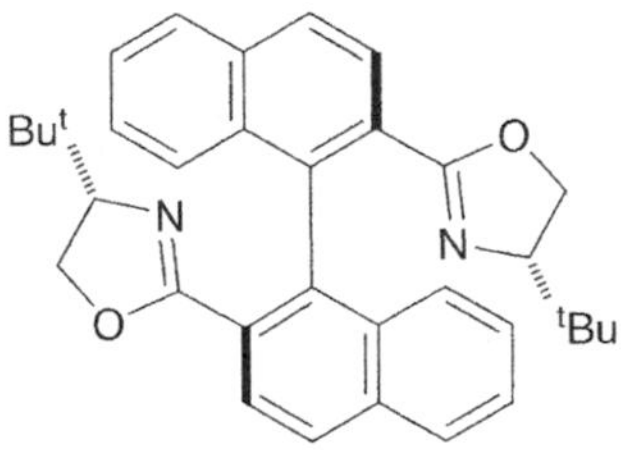

(9.14)

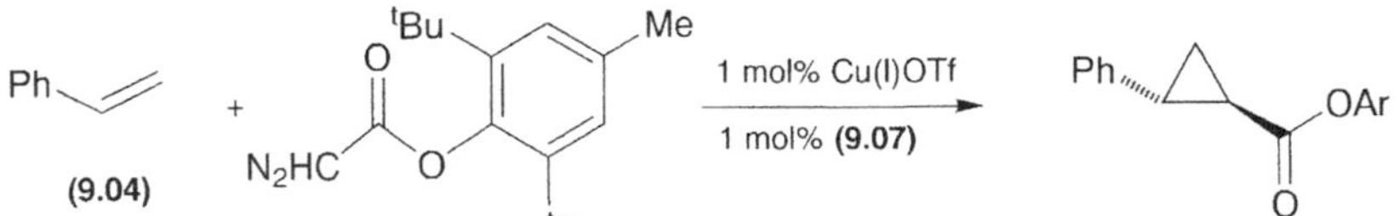

Ph (9.04)

+

(9.09)

1 mol% Cu(I)OTf

1 mol% (9.07)

(9.10) 99% ee
94:6 *trans:cis*

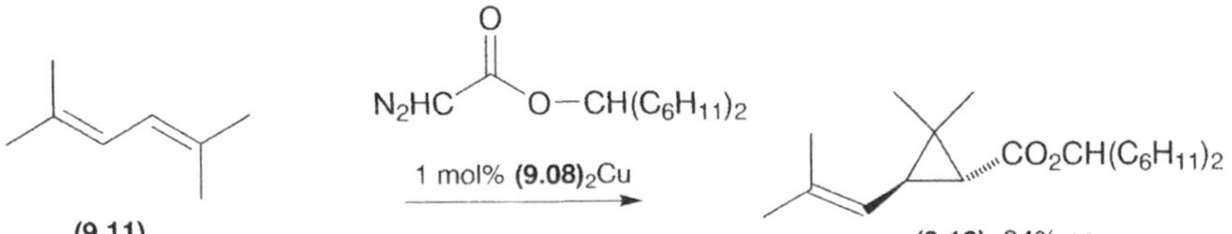

(9.11)

1 mol% (9.08)₂Cu

(9.12) 94% ee
95:5 *trans:cis*

being achieved in the cyclopropanation of styrene **(9.04)** with hindered diazoesters.

Intramolecular cyclopropanations with copper/bis-oxazoline catalysts have generally been less successful than with the Doyle rhodium catalyst (see Section 9.2). However, Shibasaki and co-workers have shown that using bis-oxazoline ligand **(9.15)**, good enantioselectivity is obtained in the cyclisation of the silyl enol ether **(9.16)**.[16, 17] Pfaltz and co-workers have shown that their original semicorrin ligand complexes **(9.02)** are best suited to the cyclisation of the diazoketone **(9.17)**. The corresponding reaction with a bis-oxazoline gave only 77% ee.[18]

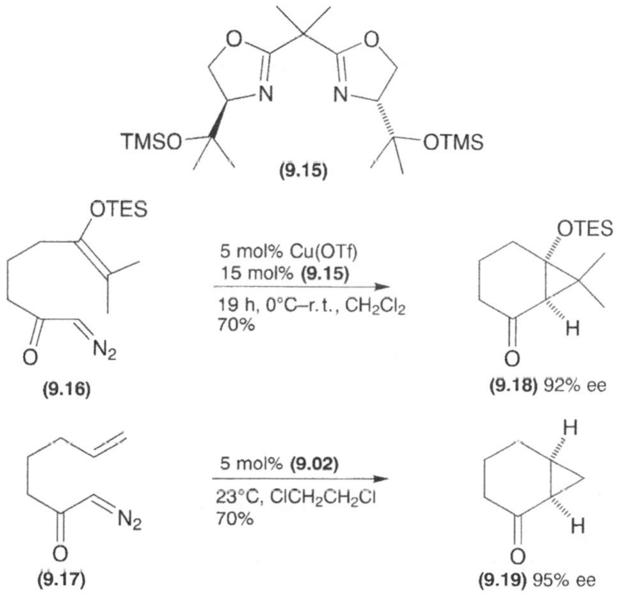

9.2 Rhodium-catalysed cyclopropanation

Enantioselective rhodium-catalysed cyclopropanation reactions have enjoyed considerable success for intramolecular cases. However, by suitable choice of catalyst, intermolecular reactions have also been highly selective in some instances. Use of the Davies/McKervey catalyst **(9.20)**[19] has been particularly successful for intermolecular cyclopropanation reactions with phenyldiazoacelates **(9.21)**[20, 21] and also with vinyl diazo-acetates, such as **(9.22)**.[22] In the latter case, Davies converted the product formed initially **(9.24)** into the cyclopropyl amino acid **(9.25)**.[23] The same product (9.24) has also been converted into the antidepressant, Sertraline (9.26), by Corey and Gant.[24]

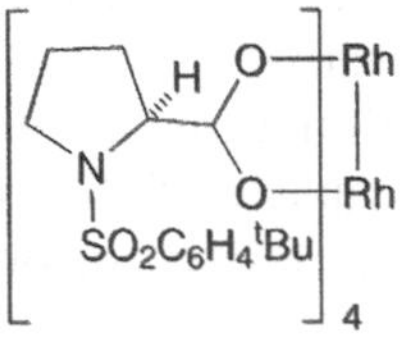

(9.20)

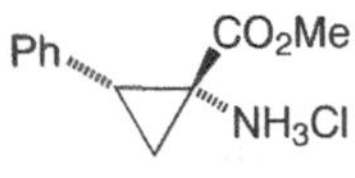

(9.25)

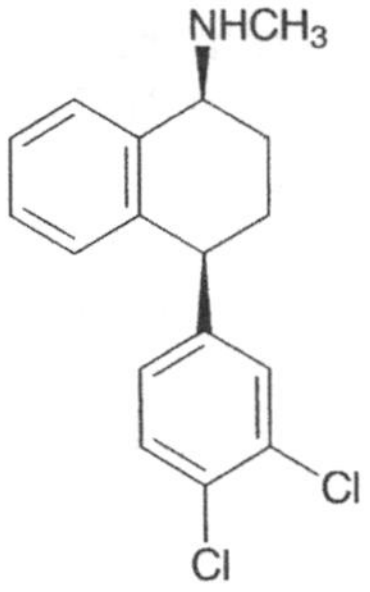

(9.26)

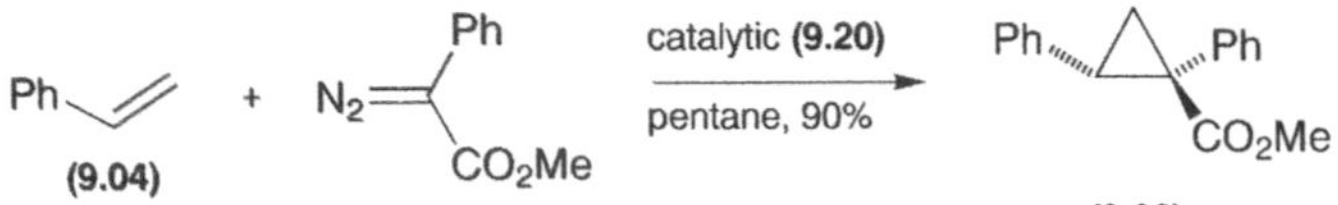

catalytic (9.20)
pentane, 90%

(9.04)

(9.21)

(9.23)
87% ee
98:2 trans:cis

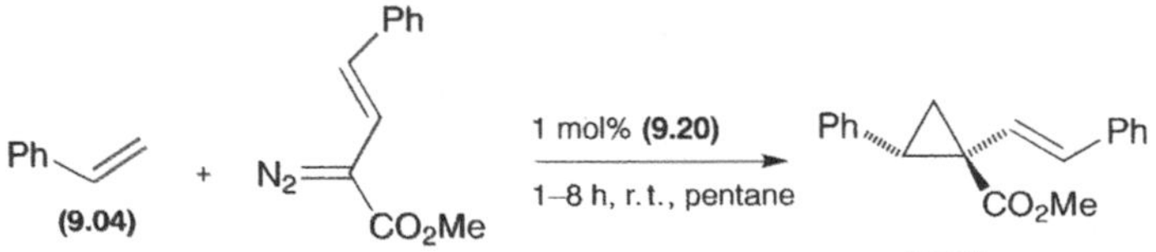

1 mol% (9.20)
1–8 h, r.t., pentane

(9.04)

(9.22)
added over 10 min

(9.24)
90% ee
>40:1 trans:cis

One of the problems with the use of other rhodium catalysts for intermolecular cyclopropanation is a lack of diastereoselectivity in the formation of product.[25] An example of this is the catalyst (9.27), which is effective for intermolecular cyclopropanation of p-chlorostyrene (9.28) with diazoester (9.29), providing up to 98% ee in the cyclopropyl products (9.30) and (9.31), but low trans/cis selectivity.[26]

However, rhodium-catalysed intramolecular cyclopropanation has been achieved for a broad range of substrates,[27] and the enantioselectivity and diastereoselectivity are often reliably high. Doyle has prepared a family of related rhodium complexes, including $Rh_2(MEPY)_4$ (9.32) and $Rh_2(MPPIM)_4$ (9.33), which are effective for various cyclopropanation reactions. Thus, the $Rh_2(MEPY)_4$ catalyst can be used for bicyclisation of the diazoesters (9.34). The selectivity is generally good, but gives poor results for the trans-substituted alkene, where much improved results are seen with the $Rh_2(4S\text{-MPPIM})_4$ catalysts.[28, 29] This is further illustrated by the reaction of the trans-substituted alkene (9.35), which undergoes a highly enantioselective cyclopropanation with the $Rh_2(MPPIM)_4$ cata-

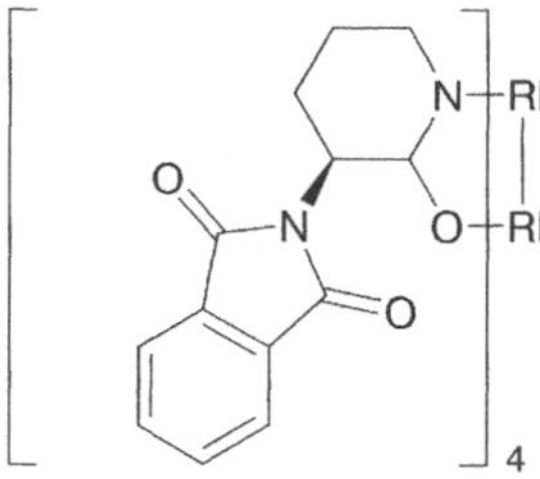

(9.27)

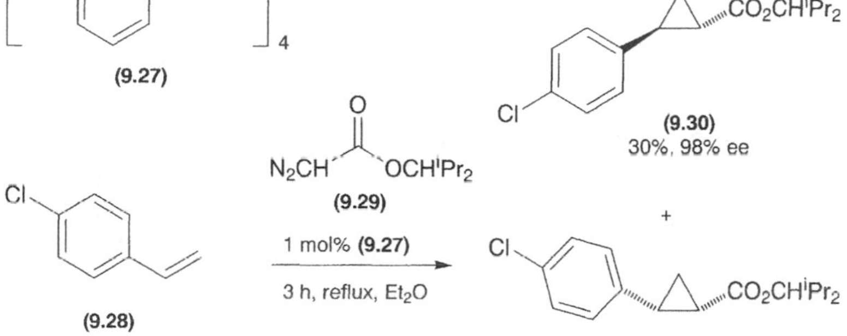

(9.29)

(9.28)

1 mol% (9.27)

3 h, reflux, Et$_2$O

(9.30)
30%, 98% ee

+

(9.31)
20%, 98% ee

lyst, but provides only 68% ee when the Rh$_2$(MEPY)$_4$ catalyst is employed.

The prochiral divinyl substrate (9.36) underwent reaction selectively with one of the enantiotopic alkene groups to give the cyclisation product (9.40), with high enantioselectivity and diastereoselectivity.[30]

Cyclisation reactions have also been performed on diazoacetamides, where the intramolecular cyclopropanation of diazoacetamide (9.37) yields the bicyclic lactam (9.41).[31]

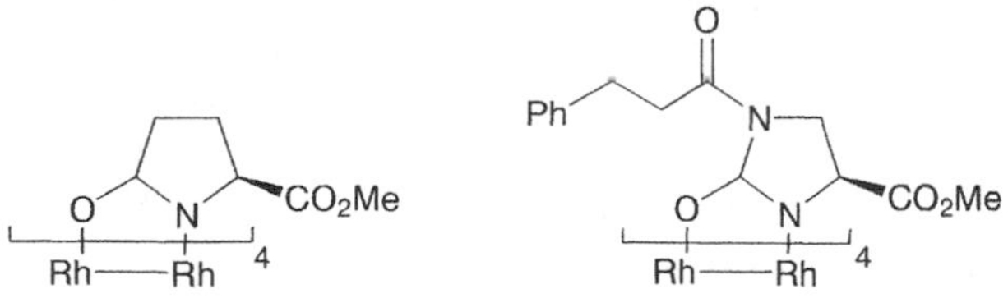

(9.32)
Rh$_2$((5S)-MEPY)

(9.33)
Rh$_2$((4S)-MPPIM)$_4$

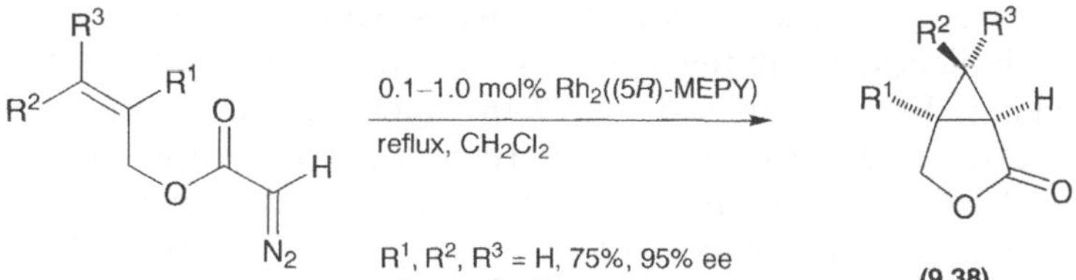

(9.34)

0.1–1.0 mol% Rh$_2$((5R)-MEPY)

reflux, CH$_2$Cl$_2$

(9.38)

R^1, R^2, R^3 = H, 75%, 95% ee
R^1= H, R^2, R^3= Me, 84%, 98% ee
R^1, R^2= H, R^3= Me, 72%, 7% ee
using Rh$_2$((4S)-MPPM)$_4$ gives 75%, 89% ee

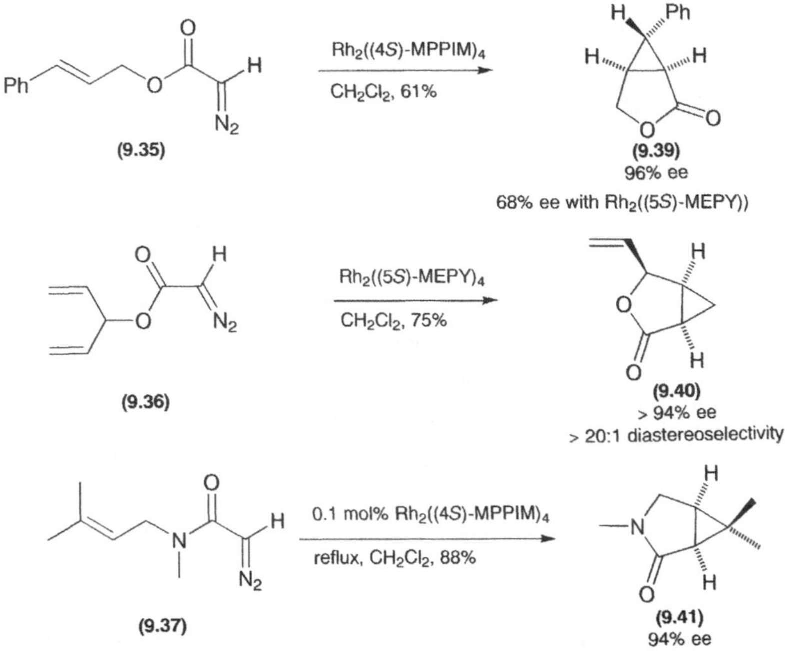

(9.35) → (9.39) 96% ee, 68% ee with Rh₂((5S)-MEPY))

(9.36) → (9.40) > 94% ee, > 20:1 diastereoselectivity

(9.37) → (9.41) 94% ee

The $Rh_2(MEPY)_4$ catalysts are also able to catalyse enantioselective cyclopropenation reactions of alkynes. The alkynes (9.42) and (9.43) are used in a tenfold excess, and the cyclopropene products (9.44) and (9.45) are somewhat unstable but are isolated in moderate to good yields.[32]

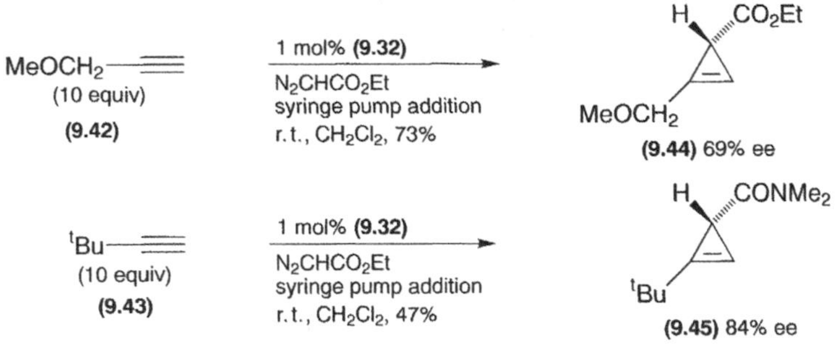

(9.42) → (9.44) 69% ee

(9.43) → (9.45) 84% ee

Davies and co-workers extended their work with catalyst (9.20) to the reaction with dienes, where the products formed initially undergo a Cope rearrangement. For example, diene (9.46) undergoes cyclopropanation with diazoester (9.22) to give an intermediate cyclopropyl derivative (9.47), which ring-expands by the Cope reaction, providing the seven-membered ring product (9.48) with good enantioselectivity.[33] Reaction with the furan (9.49) affords the bicyclic product (9.50) after rearrangement. A similar reaction sequence has also been performed in an intramolecular fashion to provide the tricyclic tremulane skeleton.[34]

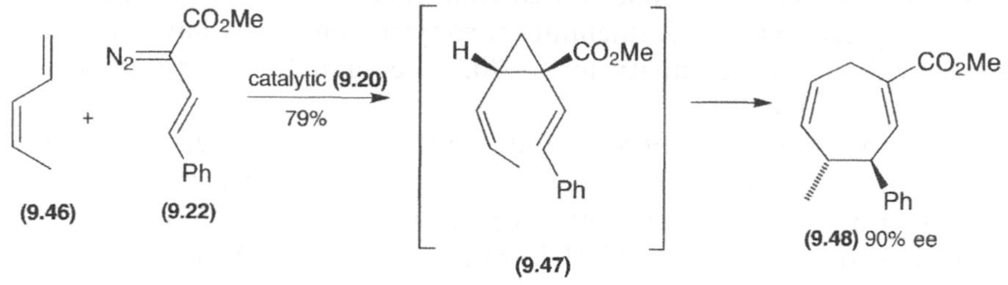

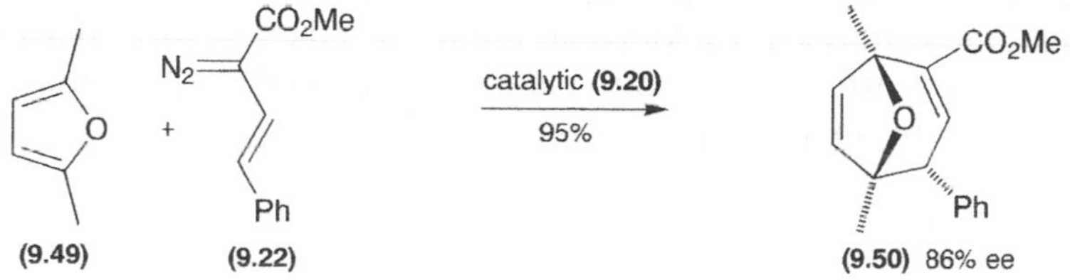

(9.46)   **(9.22)**   **(9.47)**   **(9.48)** 90% ee

(9.49)   **(9.22)**   **(9.50)** 86% ee

9.3   Simmons-Smith type cyclopropanation

The Simmons-Smith reaction involves the cyclopropanation of alkenes using CH_2I_2 and a Zn/Cu couple.[35] Simmons-Smith cyclopropanation is accelerated in the presence of catalytic amounts of ligand (**9.51**) (see also Section 6.1), although there is still some uncatalysed background reaction.[36] The reaction works well for allyl alcohols, such as cinnamyl alcohol (**9.52**), and has also been extended to vinyl silanes and vinylstannanes.[37] Denmark and co-workers improved the yields (up to 99%) and enantioselectivity (up to 89% ee) of this reaction by employing a bis-methylsulfonamide variant of the ligand and a slightly modified procedure.[38, 39]

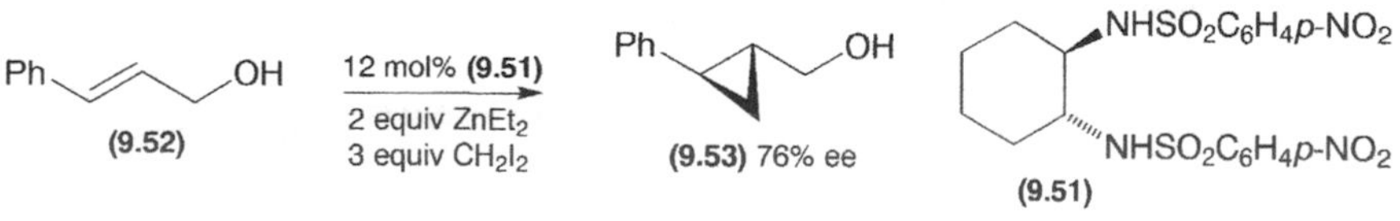

(9.52)   12 mol% (**9.51**)   **(9.53)** 76% ee   (**9.51**)
               2 equiv $ZnEt_2$
               3 equiv CH_2I_2

9.4   Other cyclopropanation catalysts

Whilst copper and rhodium catalysts have been the most widely investigated, many other metals also catalyse cyclopropanation reactions.

In particular, ruthenium and cobalt complexes have been successful for enantioselective reactions. The ruthenium/pybox complex (9.54) provides good *trans* selectivity as well as high enantiocontrol in the cyclopropanation of syrene (9.04).[40] Ruthenium porphyrin complexes have also been used, with enantioselectivity up to 90.8% ee and turnover numbers in excess of 1,000.[41]

The cobalt complexes (9.56)[42] and (9.57)[43] have been used in cyclopropanation reactions, but have given poor control of diastereoselectivity, although high enantioselectivities, up to 97% ee with (9.57), have been recorded. In contrast, the Co(III) salen complex (9.58) provides good diastereocontrol, with reasonable enantioselectivity.[44]

The palladium-catalysed cyclopropanation of alkenes with diazomethane has been attempted using (bis-oxazoline) palladium chloride complexes, but with no asymmetric induction.[45]

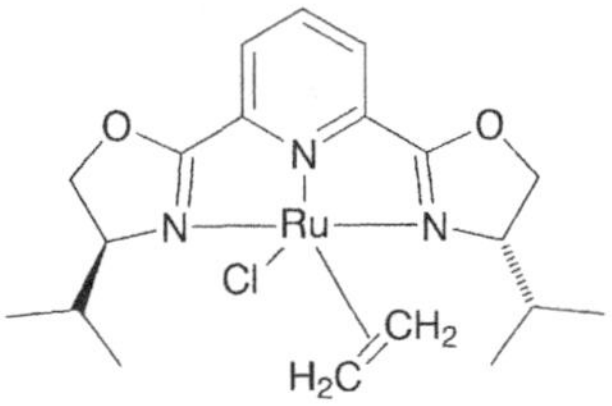

(9.54)

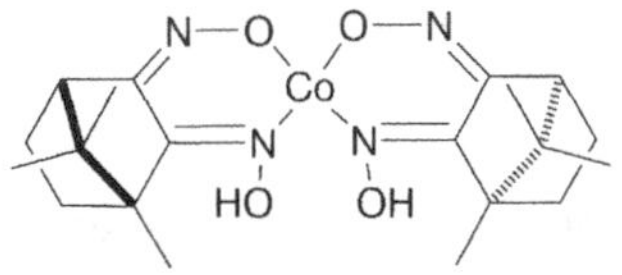

(9.56)

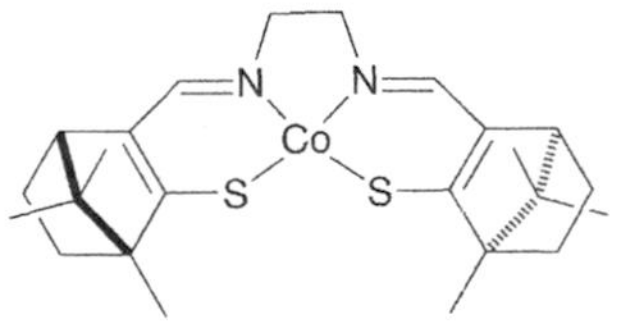

(9.57)

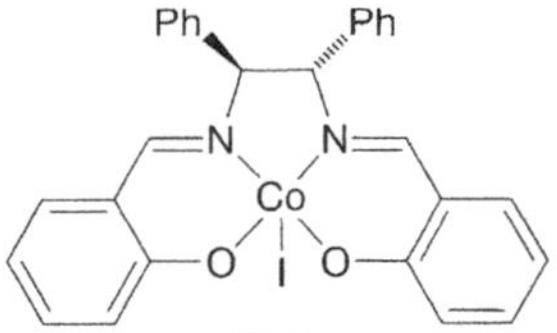

(9.58)

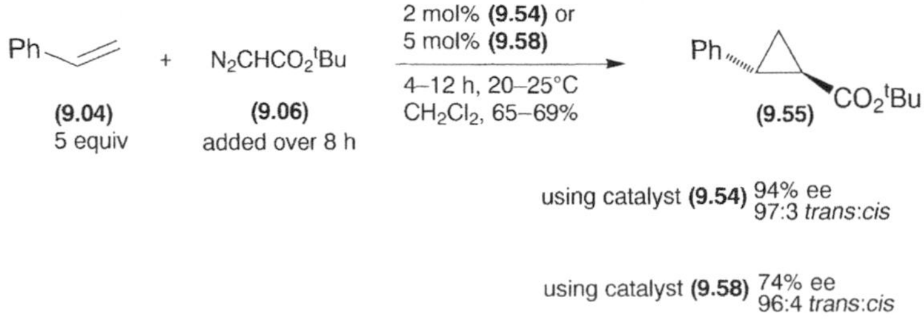

Ph⟍⟋ + N$_2$CHCO$_2$^{t}Bu

(9.04)　　　　(9.06)
5 equiv　　added over 8 h

2 mol% (9.54) or
5 mol% (9.58)
⟶
4–12 h, 20–25°C
CH$_2$Cl$_2$, 65–69%

Ph⟍△⟍CO$_2$^{t}Bu
(9.55)

using catalyst (9.54) 94% ee
97:3 *trans:cis*

using catalyst (9.58) 74% ee
96:4 *trans:cis*

9.5　Insertion reactions

Enantioselective C-H insertion reactions have been successfully performed by various rhodium catalysts over a broad range of substrates, which involve an intramolecular insertion. Of the many examples reported, a few are highlighted here. McKervey and co-workers obtained good diastereoselectivity and enantioselectivity in the C-H insertion reaction of compound (9.58) catalysed by rhodium complex (9.59).[46] Intramolecular cyclopropanation is not competitive in this case, since the alkene moiety is too remote.

Doyle and co-workers found that the Rh$_2$(4S-MACIM)$_4$ catalyst (9.61) provides higher *cis* selectivity than related 'Doyle' catalysts in the reaction of substrate (9.62), where the insertion takes place highly selectively into one of four C-H bonds, yielding essentially a single product (9.63).[47]

In an acyclic version of this reaction, substrate (9.64) was also found to undergo the C-H insertion, with high selectivity, using the Rh$_2$ (5R-MEPY)$_4$ catalyst.[48] Bicyclic β-lactones have been prepared by enantioselective C-H insertion using the Rh$_2$(5S-MEPY)$_4$ catalyst, including the formation of compound (9.67) from the diazoamide precursor (9.66).[49]

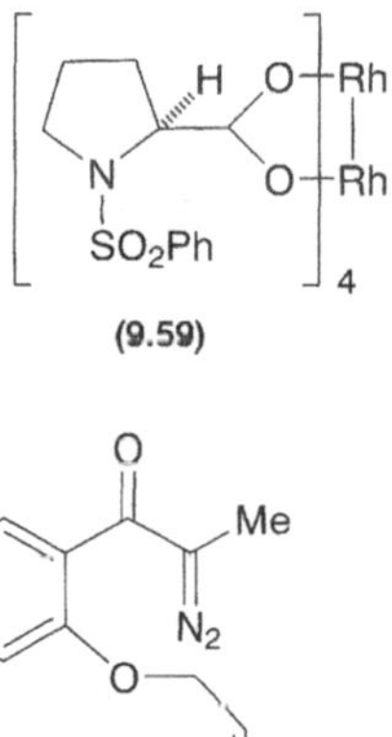

(9.59)

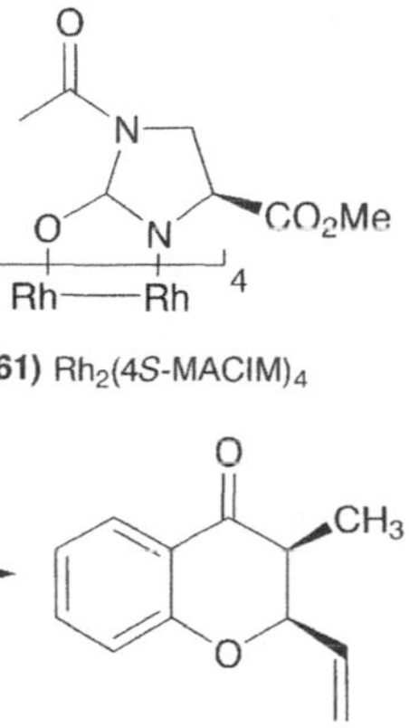

(9.61) Rh$_2$(4S-MACIM)$_4$

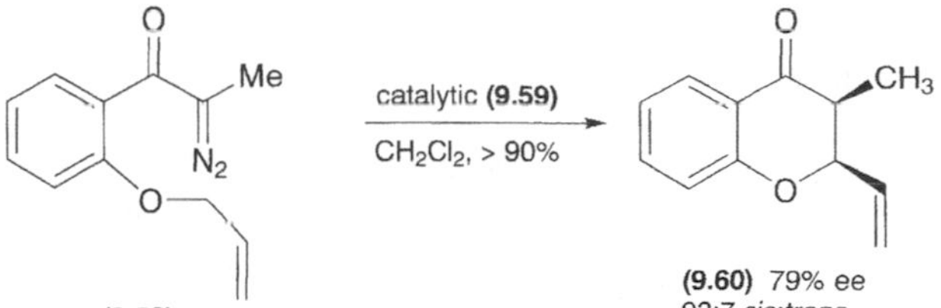

catalytic (9.59)

CH$_2$Cl$_2$, > 90%

(9.58)

(9.60) 79% ee
93:7 *cis:trans*

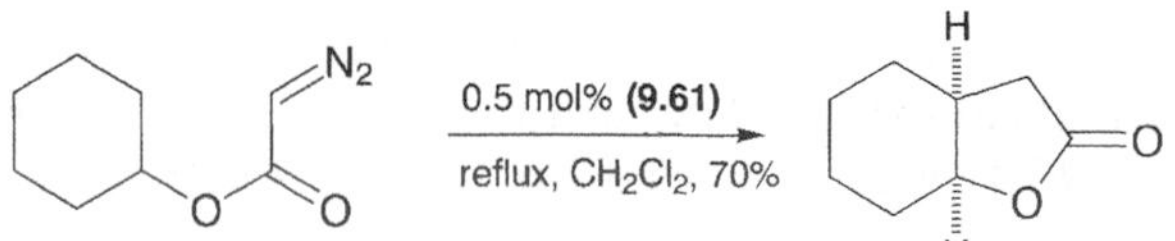

0.5 mol% (9.61)

reflux, CH$_2$Cl$_2$, 70%

(9.62)

(9.63) 97% ee
99:1 *cis:trans*

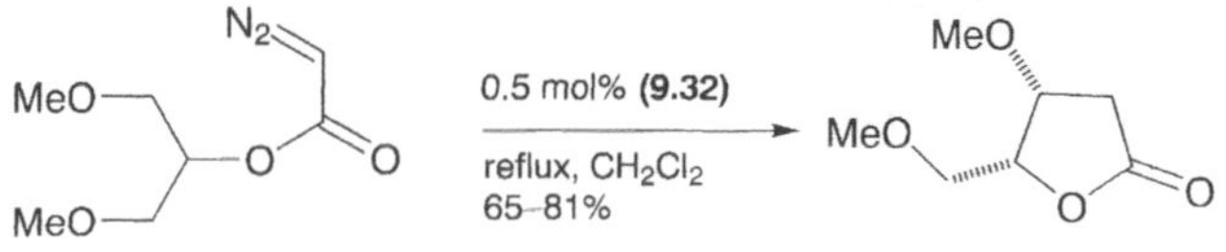

0.5 mol% (9.32)

reflux, CH$_2$Cl$_2$
65–81%

(9.64)

(9.65) 97% ee
93:7 *cis:trans*

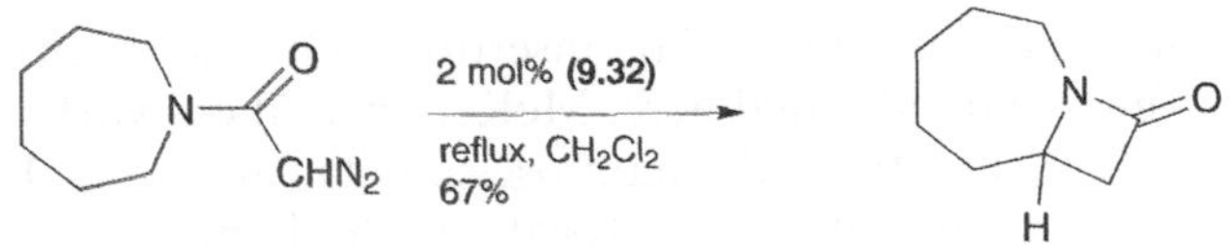

(9.66)　　　　　　　　　　　　　　　　**(9.67)** 97% ee

The Hashimoto catalyst **(9.68)** has been used to give very high stereocontrol in the selective insertion into the enantiotopic aryl C-H bonds of substrate **(9.69)**, to give the product **(9.70)** with a quaternary chiral centre in up to 98% ee.[50, 51] An asymmetric intermolecular C-H insertion reaction has been successfully achieved by Davies and Hansen using catalyst **(9.71)**.[52] Cyclopentane **(9.72)** is employed as the solvent and is converted into the product **(9.73)**, with good enantioselectivity.

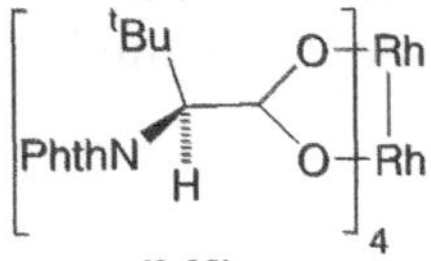

(9.68) 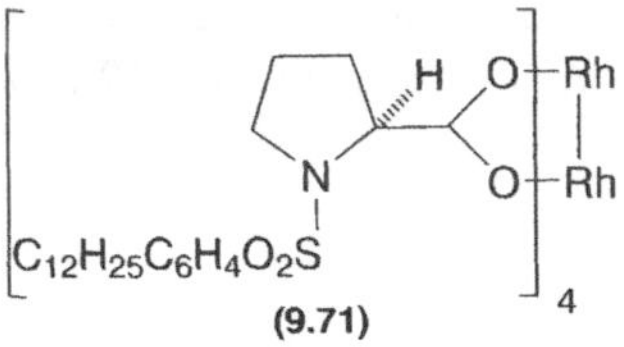　　　　　　　　　　　　**(9.71)**

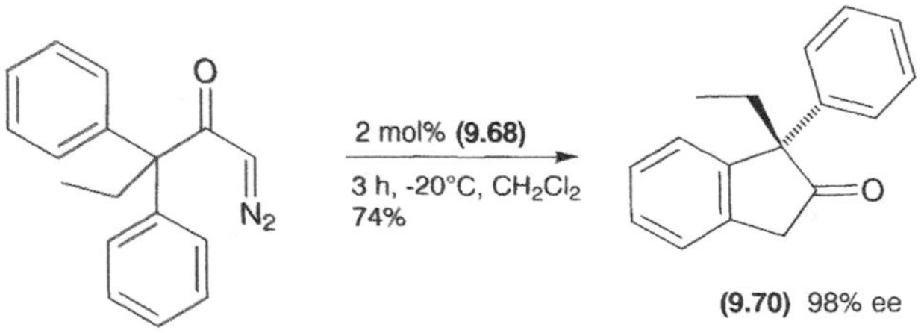

(9.70) 98% ee

(9.69)

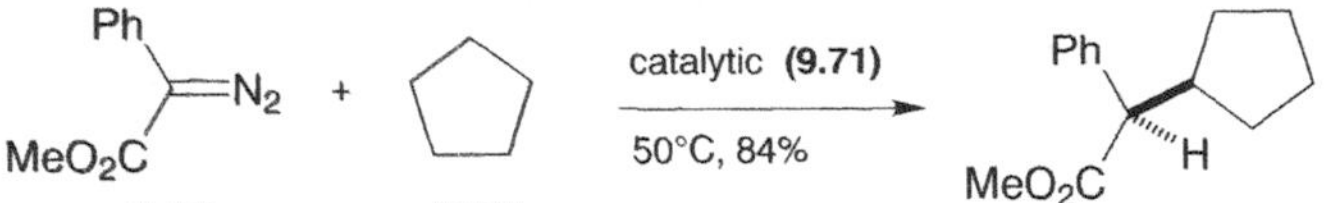

(9.21)　　　**(9.72)** solvent　　　　　　　**(9.73)** 87% ee

Enantioselective Si-H insertion reactions have also been achieved using the Rh₂(MEPY)₄ catalyst.[53] The enantioselectivity of Si-H insertion into the silane **(9.74)** was improved up to 85% by the use of catalyst **(9.71)**.[54]

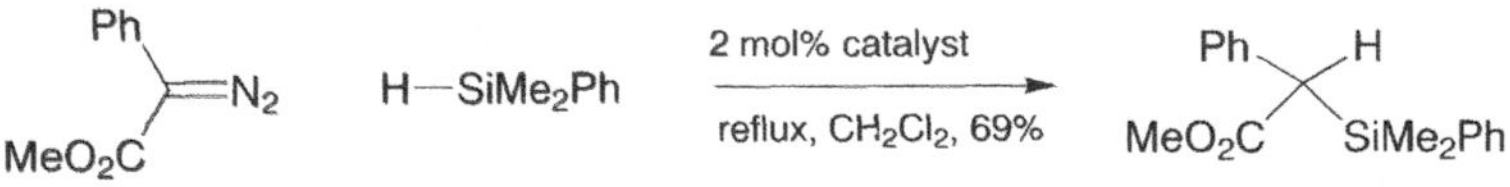

(9.21)　　　**(9.74)**　　　　　　　　　　　　**(9.75)**

with **(9.32)** 47% ee

with **(9.71)** 85% ee

Ruthenium-catalysed asymmetric insertions into S-H bonds have provided very low enantioselectivity, so far.[55] Rhodium catalysts have not given significantly better results for S-H insertion.[56] O-H insertion reactions have not been achieved with enantioselectivity using enantiomerically pure rhodium catalysts.[57] These O-H and S-H reactions may proceed in a stepwise fashion rather than by direct concerted insertion.

9.6   Asymmetric ylide reactions

Metal carbenoid complexes will react with heteroatoms to provide heteroatom ylides. On treatment with catalyst (9.68), the diazocompound (9.76) forms a intermediate oxonium ylide (9.77), which then undergoes a 2,3-sigmatropic rearrangement to give the product (9.78), with good enantioselectivity.[58] Either the rhodium catalyst remains associated to the ylide to control the enantioselectivity of the rearrangement (more likely), or the free ylide rearranges more quickly than the rate of racemisation.

A similar transformation has been reported using enantiomerically pure copper catalysts, where the diazoketone (9.79) forms an initial oxonium ylide, which then rearranges to the cyclic product (9.80), with reasonable enantioselectivity, using a copper complex of ligand (9.81).[59]

Apparent C-O insertion reactions involving decomposition of diazo-compounds mechanistically involve ylide formation and subsequent Stevens rearrangement.[60] An interesting example involves enantioselective ylide formation with one of the enantiotopic oxygens in the acetal

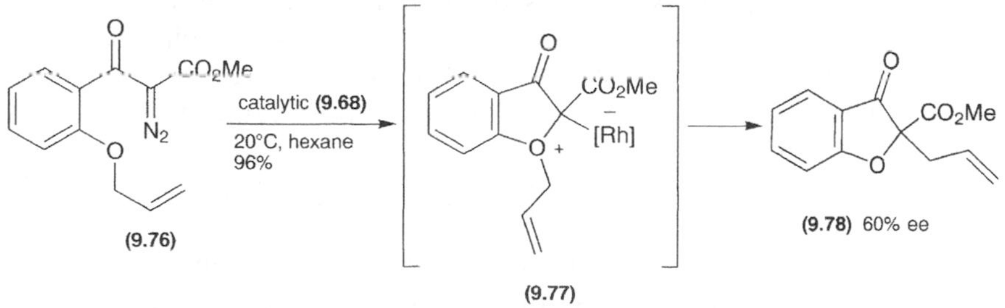

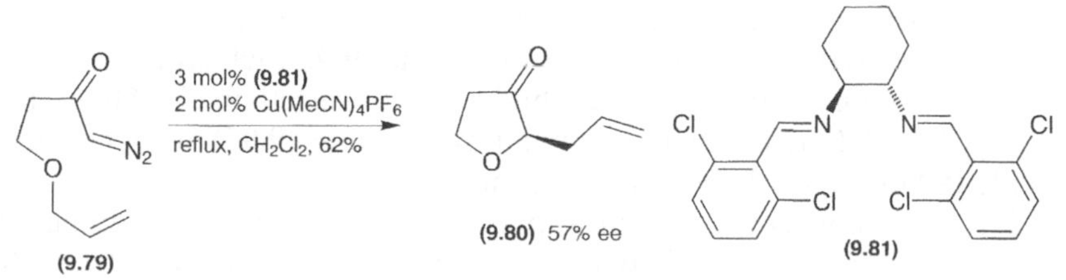

(9.82).[61] The intermediate oxonium ylide **(9.83)** rearranges to the bicyclic product **(9.84)**, which is isolated with fairly good enantiomeric excess using the $Rh_2(MPPIM)_4$ catalyst **(9.33)**.

Decomposition of diazocompounds in the presence of a carbonyl group yields carbonyl ylides, which can be trapped to give enantiomerically-enriched products.[62] For example, Hodgson and co-workers reported a tandem carbonyl ylide formation/cycloaddition process, which provides reasonable enantioselectivity.[63] The diazo precursor **(9.85)** is converted into an intermediate carbonyl ylide **(9.86)**, which undergoes an intramolecular 3 + 2 cycloaddition to provide the tricyclic product **(9.87)**.

Aggarwal and co-workers devised an unusual catalytic cycle for the conversion of carbonyl compounds into epoxides, which involves a rhodium and a sulfide catalyst.[64] A sulfide **(9.88)** is converted into a sulfur ylide **(9.90)** by reaction with a rhodium or copper catalyst and a diazo compound **(9.89)**. The sulfur ylide reacts with a carbonyl compound **(9.91)** forming the epoxide **(9.92)** and regenerating the original (catalytic) sulfide **(9.88)**.

Using an achiral rhodium or copper catalyst but an enantiomerically pure sulfide catalyst provides an interesting preparation of enantiomerically-enriched epoxides. The best sulfide for this reaction, in terms of enantioselectivity, seems to be the oxathiane **(9.93)**, which is straightforward to prepare and provides high enantioselectivity.[65] Thus, benzaldehyde **(9.94)** reacts with phenyldiazomethane **(9.95)** to give stilbene oxide **(9.96)** under these catalytic conditions.

The same concept has been applied to the conversion of imines into aziridines. Yields and selectivities are higher using stoichiometric sulfide but are still good even with sub-stoichiometric amounts.[66] For example, imine (**9.97**) is converted into the aziridine (**9.98**), with good enantios-electivity. Catalytic asymmetric cyclopropanation of enones has also been reported using similar methodology and looks very promising.[67]

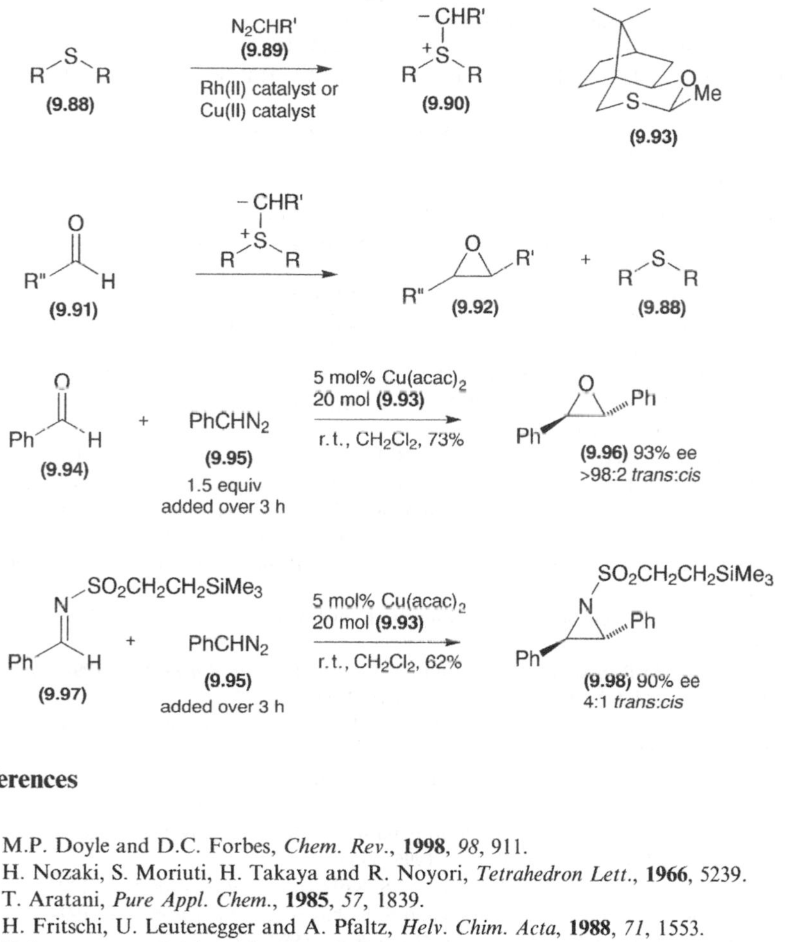

References

1. M.P. Doyle and D.C. Forbes, *Chem. Rev.*, **1998**, *98*, 911.
2. H. Nozaki, S. Moriuti, H. Takaya and R. Noyori, *Tetrahedron Lett.*, **1966**, 5239.
3. T. Aratani, *Pure Appl. Chem.*, **1985**, *57*, 1839.
4. H. Fritschi, U. Leutenegger and A. Pfaltz, *Helv. Chim. Acta*, **1988**, *71*, 1553.
5. U. Leutenegger, G. Umbricht, C. Fahrni, P. von Matt and A. Pfaltz, *Tetrahedron*, **1992**, *48*, 2143.
6. R.E. Lowenthal, A. Abiko and S. Masamune, *Tetrahedron Lett.*, **1990**, *31*, 6005.
7. D.A. Evans, K.A. Woerpel, M.M. Hinman and M.M. Faul, *J. Am. Chem. Soc.*, **1991**, *113*, 726.
8. D. Müller, G. Umbricht, B. Weber and A. Pfaltz, *Helv. Chim. Acta*, **1991**, *74*, 232.
9. R.E. Lowenthal and S. Masamune, *Tetrahedron Lett.*, **1990**, *31*, 6005.
10. A.V. Bedekar and P.G. Andersson, *Tetrahedron Lett.*, **1996**, *37*, 4073.
11. A.M. Harm, J.G. Knight and G. Stemp, *Tetrahedron Lett.*, **1996**, *37*, 6189.
12. Y. Uozumi, H. Kyota, E. Kishi, K. Kitayama and T. Hayashi, *Tetrahedron: Asymmetry*, **1996**, *7*, 1603.

13. M.T. Reetz, E. Bohres and R. Goddard, *J. Chem. Soc., Chem. Commun.*, **1998**, 935.
14. S. Kanemasa, S. Hamura, E. Harada and H. Yamamoto, *Tetrahedron Lett.*, **1994**, *35*, 7985.
15. K. Ito and T. Katsuki, *Tetrahedron Lett.*, **1993**, *34*, 2661.
16. R. Tokunoh, H. Tomiyama, M. Sodeoka and M. Shibasaki, *Tetrahedron Lett.*, **1996**, *37*, 2449.
17. For more information on asymmetric cyclopropanation of silyl enol ethers, see: R. Schumacher, F. Dammast and H.-U. Reißig, *Chem. Eur. J.*, **1997**, *3*, 614.
18. C. Piqué, B. Fähndrich and A. Pfaltz, *Synlett*, **1995**, 491.
19. M. Kennedy, M.A. McKervey, A.R. Maguire and G.H.P. Roos, *J. Chem. Soc., Chem. Commun.*, **1990**, 361.
20. H.M.L. Davies, P.R. Bruzinski and M.J. Fall, *Tetrahedron Lett.*, **1996**, *37*, 4133.
21. M.P. Doyle, Q.-L. Zhou, C. Charnsangavej and M.A. Longoria, *Tetrahedron Lett.*, **1996**, *37*, 4129.
22. H.M.L. Davies and D.K. Hutcheson, *Tetrahedron Lett.*, **1993**, *34*, 7243.
23. H.M.L. Daves, P.R. Bruzinski, D.H. Lake, N. Kong and M.J. Fall, *J. Am. Chem. Soc.*, **1996**, *118*, 6897.
24. E.J. Corey and T.G. Gant, *Tetrahedron Lett.*, **1994**, *35*, 5373.
25. For examples of rhodium-catalysted intermolecular cyclopropanation with good enantio-selectivity but poor diastereoselectivity, see: (a) H. Ishitani and K. Achiwa, *Synlett*, **1997**, 781. (b) M.P. Doyle, Q.-L. Zhou, S.H. Simonsen and V. Lynch, *Synlett*, **1996**, 697. (c) H. Ishitani and K. Achiwa, *Synlett*, **1997**, 781. (d) M.P. Doyle, Q.-L. Zhou, S.H. Simonsen and V. Lynch, *Synlett*, **1996**, 697.
26. S. Kitagaki, H. Matsuda, N. Watanabe, S.-I. Hashimoto, *Synlett*, **1997**, 1171.
27. M.P. Doyle, R.E. Austin, A. Scott Bailey, M.P. Dwyer, A.B. Dyatkin, A.V. Kalinin, M.M.Y. Kwan, S. Liras, C.J. Oalmann, R.J. Pieters, M.N. Protopopova, C.E. Raab, G.H.P. Roos, Q.-L. Zhou and S.F. Martin, *J. Am. Chem. Soc.*, **1995**, *117*, 5763.
28. M.P. Doyle, Q.-L. Zhou, A.B. Dyatkin and D.A. Ruppar, *Tetrahedron Lett.*, **1995**, *36*, 7569.
29. M.P. Doyle, C.S. Peterson, Q.-L. Zhou and H. Nishiyama, *J. Chem. Soc., Chem. Commun.*, **1997**, 211.
30. S.F. Martin, M.R. Spallar, S. Liras and B. Hartmann, *J. Am. Chem. Soc.*, **1994**, *116*, 4493.
31. M.P. Doyle and A.V. Kalinin, *J. Org. Chem.*, **1996**, *61*, 2179.
32. M.P. Doyle, M. Protopopova, P. Müller, D. Ene and E.A. Shapiro, *J. Am. Chem. Soc.*, **1994**, *116*, 8492.
33. H.M.L. Davies, Z.-Q. Peng and J.H. Houser, *Tetrahedron Lett.*, **1994**, *35*, 8939.
34. H.M.L. Davies and B.D. Doan, *Tetrahedron Lett.*, **1996**, *37*, 3967.
35. For a review on asymmetric cyclopropanation with iodomethylzinc reagents see: A.B. Charette and J.-F. Marcoux, *Synlett*, **1995**, 1197.
36. H. Takahashi, M. Yoshioka, M. Ohno and S. Kobayashi, *Tetrahedron Lett.*, **1992**, *33*, 2575.
37. N. Imai, K. Sakamoto, H. Takahashi and S. Kobayashi, *Tetrahedron Lett.*, **1994**, *35*, 7045.
38. S.E. Denmark and S.P. O'Connor, *J. Org. Chem.*, **1997**, *62*, 584.
39. S.E. Denmark, S.P. O'Connor and S.R. Wilson, *Angew. Chem., Int. Ed. Engl.*, **1998**, *37*, 1149.
40. H. Nishiyama, Y. Itoh, H. Matsumoto, S.-B. Park and K. Itoh, *J. Am. Chem. Soc.*, **1994**, *116*, 2223.
41. W.-C. Lo, C.-M. Che, K.-F. Cheng and T.C.W. Mak, *J. Chem. Soc., Chem. Commun.*, **1997**, 1205.
42. A. Nakamura, *Pure Appl. Chem.*, **1978**, *50*, 37.
43. G. Jommi, R. Pagliarin, G. Rizzi and M. Sisti, *Synlett*, **1993**, 833.
44. T. Fukuda and T. Katsuki, *Synlett*, **1995**, 825.

45. S.E. Denmark, R.A. Stavenger, A.-M. Faucher and J.P. Edwards, *J. Org. Chem.*, **1997**, *62*, 3375.
46. T. Ye, C.F. Garcia and M.A. McKervey, *J. Chem. Soc., Perkin Trans. 1*, **1995**, 1373.
47. M.P. Doyle, A.B. Dyatkin, G.H.P. Roos, F. Cañas, D.A. Pierson, A. van Basten, P. Müller and P. Polleux, *J. Am. Chem. Soc.*, **1994**, *116*, 4507.
48. M.P. Doyle, A.B. Dyatkin and J.S. Tedrow, *Tetrahedron Lett.*, **1994**, *35*, 3853.
49. M.P. Doyle and A.V. Kalinin, *Synlett*, **1995**, 1075.
50. N. Watanbe, T. Ogawa, Y. Ohtake, S. Ikegami and S.-I. Hashimoto, *Synlett*, **1996**, 85.
51. N. Watanabe, Y. Ohtake, S.-I. Hashimoto, M. Shiro and S. Ikegami, *Tetrahedron Lett.*, **1995**, *36*, 1491.
52. H.M.L. Davies and T. Hansen, *J. Am. Chem. Soc.*, **1997**, *119*, 9075.
53. R.T. Buck, M.P. Doyle, M.J. Drysdale, L. Ferris, D.C. Forbes, D. Haigh, C.J. Moody, N.D. Pearson and Q.-L. Zhou, *Tetrahedron Lett.*, **1996**, *37*, 7631.
54. H.M.L. Davies, T. Hansen, J. Rutberg and P.R. Bruzinski, *Tetrahedron Lett.*, **1997**, *38*, 1741.
55. E. Galardon, S. Roué, P. Le Maux and G. Simonneaux, *Tetrahedron Lett.*, **1998**, *39*, 2333.
56. H. Brunner, K. Wutz and M.P. Doyle, *Monatsh. Chem.*, **1990**, *121*, 755.
57. L. Ferris, D. Haigh and C.J. Moody, *Tetrahedron Lett.*, **1996**, *37*, 107.
58. N. Pierson, C. Fernández-García and M.A. McKervey, *Tetrahedron Lett.*, **1997**, *38*, 4705.
59. J.S. Clark, M.Fretwell, G.A. Whitlock, C.J. Burns and D.N.A. Fox, *Tetrahedron Lett.*, **1998**, *39*, 97.
60. K. Ito and T. Katsuki, *Chem. Lett.*, **1994**, 1857.
61. M.P. Doyle, D.G. Ene, D.C. Forbes and J.S. Tedrow, *Tetrahedron Lett.*, **1997**, *38*, 4367.
62. M.P. Doyle, D.C. Forbes, M.N. Protopopova, S.A. Stanley, M.M. Vasbinder and K.R. Xavier, *J. Org. Chem.*, **1997**, *62*, 7210.
63. D.M. Hodgson, P.A. Stupple and C. Johnstone, *Tetrahedron Lett.*, **1997**, *138*, 6471.
64. V.K. Aggarwal, *Synlett*, **1998**, 329.
65. V.K. Aggarwal, J.G. Ford, A. Thompson, R.V.H. Jones and M. Standen, *J. Am. Chem. Soc.*, **1996**, *118*, 7004.
66. V.K. Aggarwal, A. Thompson, R.V.H. Jones and M.C.H. Standen, *J. Org. Chem.*, **1996**, *61*, 8368.
67. V.K. Aggarwal, H.W. Smith, R.V.H. Jones and R. Fieldhouse, *J. Chem. Soc., Chem. Commun.*, **1997**, 1785.

10   Catalytic carbon-carbon bond-forming reactions

This chapter describes carbon-carbon bond-forming reactions, which are metal-catalysed. The reactions discussed include simple cross-coupling reactions and variations on this theme. Where appropriate, related C-X and C-H bond-forming reactions are also considered, when they are mechanistically related. Of course, there are many other catalytic reactions which involve the formation of carbon-carbon bonds, such as the aldol reaction, cyclopropanation and conjugate addition, but these are detailed in other chapters.

10.1   Cross-coupling reactions

The cross-coupling of organometallic reagents with organohalides using a transition metal catalyst is a very powerful method for $C-C$ bond formation.[1] Typically, the reactions do not involve coupling of sp^3 hybridised carbon, limiting the options for asymmetric induction.

However, the coupling of racemic sp^3-hybridised organometallics to sp^2-hybridised organohalides has been achieved with asymmetric induction, and the area was reviewed in 1993.[2] Typical examples include the coupling of vinyl bromide (**10.01**) with racemic Grignard reagent (**10.02**), using either a palladium- or nickel-based catalyst.[3,4] The ligands (**10.07–10.09**) are amongst the best for catalysing these reactions with good enantioselectivity. The coupling of silylated Grignard reagents (**10.03**) has provided a route to enantiomerically-enriched allylsilanes (**10.06**) as the coupled products.[5]

Addition of zinc chloride causes transmetallation of the Grignard reagent prior to the coupling reaction. Either the enantiomerically pure catalyst selectively picks out one enantiomer of the organometallic reagent to react with, or the intermediate palladium/nickel alkyl complex undergoes epimerisation before reductive elimination. In any event, the

(10.07)   (10.08)   (10.09)

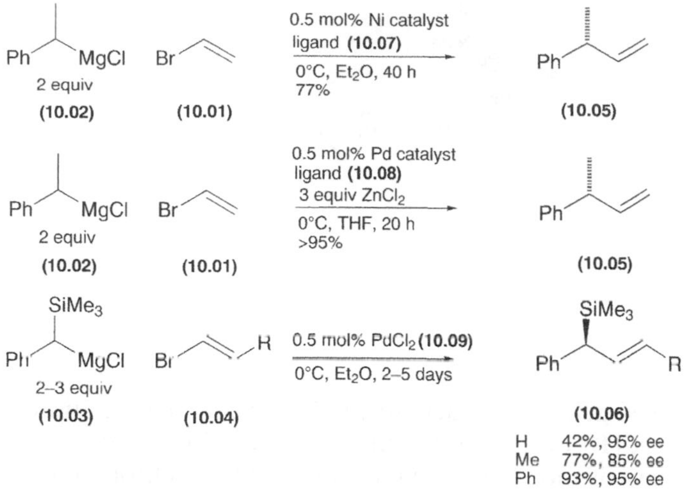

mechanism of enantioselection in these reactions is not well understood and advances in this area have been sporadic over the past few years.[6]

Asymmetric carbon-carbon bond-forming catalysis has been used to generate selectivity in the axial chirality of binaphthyls. In one impressive example, coupling of the two achiral components (10.10) and (10.11) was achieved, with high yield and enantioselectivity.[7] The methoxy group in ligand (10.16) is not considered to bind to the transition metal catalyst during the reaction.

Hayashi and co-workers induced axial chirality by the use of enantio-position selective cross-coupling reactions. Selective replacement of one of the triflate groups in substrate (10.12) leads to the axially chiral product (10.14), with good enantioselectivity.[8] In the case of alkynyl Grignard reagents, such as compound (10.13), the selectivity is particularly high.[9] To some extent, the reaction is self-correcting. When the initial coupling reaction affords the wrong enantiomer of product, this is more likely to undergo a second cross-coupling step to afford the

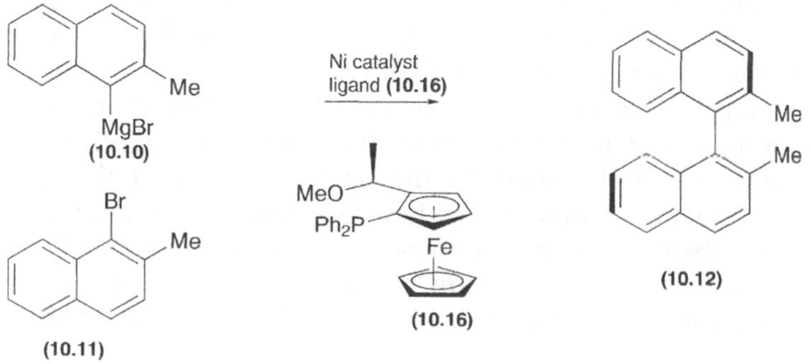

(10.12)   (10.14)   (10.15)

4 h   91% (88% ee)   0%
17 h   53% (>99% ee)   43%

disubstituted achiral product (**10.15**). Therefore, the enantioselectivity of the product (**10.14**) increases with longer reaction times.

Desymmetrisation of *ortho*-dichlorobenzene chromium tricarbonyl complex has been achieved using enantiomerically pure palladium complexes, which catalyse cross-coupling reactions, including the Suzuki coupling of alkenylboronic acids.[10]

Buchwald and co-workers recently described the cross-coupling of ketone enolates with aryl bromides, using enantiomerically pure palladium/ BINAP complexes.[11] The ketone (**10.18**) was shown to be a particularly suitable substrate, yielding products (**10.19**), with high enantioselectivity.

(10.18)   (10.19) 94–98% ee

10.2   Metal-catalysed allylic substitution

Enantioselective metal-catalysed allylic substitution reactions have attracted considerable attention, especially in recent years; this topic was reviewed in 1996.[12] The metal which has been most widely investigated for allylic substitution reactions is palladium. Palladium-catalysed allylic substitution typically involves a double inversion mechanism, resulting in overall retention of relative stereochemistry.[13] So, if the stereochemistry of the product is simply based on the stereochemistry of the starting material, how can an asymmetric synthesis be possible? The answer lies in the choice of substrate for the enantioselective version of the palladium-catalysed allylic substitution reaction. For example, the substrate (**10.20**) proceeds *via* a *meso* intermediate complex (**10.21**). The end of the allyl group to which the nucleophile adds dictates the enantiomer of product that is formed: **(10.22)** or **ent (10.22)**.

The most commonly employed test reaction for enantioselective palladium-catalysed allylic substitution reactions is the allyl acetate **(10.23)**, using dimethylmalonate anion as the incoming nucleophile, to provide the substitution product **(10.24)**.

Many ligands have been used for this reaction, and enantioselectivities in excess of 90% ee have often been achieved. The first ligands to work with this level of selectivity were designed by Hayashi and co-workers: it is anticipated that the ligand **(10.25)** will bind to the palladium and the 'arm' on the ligand is able to reach round to the other side of the allyl group to direct the approach of the incoming nucleophile, as indicated by structure **(10.26)**.[14, 15]

Indeed, directing the approach of the nucleophile to one end of the allyl group appears to be a formidable task, since the ligand is on the wrong side of the allyl group. However, the ligand can perturb the symmetry of the allyl group by more direct steric interactions, such that one end of the allyl group is pushed away from the metal. The incoming nucleophile is expected to add to the allyl terminus further from the palladium, represented by structure **(10.28)**. Bidentate nitrogen ligands, including sparteine,[16] bis-oxazoline **(10.27)**[17] and bis-aziridines,[18] as well as bidentate phosphine ligands,[19, 20] have all been used in allylic substitution reactions.

An alternative concept for disrupting the geometry of the allyl palladium intermediate involves using ligands with two different donor atoms. If one of the donor atoms is a better π-acceptor, then it is expected that the nucleophile will approach *trans* to that atom, as shown in structure **(10.32)**. The use of the mixed phosphine/oxazoline ligands **(10.29)** demonstrates this principle well.[21, 22, 23] Other ligands with various combinations of donor atoms have also been used successfully. These ligands include the QUINAP ligand of Brown and co-workers (**10.30**),[24] Josiphos (**10.31**), which has two electronically different phosphorus donor atoms,[25] and other ligand designs.[26, 27]

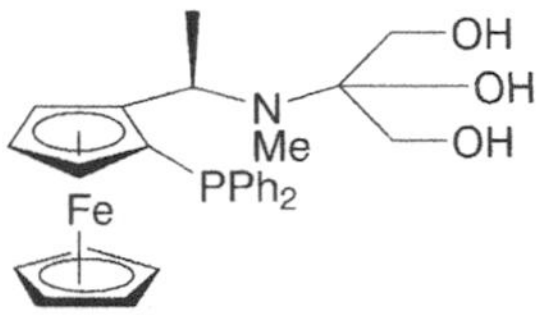

(10.25)

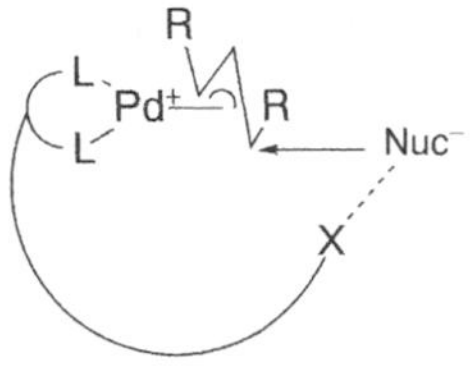

(10.26)

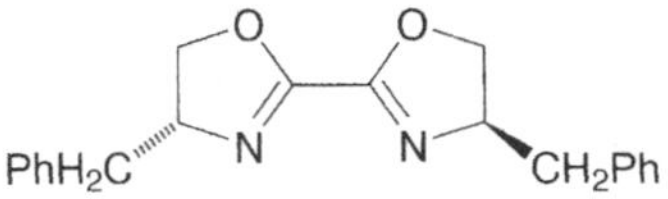

(10.27)

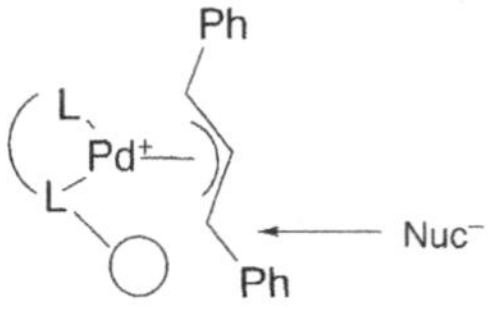

(10.28)

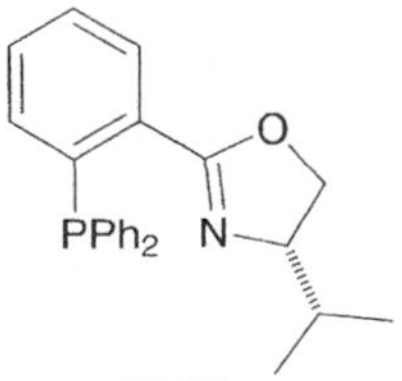

(10.29)

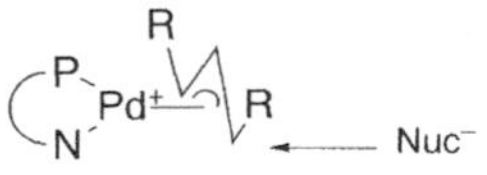

(10.32)

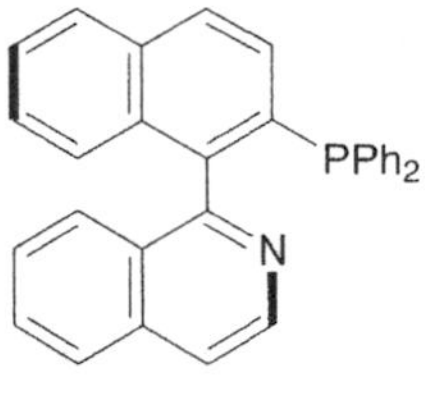

(10.30)

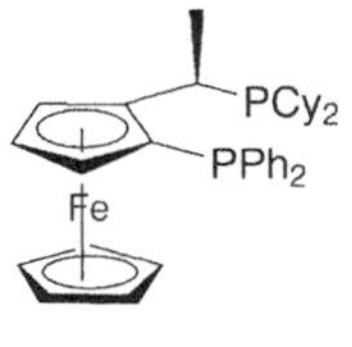

(10.31)

The substrate (**10.23**) does not give a good indication of how useful particular ligands will be across a broad range of different substrates. In particular, most of these ligands give lower enantioselectivities with cyclic substrates. However, the ligands (**10.33**), developed by Trost and Bunt seem to be generally applicable to cyclic substrates to provide excellent asymmetric induction.[28] Thus, cyclohexenyl acetate (**10.34**) is converted into the substitution product (**10.35**), with excellent selectivity. Helmchen and co-workers have used enantiomerically pure ligands containing a phosphorus donor atom tethered to a carboxylic acid, which also provided excellent enantioselectivity in the same reaction.[29]

(10.33)

OAc

2.5 mol% [Pd(allyl)Cl]$_2$
7.5 mol% (10.33)
NaCH(CO$_2$Me)$_2$
(C$_6$H$_{13}$)$_4$NBr
0°C, CH$_2$Cl$_2$, 86%

(10.34)

CO$_2$Me
CO$_2$Me

(10.35) >98% ee

Substrates which do not proceed *via* symmetrical allyl intermediates, e.g. compound (10.21), can also be subject to enantioselective allylic substitution reactions in some circumstances. Substrates which can equilibrate through the π-σ-π mechanism provide one option, where the intermediate π-allyl group must contain two identical groups at one terminus for racemisation/epimerisation to occur. Thus, the racemic compound (10.36) has been used with bidentate phosphines[30] and with the phosphine/oxazoline ligand (10.29) to give very high enantioselectivity.[31] The product (10.37) could be converted into a range of substituted β-amino acids, succinic acids and other products.[32]

Ph

Ph

OAc Ph

(10.36)

2.5 mol% [Pd(allyl)Cl]$_2$
5 mol% (10.29)
NaCH(CO$_2$Me)$_2$
r.t., THF, 24 h, 95%

Ph

Ph

(McO$_2$C)$_2$CH Ph

(10.37) 97% ee

Substrates which possess enantiotopic leaving groups provide another opportunity for an asymmetric reaction. The dibenzoate (10.38) undergoes selective substitution of one of the enantiotopic benzoate groups using the Trost ligand (10.33), with diketone (10.39) as the incoming nucleophile.[33] The same ligand has also been used to control enantioselectivity in the displacement of one of the enantiotopic acetates of the *gem* diacetate (10.40).[34] Interestingly, the same substrate (10.40) has been shown to react with the prochiral nucleophile (10.41), with excellent control of diastereoselectivity and enantioselectivity.[35] In fact, the control of stereochemistry in the prochiral nucleophile is not easily achieved and enantioselectivity has been low, with only a few exceptions.[36]

(10.38) → Pd cat, ligand **(10.33)**, **(10.39)**, DBU, 84% → **(10.42)** 98% ee

(10.40) → 2.5 mol% [Pd(allyl)Cl]₂, 7.5 mol% **(10.33)**, MeCH(CO₂Me)₂ → **(10.43)**

(10.40) + **(10.41)** → 2.5 mol% [Pd(allyl)Cl]₂, 7.5 mol% **(10.33)**, 0–5°C, 88% → **(10.44)** 99% ee, 90% de

The use of heteroatom nucleophiles has provided the opportunity to extend the synthetic repertoire of allylic substitution reactions. Nitrogen nucleophiles have been particularly popular. The phosphino-oxazoline ligand **(10.45)**[37] has given slightly better results than other phosphino-oxazoline ligands in the amination of substrate **(10.23)** with potassium phthalimide **(10.46)**.[38, 39] The ferrocenyl pyrazole ligand of Togni and co-workers **(10.48)** also provides excellent enantioselectivity in the amination of the allylic carbonate **(10.49)**.[40]

Trost and Bunt reported high selectivities for allylic amination of cyclic substrates using their ligands.[28] For example, cycloheptenyl acetate **(10.50)** is converted into the allylic amine derivative **(10.51)**, with excellent control of enantioselectivity.

(10.45)   **(10.48)**

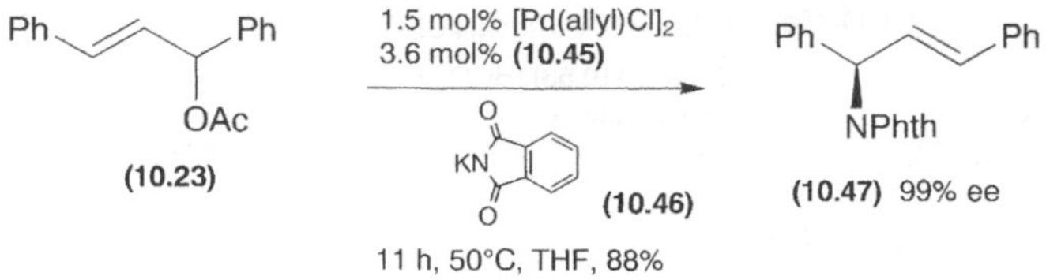

Ph⁀⁀Ph
|
OAc

(10.23)

1.5 mol% [Pd(allyl)Cl]₂
3.6 mol% **(10.45)**
———————————————→
KN-phthalimide **(10.46)**
11 h, 50°C, THF, 88%

Ph⁀⁀Ph
|
NPhth

(10.47) 99% ee

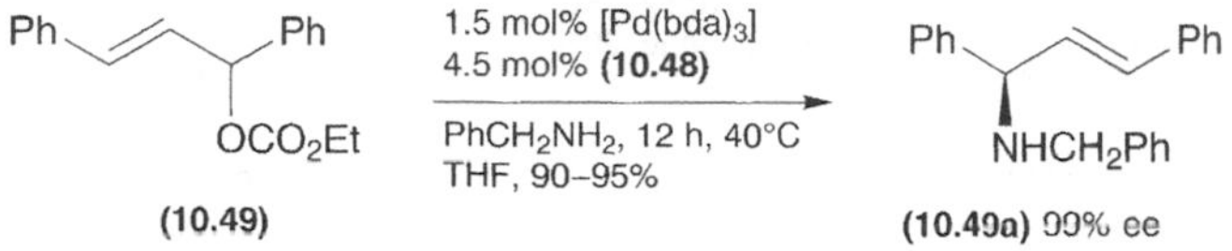

Ph⁀⁀Ph
|
OCO₂Et

(10.49)

1.5 mol% [Pd(bda)₃]
4.5 mol% **(10.48)**
———————————————→
PhCH₂NH₂, 12 h, 40°C
THF, 90–95%

Ph⁀⁀Ph
|
NHCH₂Ph

(10.49a) 99% ee

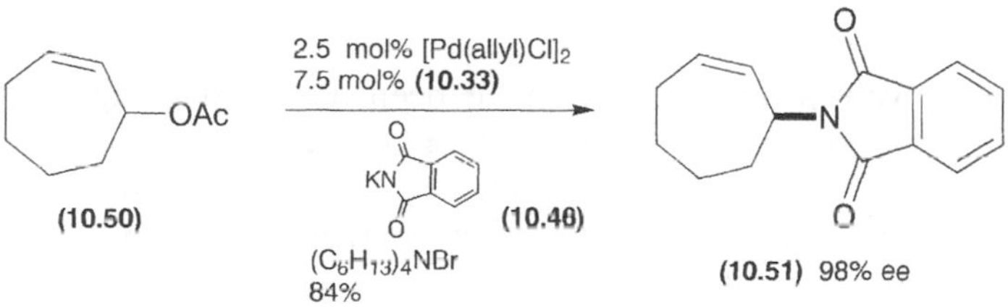

(cycloheptenyl)—OAc

(10.50)

2.5 mol% [Pd(allyl)Cl]₂
7.5 mol% **(10.33)**
———————————————→
KN-phthalimide **(10.46)**
(C₆H₁₃)₄NBr
84%

(cycloheptenyl)—N-phthalimide

(10.51) 98% ee

Even oxygen nucleophiles have been introduced, with good enantio-selectivity. The conditions of the reaction need to be sufficiently mild that the product does not become a substrate for the allylic substitution, since this will ultimately lead to racemisation. Pivalate $(tBuCO_2^-)$[41] and phenols have been used as nucleophiles, with good results.[42] Sulfur nucleophiles have also been used in enantioselective allylic substitution reactions.[43]

The reduction of allylic esters has been achieved using formate as the hydride source. Although not widely investigated with a variety of ligands, Hayashi and co-workers have demonstrated that the mono dentate MOP ligands **(10.52)** and **(10.53)** are certainly effective.[44] The reduction of substrates **(10.54)** and **(10.55)** has been achieved, with good control of enantioselectivity.

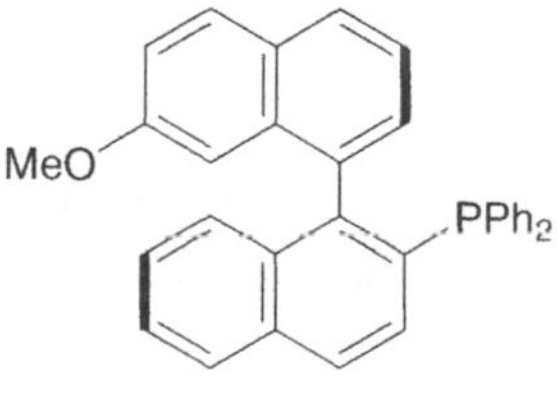

MeO—(binaphthyl)—PPh₂

(10.52)

MeO—(binaphthyl)—PPh₂

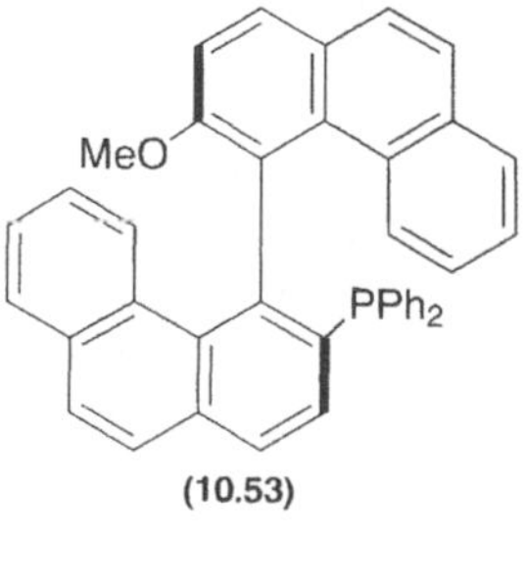

(10.53)

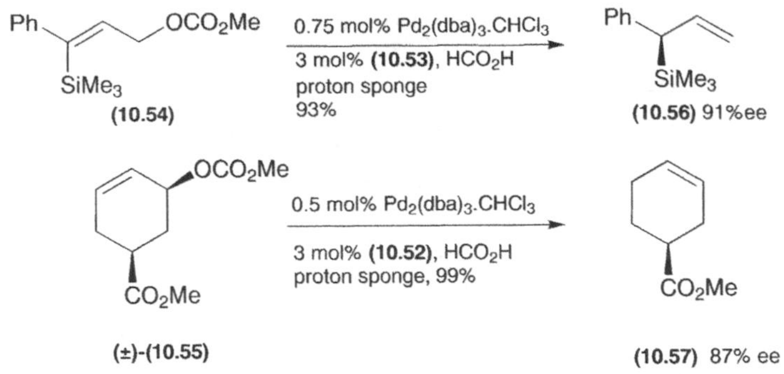

Ph $\diagdown$ ⟋⟍⟋ OCO₂Me
SiMe₃
(10.54)

0.75 mol% Pd₂(dba)₃.CHCl₃
3 mol% (10.53), HCO₂H
proton sponge
93%
→

Ph $\diagdown$
SiMe₃
(10.56) 91%ee

OCO₂Me

CO₂Me
(±)-(10.55)

0.5 mol% Pd₂(dba)₃.CHCl₃
3 mol% (10.52), HCO₂H
proton sponge, 99%
→

CO₂Me
(10.57) 87% ee

proton sponge = 1,8-bis(dimethylamino)naphthalene

Various cyclisation reactions have been achieved using enantiomeri-cally pure palladium catalyst through an allylic substitution process. One of the first reported examples of such a cyclisation, with substrate (10.58), is also one of the most selective, with control of both enantioselectivity and diastereoselectivity.[45] Enantioselective cyclisation reactions using phosphino-oxazoline ligands have been reported by Pfaltz and Koch, although the enantioselectivities (up to 87% ee) are lower than for most intermolecular reactions.[46] The *meso* diol (10.60) forms an intermediate (10.61), which undergoes cyclisation with high enantioselectivity to give the product (10.62). The use of triethylamine was crucial to this selectivity.[47]

Reactions involving a Wacker-type oxidative cyclisation have been reported, with very high enantioselectivity in some cases.[48]

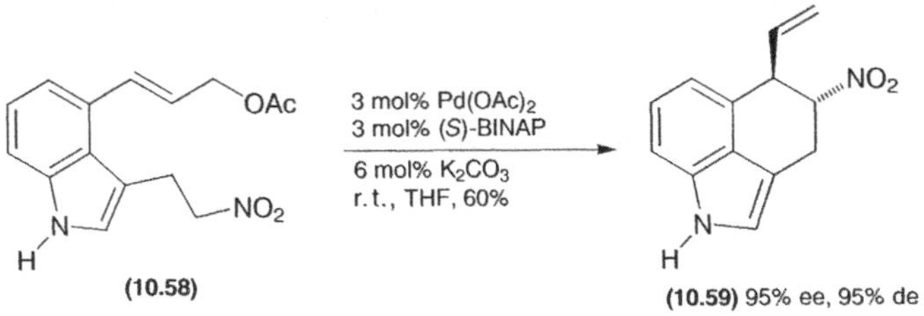

OAc

N
H
NO₂
(10.58)

3 mol% Pd(OAc)₂
3 mol% (S)-BINAP
6 mol% K₂CO₃
r.t., THF, 60%
→

NO₂
N
H
(10.59) 95% ee, 95% de

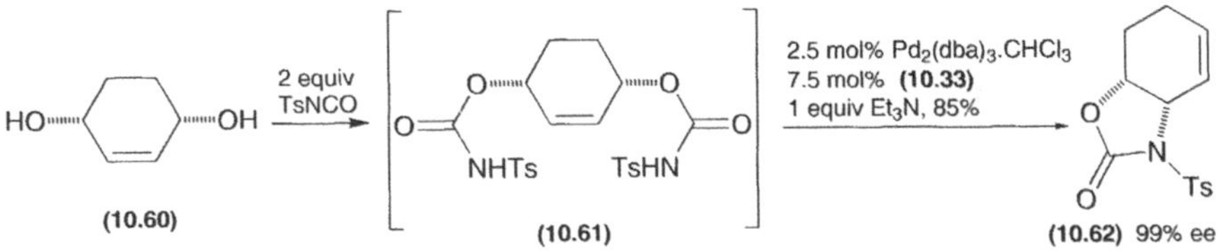

HO⟍ ⟍OH
(10.60)

2 equiv
TsNCO
→

O
O
NHTs
O
O
TsHN
(10.61)

2.5 mol% Pd₂(dba)₃.CHCl₃
7.5 mol% (10.33)
1 equiv Et₃N, 85%
→

O
N
O
Ts
(10.62) 99% ee

Whilst palladium has been examined more than any other metal for the allylic substitution reaction, other transition metals have also been used successfully, albeit with fewer examples, so far.

One of the advantages of using an alternative metal is that the regiochemistry of the substitution process is often biased towards attack at the more substituted terminus. For palladium-catalysed allylic substitution, the regiochemical outcome is dependent upon the electronic nature of the ligand.[49]

The tungsten complex (10.63)[50] and the nickel complex (10.64)[51] have been used in allylic substitution reactions, providing over 90% ee in the formation of some products. By appropriate choice of ligand, Janssen and Helmchen have begun to develop a remarkable iridium-catalysed enantioselective allylic substitution process.[52] The allylic acetate (10.65) undergoes a highly regioselective reaction to give the substitution product (10.67), with very good enantioselectivity, using the ligand (10.69).

The first reports of asymmetric molybdenum-catalysed allylic substitution reactions have also provided very high selectivities.[53] The allylic carbonate (10.66) is converted into the substitution product (10.70), with excellent enantioselectivity and good regioselectivity, using the pyridine-containing ligand (10.72).

The use of metals other than palladium is, thus, set to become an interesting area for future research.

(10.63)

(10.64)

(10.65)

4 mol% [Ir(cod)Cl]$_2$
4 mol% (10.69)
NaCH(CO$_2$Me)$_2$
24 h, reflux, THF
98%
(10.67):(10.68) 99:1

CH(CO$_2$Me)$_2$

(10.67) 95% ee

+

CH(CO$_2$Me)$_2$

(10.68)

P(C$_6$H$_4$-pCF$_3$)$_2$

(10.69)

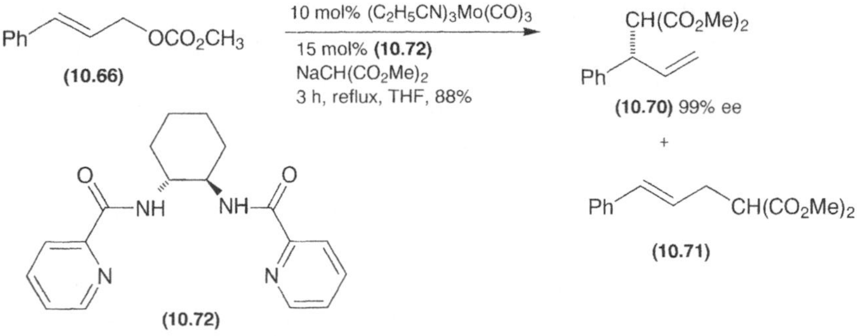

(10.72)

10.3   Heck reactions

In general, the Heck reaction involves the alkylation or arylation of an alkene to give the more substituted alkene product. In order for an asymmetric Heck reaction to take place, it must either take place on one of two enantiotopic alkenes or the standard reaction mechanism must be diverted. The reaction pathway can be altered if the palladium hydride elimination step cannot be achieved, which occurs when the palladium and β-hydride are unable to align in a *syn* fashion. This is true for cyclic substrates and in certain other cases. For example, dihydrofuran (**10.73**) forms the initial palladium alkyl adduct (**10.74**). The hydride next to the R group is unable to undergo β-hydride elimination. However, elimination to give the palladium alkene complex (**10.75**) is possible. This species can either decomplex directly, to give product (**10.76**) or undergoes further β-hydride rearrangements to give the regioisomer (**10.77**). The asymmetric Heck reaction was reviewed in 1997.[54]

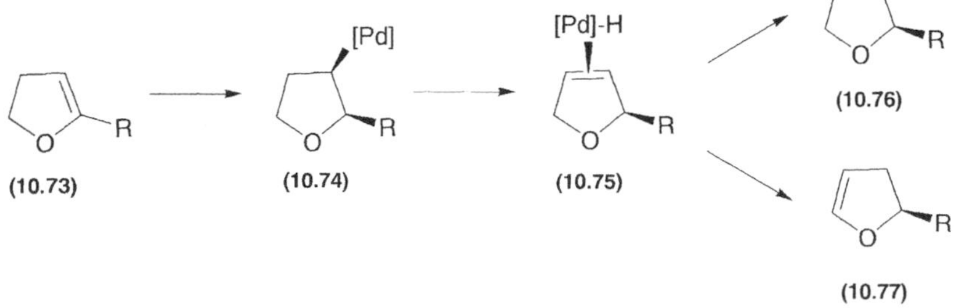

The first asymmetric Heck reactions were reported by Shibasaki and co-workers in 1989.[55] Typical examples from this group include the cyclisation of vinyl iodide (**10.78**). The silver salt is required to generate a cationic palladium intermediate (which then contains no chloride).[56]

In some cases, pinacol has been found to have a beneficial effect on yield and enantiomeric excess of the product, for example in the formation of the decalin (**10.81**). In the absence of pinacol, only 6% yield was obtained after 106 h, with 92% ee. When triflates are used as starting materials, silver salts are not required, since the triflate is more readily dissociated from the palladium intermediates.

(10.78) (10.80) 44% ee

(10.79) (10.81) 95% ee

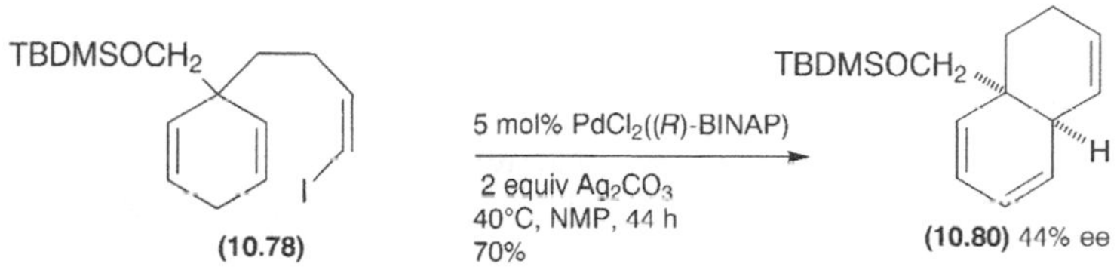

Ashimori and Overman have shown how the use of silver salts can change the sense of asymmetric induction of the cyclised product.[57] Thus, the iodide (**10.82**) can be converted into the product (**10.83**) with the (*S*)-enantiomer predominating, when the reaction is run in the presence of silver salts. In the absence of silver salts, the (*R*)-enantiomer is the major product.

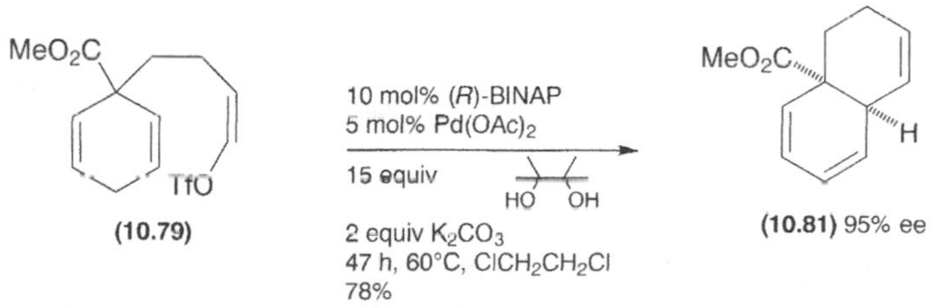

Overman and Poon have shown that the mechanism without silver salts, indeed even with deliberately added halide salt (e.g. Bu_4NBr), can give high enantioselectivities, as demonstrated by the conversion of triflate (**10.84**) into the cyclised product (**10.85**). This shows that the cationic mechanism is not a requirement for a successful outcome, and the authors suggest an intermediate, in which halide, bidentate phosphine, alkene and aryl are all coordinated to the palladium simultaneously (i.e. penta-coordinate).[58] In the absence of Bn_4NBr, a 72% yield was achieved but with only 43% ee.

(10.82) (*S*)-(10.83) 71% ee

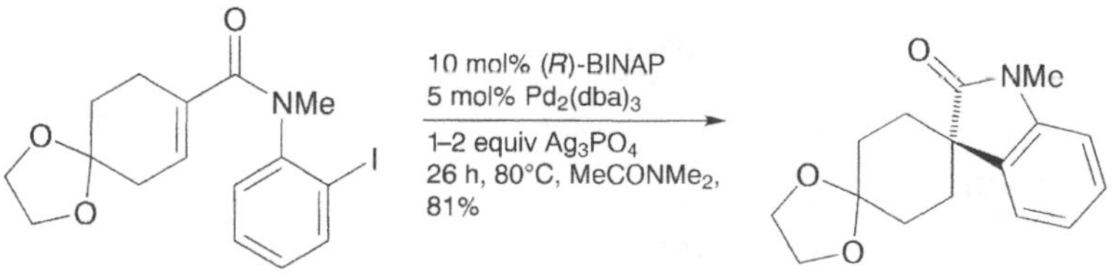

(10.82)

10 mol% (*R*)-BINAP
5 mol% Pd$_2$(dba)$_3$

5 equiv
Me — Me
Me — N — Me
Me

140 h, 80°C, MeCONMe$_2$
77%

(*R*)-(10.83) 66% ee

(10.84)

5 mol% [Pd$_2$(dba)$_3$].CHCl$_3$
10 mol% (*R*)-BINAP

1 equiv Bu$_4$NBr
4 equiv
Me — Me
Me — N — Me
Me

23 h, 100°C, MeCONMe$_2$
then, HCl, H$_2$O, NaBH$_4$, THF
59%

(10.85) 93% ee

Dihydrofuran **(10.73)** has also proved to be a popular substrate for the asymmetric Heck reaction. Hayashi and co-workers reported that, using a Pd/BINAP catalyst, not only is the initial addition enantioselective but the diastereomeric intermediates, i.e. of structure **(10.75)**, preferentially give different regioisomeric products **(10.86)** and **(10.87)**. This effect is similar to that of a kinetic resolution (see Section 4.1).[59]

Other ligands which work well in this reaction include the bidentate phosphine **(10.88)**, where the 3,5-dialkyl substituents are important to enantioselectivity (the 3,5-dialkyl *meta* effect).[60] Phosphino-oxazoline ligands, such as compound **(10.89)** have also been used to great effect.[61]

Interestingly, these last two examples show how the regiochemistry of the reaction is controlled by the ligand and reaction conditions. The same phosphino-oxazoline ligand **(10.89)** has also been applied to other enantioselective Heck reactions, including the coupling of triflates to give the products **(10.93)** and **(10.94)**, with good to excellent stereocontrol. This ligand has also been employed in the Heck arylation of cyclic enamides.[62]

(10.73)
5 equiv

3 mol% Pd(OAc)$_2$
6 mol% (*R*)-BINAP

PhOTf
3 equiv ^{i}Pr$_2$NEt
24 h, 40°C, benzene
(10.86):(10.87) 98:2

(10.86) 82% ee   **(10.87)** 60% ee

(10.88)

Ar = 3,5-(^{t}Bu)$_2$C$_6$H$_3$-

provides >98% ee
of regiosiomer (ent-10.86)
(selectivity 95:5)

(10.89)

provides 97% ee
of regiosiomer (ent-10.87)
(selectivity >99:1)

(10.73) + **(10.91)**

3 mol% Pd(dba)$_2$
6 mol% (10.89)
──────────────→
^{i}Pr$_2$NEt
3 days, 30°C, benzene
92%

(10.93) 99% ee

(10.90) + **(10.92)**

3 mol% Pd(dba)$_2$
6 mol% (10.89)
──────────────→
^{i}Pr$_2$NEt
7 days, 70°C, THF
70%

(10.94) 92% ee

In the presence of nucleophiles, Shibasaki and co-workers have extended their methodology to a Heck reaction/carbanion capture sequence, which gives good enantiomeric excess. For example, using nucleophile **(10.95)** provided the highest enantioselectivity in the cyclisation/nucleophilic capture of triflate **(10.96)**.[63] Tietze and Raschke used a 'silane-terminated' Heck reaction, in which the substrate **(10.98)** loses the silyl group to give the cyclised product **(10.99)**[64]. An asymmetric polyene cyclisation (asymmetric tandem Heck) has also been reported in the synthesis of the pentacyclic polyketide (+)-xestoquinone.[65]

The use of a hydride source affords a hydroarylation of alkenes, which takes place when the alkene adduct formed initially is unable to undergo cis-β-hydride elimination.[66] For example, the Heck reaction of norbornene, **(10.100)** with triflate **(10.101)** fails because of the geometry of the palladium alkyl intermediate initially formed. However, in the presence of a hydride source (Et$_3$N/HCO$_2$H), reductive elimination to give the product **(10.102)** takes place enantioselectively using ligand **(10.103)**.

(10.96)

+

NaCH(SO₂Ph)₂
(10.95)
2 equiv

2.5 mol% [Pd(allyl)Cl]₂
6.3 mol% (S)-BINAP
────────────────
2 equiv NaBr
r.t., DMSO
83%

(10.97) 94%ee

2.5 mol% Pd₂(dba)₃.CHCl₃
7.0 mol% (R)-BINAP
────────────────
1.1 equiv Ag₃PO₄
48 h, 80°C, 91%

(10.98)

(10.99) 92% ee

1.2 mol% Pd(OAc)₂
2.4 mol% (10.103)
────────────────
1.5 equiv PhOTf (10.101)
3.5 equiv Et₃N
3 equiv HCO₂H
20 h, DMSO, 65°C
62%

(10.100)

(10.102) 70% ee

(10.103)

10.4 Alkylmetalation of alkenes

The asymmetric alkylmagnesiation of alkenes has been achieved with enantiomerically pure zirconocene catalysts.[67, 68] The reaction with allyl ethers is a useful procedure, typified by the reaction of the cyclic allylic ether (10.104), catalysed by the Brintzinger complex (10.105). The reaction is believed to proceed *via* the alkylmagnesium intermediate (10.106), which undergoes elimination to afford the product (10.107) with high enantioselectivity. The reaction has been employed for the ethylmagnesiation of larger ring systems,[69] as well in an efficient kinetic resolution, as demonstrated by the reaction of racemic dihydropyran (10.108).[70] In some cases, the starting material could be prepared *in situ* by ruthenium-catalysed ring-closing metathesis.

The enantioselective zirconium-catalysed alkylalumination of alkenes has been reported by Kondakov and Negishi.[71] The intermediate organoaluminium species can be quenched in a number of ways, including dioxygen quench to provide an alcohol. For example, hex-l-ene (10.110) undergoes ethylalumination/oxidation with good enantioselectivity in the product alcohol (10.111), using the catalyst (10.112).[72] Lautens and co-workers have shown that the nickel-catalysed hydroalumination of

(10.105)

(10.112)

(10.104)

10 mol% (10.105)
EtMgBr
25°C, 6–12 h, THF
65%

ClMg

H [Zr]

(10.106)

Et

OH

(10.107) >97% ee

(10.108)

10 mol% (10.105)
EtMgCl
70°C, 6–12 h, THF
60% conversion

C_3H_7

(10.108) >99% ee

+

Et

OH C_3H_7

(10.109) 94% ee

Bu

(10.110)

8 mol% (10.112)
Et$_3$Al
0°C, 24 h, CH$_3$CHCl$_2$
O$_2$ quench, 74%

Bu

OH

Et

(10.111) 93% ee

certain alkenes can also be achieved, with high enantioselectivity.[73] The mechanism of these reactions has been found to be dependent upon a number of factors, and does not necessarily involve an organoalane intermediate.[74]

References

1. J. Tsuji, in *Palladium Reagents and Catalysts*, John Wiley and Sons, Chichester, **1995**.
2. T. Hayashi, in *Catalytic Asymmetric Synthesis*, (I. Ojima, ed.) VCH, New York, **1993**, 325.
3. T. Hayashi, M. Konishi, M. Fukushima, K. Kanehira, T. Hioki and M. Kumada, *J. Org. Chem.*, **1983**, *48*, 2195.
4. T. Hayashi, A. Yamamoto, M. Hojo and Y. Ito, *J. Chem. Soc., Chem. Commun.*, **1989**, 495.
5. T. Hayashi, M. Konishi, Y. Okamoto, K. Kabeta and M. Kumada, *J. Org. Chem.*, **1986**, *51*, 3772.
6. For example, see: G.C. Lloyd-Jones and C.P. Butts, *Tetrahedron*, **1998**, *54*, 901.
7. T. Hayashi, K. Hayashizaki, T. Kiyoi and Y. Ito, *J. Am. Chem. Soc.*, **1988**, *110*, 8153.
8. T. Hayashi, S. Niizuma, T. Kamikawa, N. Suzuki and Y. Uozumi, *J. Am. Chem. Soc.*, **1995**, *117*, 9101.

9. T. Kamikawa, Y. Uozumi and T. Hayashi, *Tetrahedron Lett.*, **1996**, *37*, 3161.
10. M. Uemura and H. Nishimura, *J. Organomet. Chem.*, **1994**, *473*, 129.
11. J. Åhman, J.P. Wolfe, M.V. Troutman, M. Palucki and S.L. Buchwald, *J. Am. Chem. Soc.*, **1998**, *120*, 1918.
12. B.M. Trost, *Chem. Rev.*, **1996**, *96*, 395.
13. C.G. Frost, J. Howarth and J.M.J. Williams, *Tetrahedron: Asymmetry*, **1992**, *3*, 1089.
14. T. Hayashi, A. Yamamoto, T. Hagihara and Y. Ito, *Tetrahedron Lett.*, **1986**, *27*, 191.
15. T. Hayashi, *Pure Appl. Chem.*, **1998**, *60*, 7.
16. A. Togni, *Tetrahedron: Asymmetry*, **1991**, *2*, 683.
17. A. Pfaltz, *Acc. Chem. Res.*, **1993**, *26*, 339.
18. D. Tanner, P.G. Andersson, A. Harden and P. Somfai, *Tetrahedron Lett.*, **1994**, *35*, 4631.
19. C. Bolm, D. Kaufman, S. Gessler and K. Harms, *J. Organomet. Chem.*, **1995**, *502*, 47.
20. G. Zhu, M. Terry and X. Zhang, *Tetrahedron Lett.*, **1996**, *37*, 4475.
21. J. Sprinz and G. Helmchen, *Tetrahedron Lett.*, **1993**, *34*, 1769.
22. G.J. Dawson, C.G. Frost, J.M.J. Williams and S.J. Coote, *Tetrahedron Lett.*, **1993**, *34*, 3149.
23. P. von Matt and A. Pfaltz, *Angew. Chem., Int. Ed. Engl.*, **1993**, *32*, 566.
24. J.M. Brown, D.I. Hulmes and P.J. Guiry, *Tetrahedron*, **1994**, *50*, 4493.
25. H.C.L. Abbenhuis, U. Burckhardt, V. Gramlich, C. Kollner, P.S. Pregosin, R. Salzman and A. Togni, *Organometallics*, **1995**, *14*, 759.
26. J.V. Allen, S.J. Coote, G.J. Dawson, C.G. Frost, C.J. Martin and J.M.J. Williams, *J. Chem. Soc., Perkin Trans. 1*, **1994**, 2065.
27. H. Kubota and K. Koga, *Tetrahedron Lett.*, **1994**, *35*, 6689.
28. B.M. Trost and R.C. Bunt, *J. Am. Chem. Soc.*, **1994**, *116*, 4089.
29. G. Knühl, P. Sennhenn and G. Helmchen, *J. Chem. Soc., Chem. Commun.*, **1995**, 1845.
30. P.R. Auburn, P.B. Mackenzie and B. Bosnich, *J. Am. Chem. Soc.*, **1985**, *107*, 2033.
31. G.J. Dawson, J.M.J. Williams and S.J. Coote, *Tetrahedron Lett.*, **1995**, *36*, 461.
32. J.M.J. Williams, *Synlett*, **1996**, 705.
33. B.M. Trost, D.L. VanVranken and C. Bingel, *J. Am. Chem. Soc.*, **1992**, *114*, 9327.
34. B.M. Trost, C.B. Lee and J.M. Weiss, *J. Am. Chem. Soc.*, **1995**, *117*, 7247.
35. B.M. Trost and X. Ariza, *Angew. Chem., Int. Ed. Engl.*, **1997**, *36*, 2635.
36. B.M. Trost, R. Radinov and E.M. Grenzer, *J. Am. Chem. Soc.*, **1997**, *119*, 7879.
37. A. Sudo and K. Saigo, *J. Org. Chem.*, **1997**, *62*, 5508.
38. P. von Matt, O. Loiseleur, G. Koch and A. Pfaltz, *Tetrahedron: Asymmetry*, **1994**, *5*, 573.
39. R. Jumnah, A.C. Williams and J.M.J. Williams, *Synlett*, **1995**, 821.
40. A. Togni, U. Burckhardt, V. Gramlich, P.S. Pregosin and R. Salzmann, *J. Am. Chem. Soc.*, **1996**, *118*, 1031.
41. B.M. Trost and M.G. Organ, *J. Am. Chem. Soc.*, **1994**, *116*, 10320.
42. B.M. Trost and F.D. Toste, *J. Am. Chem. Soc.*, **1998**, *120*, 815.
43. H. Eichelmann and H.-J. Gais, *Tetrahedron Asymmetry*, **1995**, *6*, 643.
44. T. Hayashi, H. Iwamura, Y. Uozumi, Y. Matsumoto and F. Ozawa, *Synthesis*, **1994**, 526.
45. N. Kardos and J.-P. Genêt, *Tetrahedron: Asymmetry*, **1994**, *5*, 1525.
46. G. Koch and A. Pfaltz, *Tetrahedron: Asymmetry*, **1994**, *5*, 1525.
47. B.M. Trost and D.E. Patterson, *J. Org. Chem.*, **1998**, *63*, 1339.
48. Y. Uozumi, K. Kato and T. Hayashi, *J. Am. Chem. Soc.*, **1997**, *119*, 5063.
49. R. Prétôt and A. Pfaltz, *Angew. Chem., Int. Ed. Engl.*, **1998**, *37*, 323.
50. G.C. Lloyd Jones and A. Pfaltz, *Angew. Chem., Int. Ed. Engl.*, **1995**, *34*, 462.
51. G. Consiglio and A. Indolese, *Organometallics*, **1991**, *10*, 3425.
52. J.P. Janssen and G. Helmchen, *Tetrahedron Lett.*, **1997**, *37*, 8025.
53. B.M. Trost and I. Hachiya, *J. Am. Chem. Soc.*, **1998**, *120*, 1104.
54. M. Shibasaki, C.D.J. Boden and A. Kojima, *Tetrahedron*, **1997**, *53*, 7371.

55. Y. Sato, M. Sodeoka and M. Shibasaki, *J. Org. Chem.*, **1989**, *54*, 4738.
56. K. Ohrai, K. Kondo, M. Sodeoka and M. Shibasaki, *J. Am. Chem. Soc.*, **1994**, *116*, 11737.
57. A. Ashimori and L.E. Overman, *J. Org. Chem.*, **1992**, *57*, 4571.
58. L.E. Overman and D.J. Poon, *Angew. Chem., Int. Ed. Engl.*, **1997**, *36*, 518.
59. F. Ozawa, A. Kubo, Y. Matsumoto and T. Hayashi, *Organometallics*, **1993**, *12*, 4188.
60. G. Trabesinger, A. Albinati, N. Feiken, R.W. Kunz, P.S. Pregosin and M. Tschoerner, *J. Am. Chem. Soc.*, **1997**, *119*, 6315.
61. O. Loiseleur, M. Hayashi, N. Schmees and A. Pfaltz, *Synthesis*, **1997**, 1338.
62. L. Ripa and A. Hallberg, *J. Org. Chem.*, **1997**, *62*, 595.
63. T. Ohshima, K. Kagechika, M. Adachi, M. Sodeoka and M. Shibasaki, *J. Am. Chem. Soc.*, **1996**, *118*, 7108.
64. L.F. Tietze and T. Raschke, *Synlett*, **1995**, 597.
65. S.P. Maddaford, N.G. Andersen, W.A. Cristofoli and B.A. Keay, *J. Am. Chem. Soc.*, **1996**, *118*, 10766.
66. S. Sakuraba, K. Awano and K. Achiwa, *Synlett*, **1994**, 291.
67. L. Bell, R.J. Whitby, R.V.H. Jones and M.C.H. Standen, *Tetrahedron Lett.*, **1998**, *39*, 7139.
68. J.P. Morken, M.T. Didiuk and A.H. Hoveyda, *J. Am. Chem. Soc.*, **1993**, *115*, 6997.
69. M.S. Visser, N.M. Heron, M.T. Didiuk, J.F. Sagal and A.H. Hoveyda, *J. Am. Chem. Soc.*, **1996**, *118*, 4291.
70. M.S. Visser and A.H. Hoveyda, *Tetrahedron*, **1995**, *51*, 4383.
71. D.Y. Kondakov and E. Negishi, *J. Am. Chem. Soc.*, **1995**, *117*, 10771.
72. D.Y. Kondakov and E.-I. Negishi, *J. Am. Chem. Soc.*, **1996**, *118*, 1577.
73. M. Lautens, P. Chiu, S. Ma and T. Rovis, *J. Am. Chem. Soc.*, **1995**, *117*, 532.
74. M. Lautens and T. Rovis, *J. Am. Chem. Soc.*, **1997**, *119*, 11090.

11 Conjugate addition reactions

The 1,4-addition of nucleophiles to alkenes attached to electron-withdrawing groups (ketones, aldehydes, esters, nitriles, etc.) is often referred to as conjugate addition. Nucleophiles which undergo conjugate addition rather than a direct nucleophilic attack on the electron-withdrawing group include enolates, thiolates and cuprates (or copper catalysed addition of other organometallic reagents).

11.1 Conjugate addition of enolates

The Michael addition reaction involves the conjugate addition of stabilised enolates to α,β-unsaturated carbonyl compounds. The reaction can be catalysed by activation of either the nucleophile or the Michael acceptor. In some cases, the activation of both partners probably occurs.

Rhodium complexes of the *trans*-chelating phosphine ligand, TRAP (11.01) probably coordinate to the nitrile to facilitate enolisation of the nucleophile (11.02), and control the subsequent stereochemistry with good enantiocontrol in additions to the Michael acceptors (11.03) and (11.04).[1] α-Cyano Weinreb amides also work well as nucleophiles in this reaction, and the products are readily converted into the corresponding

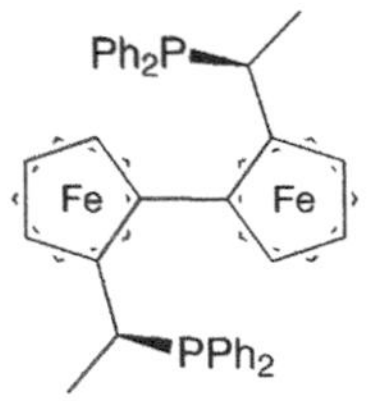

(11.01)

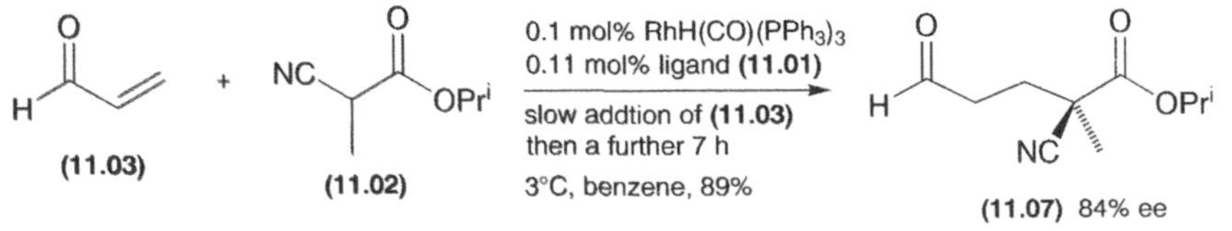

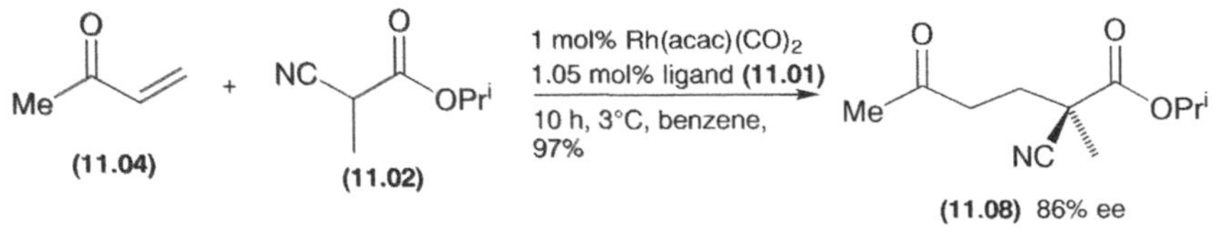

aldehydes and ketones.[2] For example, the Weinreb amide (**11.05**) undergoes Michael addition with the unsaturated ketone (**11.06**) to give the product (**11.09**), with excellent yield and good enantioselectivity.

(**11.06**) (**11.05**) (**11.09**) 86% ee

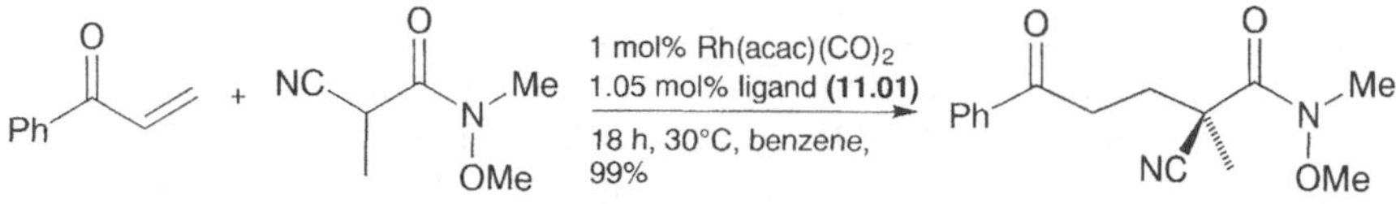

Enantioselective Michael addition reactions are certainly not restricted to nitrile-containing nucleophiles. The cyclic β-ketoester (**11.10**) undergoes an enantioselective Michael addition catalysed by the copper(II) complex (**11.11**).[3] The same reaction has also been catalysed by a potassium t-butoxide/enantiomerically pure crown ether combination, with up to 99% ee in the formation of the product (**11.12**).[4]

In fact, enantiomerically pure crown ethers have been used to catalyse other Michael reactions,[5] including the use of crown ether (**11.13**) in the conjugate addition reaction between the ester (**11.14**) and Michael acceptor (**11.15**).[6] The reaction is remarkably rapid (1 min at −78°C).

A catalytic enantioselective Michael addition of diisopropyl malonate to α,β-unsaturated ketones and aldehydes has been achieved using 5 mol% of the rubidium salt of L-proline. Reminiscent of the proline-catalysed Hajos-Wiechert aldol reaction (see Section 7.1), this reaction may proceed *via* transient formation of an iminium ion.[7]

(**11.11**)

(**11.13**)

(**11.10**) (**11.04**) (**11.12**) 75% ee

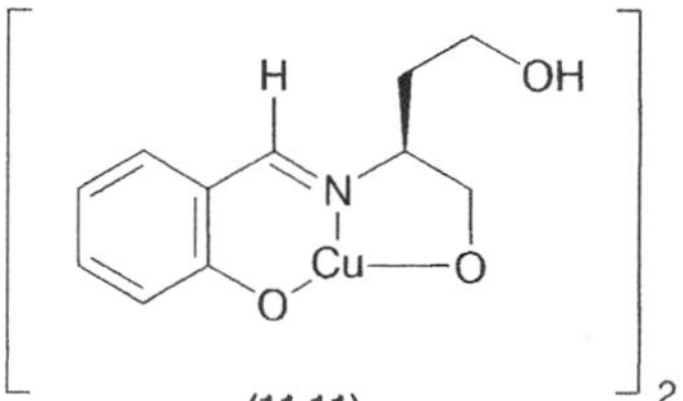

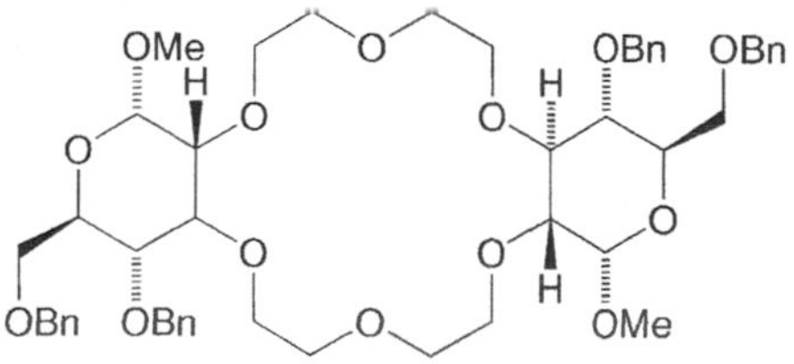

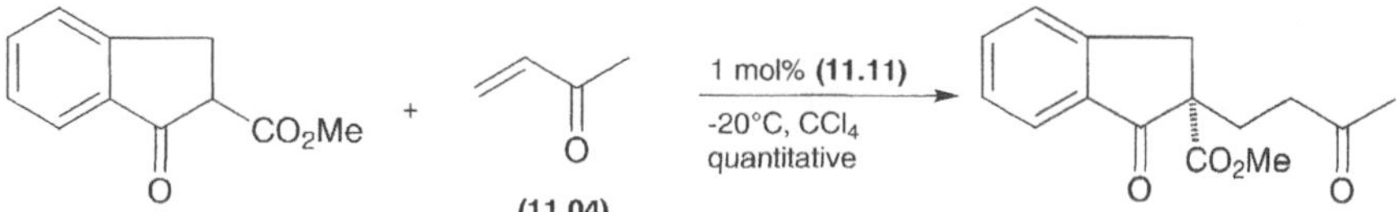

(11.14) (11.15) 1.3 equiv (11.16) 84.4% ee

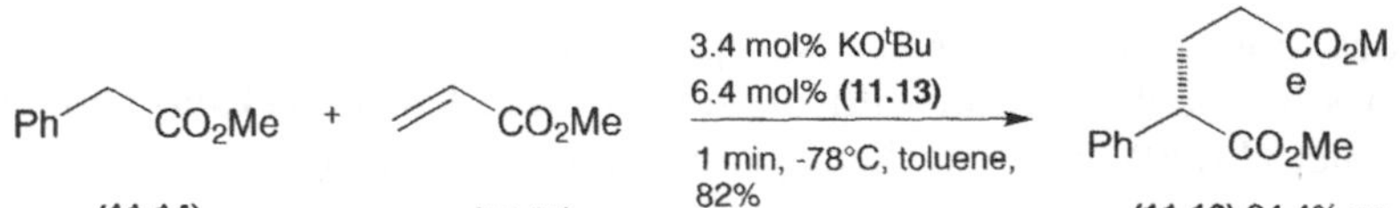

Shibasaki's heterobimetallic complexes provide the opportunity to activate both the nucleophile and the Michael acceptor. Whilst the aluminium lithium bis-BINOL complex (ALB) (**11.17**) does not catalyse conjugate addition of α-phosphonate ester (**11.18**) with cyclopentenone (**11.19**) by itself, addition of sodium tert-butoxide allows a highly enantioselective reaction to take place.[8] Heterobimetallic complexes have also been used to catalyse the addition of α-nitroesters[9] and malonates[10] to Michael acceptors. The catalytic asymmetric Michael reaction using silyl enol esters (Mukaiyama-Michael reaction) as the pronucleophiles has been reported using a titanium/BINOL catalyst (with up to 90% ee). Considering furan (**11.21**) as a silyl enol ether, this has been shown to undergo nucleophilic addition to the Michael acceptor (**11.22**). The product (**11.23**) can be obtained with excellent diastereocontrol with the scandium complex of ligand (**11.24**), or with excellent enantiocontrol using the copper complex of bis-oxazoline (**11.25**).[11,12,13]

(11.17)

(11.24)

(11.25)

(11.19) (11.18) (11.20) 95% ee

(11.21) (11.22) (11.23)

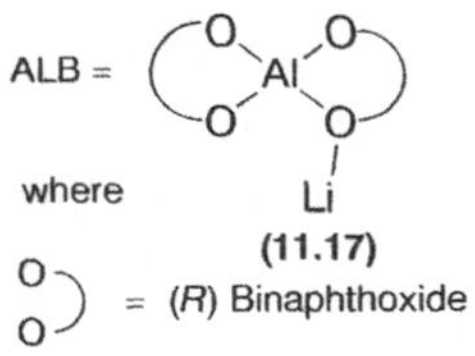

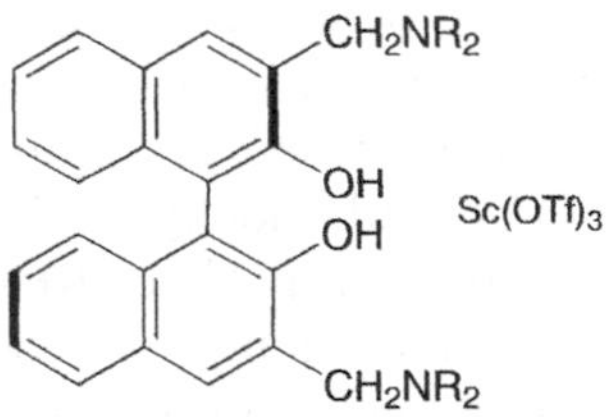

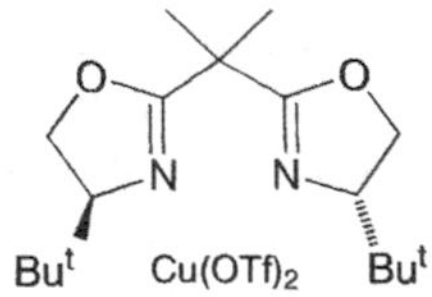

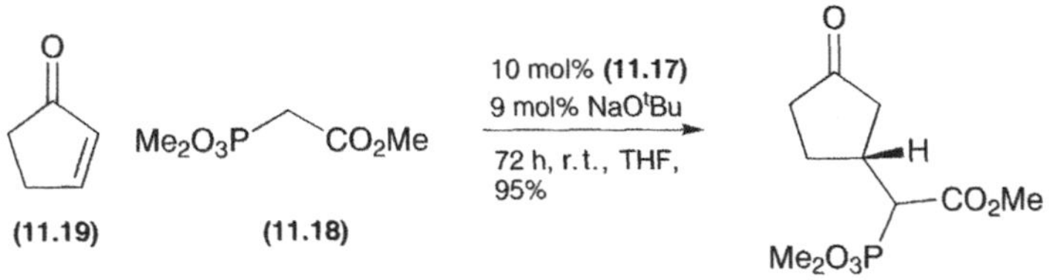

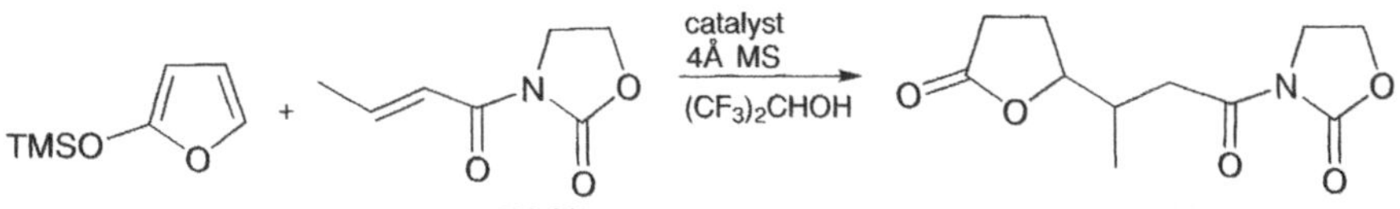

(11.24) provides;
68% ee (*anti*:*syn* >50:1)

(11.25) provides;
95% ee (*anti*:*syn* >8.5:1)

The intermediate enolate formed in a Michael reaction normally undergoes protonation to give the ketone product. However, in the presence of an aldehyde, a Michael/Aldol cascade can occur. Using the racemic Michael acceptor (11.26), Shibasaki and co-workers demonstrated an enantioselective and diastereoselective Michael/Aldol cascade involving the malonate nucleophile (11.27) and the aldehyde (11.28). The process also involves a kinetic resolution of the starting material.[14] Not only is this a remarkable example of several aspects of stereoselectivity, but also the product (11.29) is a useful prostaglandin precursor.

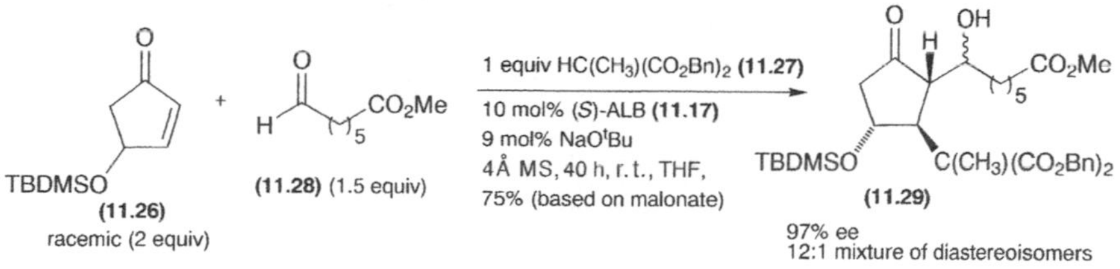

11.2 Conjugate addition of sulfur nucleophiles

Sulfur nucleophiles are soft and preferentially react by conjugate addition with α,β-unsaturated carbonyl compounds. Only catalytic amounts of the lithium thiolate (11.30) are required, since addition to the enone (11.31) generates the enolate (11.32). The enolate is then able to deprotonate thiol (11.33), regenerating thiolate (11.30) with formation of the product (11.34).

Since the lithium will coordinate to an appropriate enantiomerically pure ligand, an asymmetric catalytic reaction can be achieved. Tomioka and co-workers have established this principle using ligand (11.35).[15]

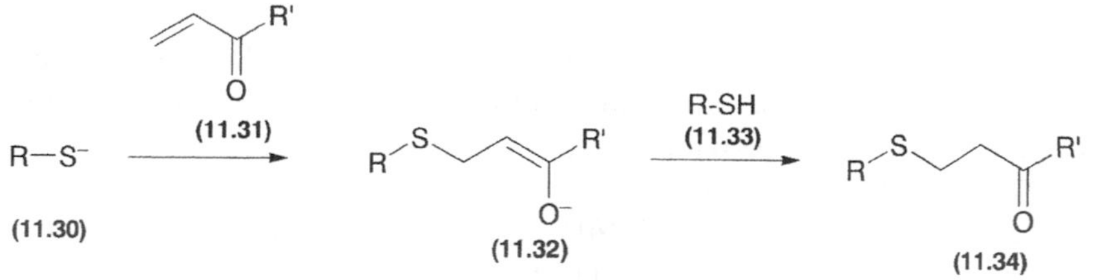

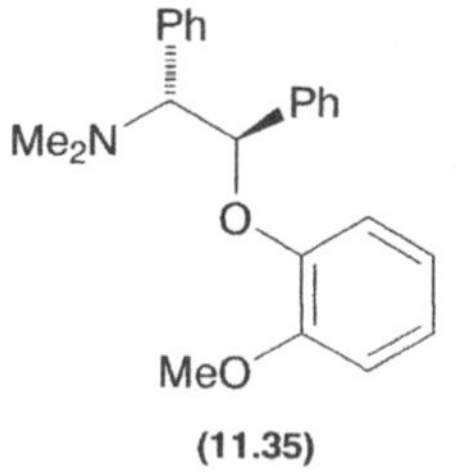

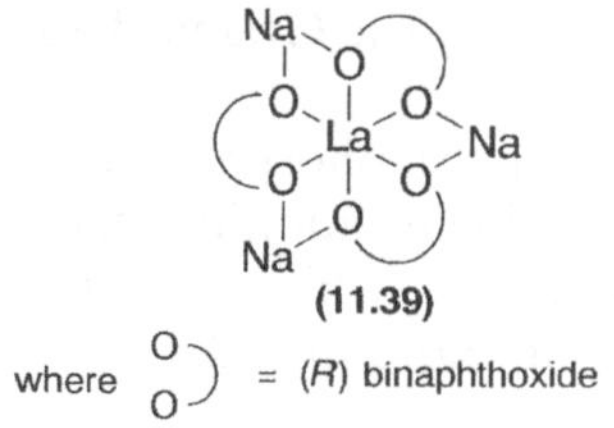

Methyl crotonate (11.36) and thiophenol (11.37) afford the conjugate addition product (11.38), with fairly good enantioselectivity. Improved enantioselectivities were obtained using 2-(trimethylsilyl)thiophenol as the incoming nucleophile (up to 97% ee).[16] A related concept using Shibasaki's heterobimetallic complexes, such as lanthanum sodium BINOL (LSB) (11.39), provides good enantiocontrol in the conjugate addition of benzylthiol (11.40) to cyclohexenone (11.41).[17]

11.3 Conjugate addition of non-stabilised carbanions

The conjugate addition of organocuprates of α,β-unsaturated enones and enoates is an important synthetic procedure. Whilst there are several examples of enantiomerically pure ligands which work well when used stoichiometrically,[18] it is only fairly recently that significant advances have been made for catalytic reactions. In order to achieve a catalytic reaction, the ligand design must be such that there will be preferential binding to the copper salt rather than to the metal counterion of the carbanion.

In 1993, Tanaka and co-workers used a sub-stoichiometric quality of ligand (11.43) in the conjugate addition reaction of methyllithium to enone (11.44), which provides the natural product muscone (11.45), with excellent enantioselectivity.[19]

The copper complexes (11.46)[20] and (11.47)[21] have been used to catalyse conjugate addition reactions, with good yields and enantio-selectivities, as shown by the Grignard additions to substrates (11.48) and (11.41).

Enantiomerically pure phosphine ligands have also been used with copper-catalysed conjugate addition reactions. In particular, ligand (11.51) gives good selectivities,[22] which were further improved (over 90% ee in some cases) by using more ligand.

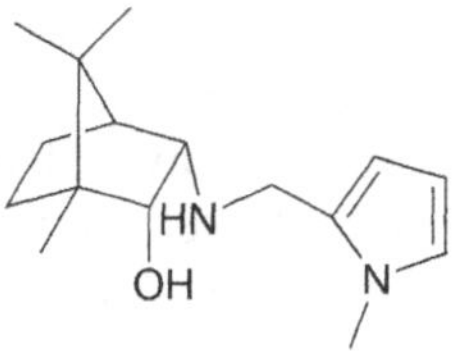

(11.43) 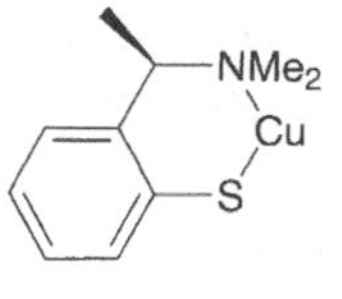   **(11.46)** 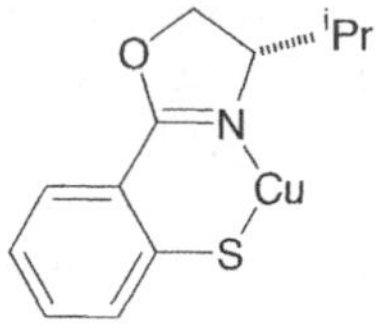   **(11.47)** 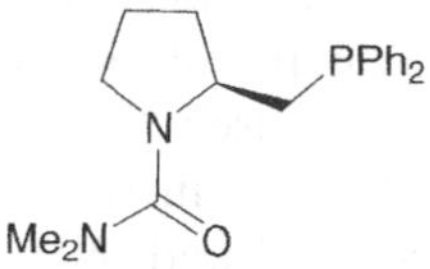   **(11.51)**

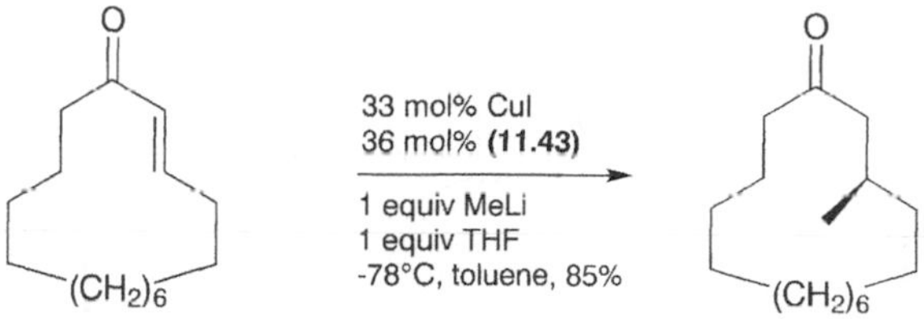

(11.44)

33 mol% CuI
36 mol% **(11.43)**

1 equiv MeLi
1 equiv THF
-78°C, toluene, 85%

(CH₂)₆
(11.45)
99% ee

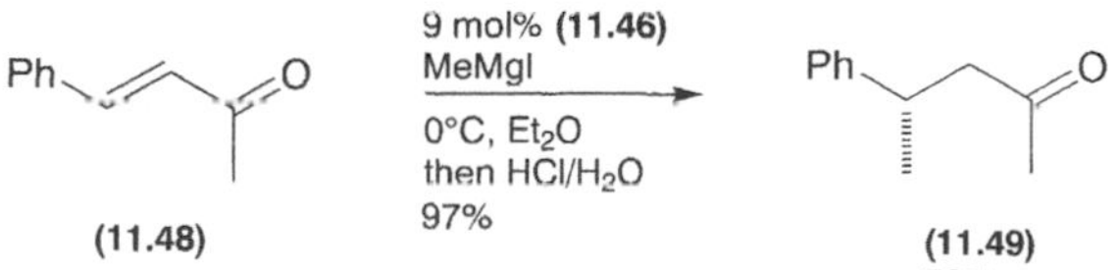

Ph
(11.48)

9 mol% **(11.46)**
MeMgI

0°C, Et₂O
then HCl/H₂O
97%

Ph
(11.49)
76% ee

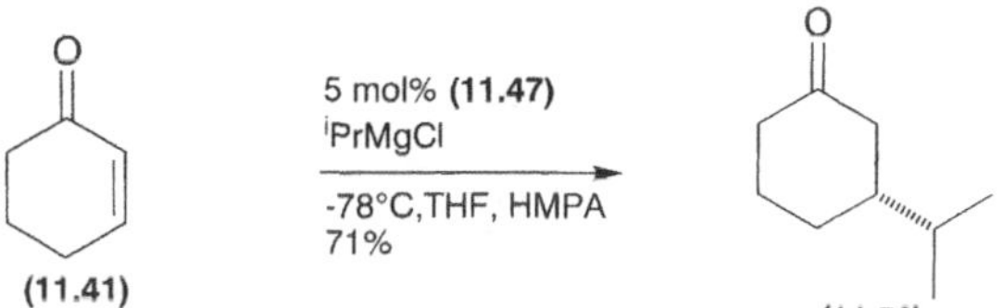

(11.41)

5 mol% **(11.47)**
ᶦPrMgCl

-78°C,THF, HMPA
71%

(11.50)
72% ee

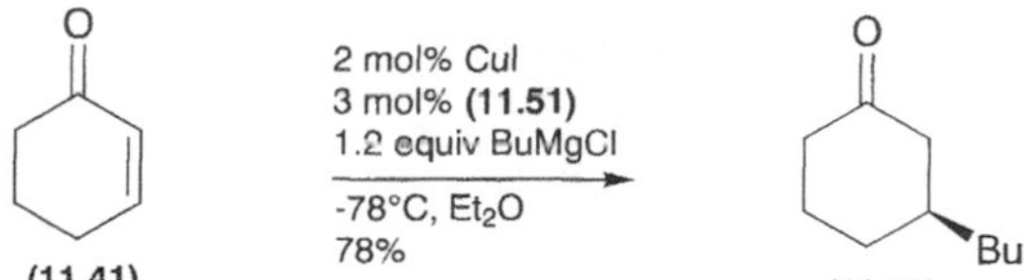

(11.41)

2 mol% CuI
3 mol% **(11.51)**
1.2 equiv BuMgCl

-78°C, Et₂O
78%

Bu
(11.52)
87% ee

The conjugate addition of diethylzinc to enones has received special attention.[23] Interestingly, this reaction is usually catalysed by copper(II) salts, rather than the copper(I) salts previously mentioned, although it is not really clear which oxidation state is required in the catalytic cycle. In fact, nickel(II) salts can also be effective, for example, in the conversion of chalcone **(11.53)** into the addition product **(11.54)**, using the diamine ligand **(11.55)**.[24]

The ligands **(11.56)**[25] and **(11.57)**[26] designed by Pfaltz and co-workers and Feringa and co-workers, respectively, display exceptional selectivity, although the range of substrates is limited to cyclohexenone (**11.41**) and similar molecules, including enone (**11.59**), for the highest selectivities.[27] Functionalised zinc reagents, such as $Zn[(CH_2)_3CH(OEt)_2]_2$ have also been used by Feringa and co-workers to give high enantioselectivity (97% ee) in the catalysed conjugate addition to cyclohexenone (**11.41**). Sewald and Wendisch used the Feringa catalysts in conjugate addition reactions to nitroalkenes.[28]

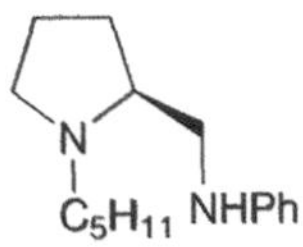

(11.55)

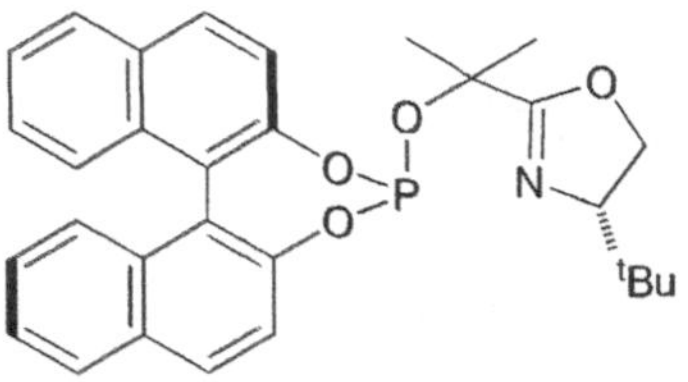

(11.56)

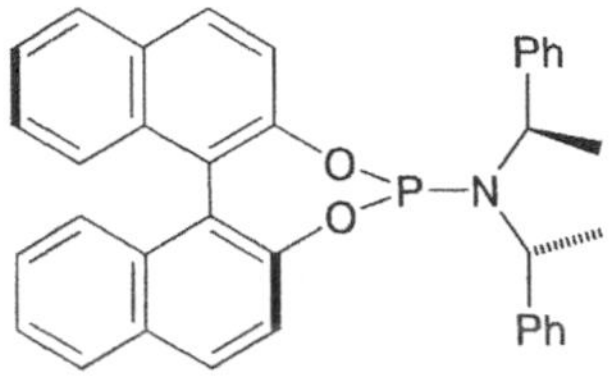

(11.57)

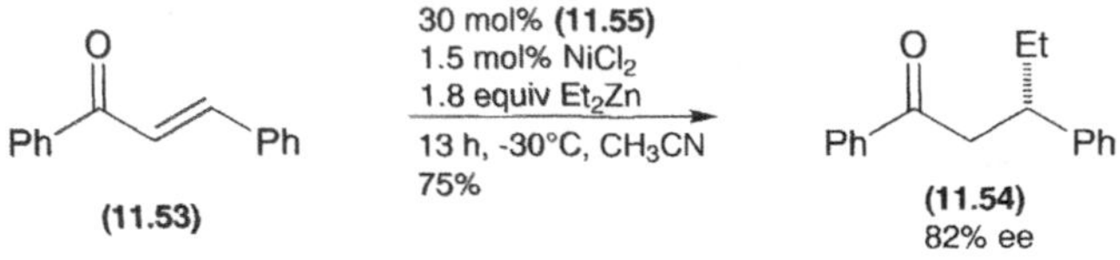

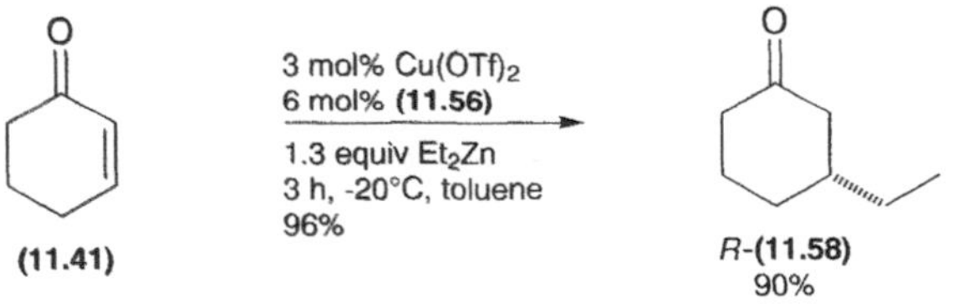

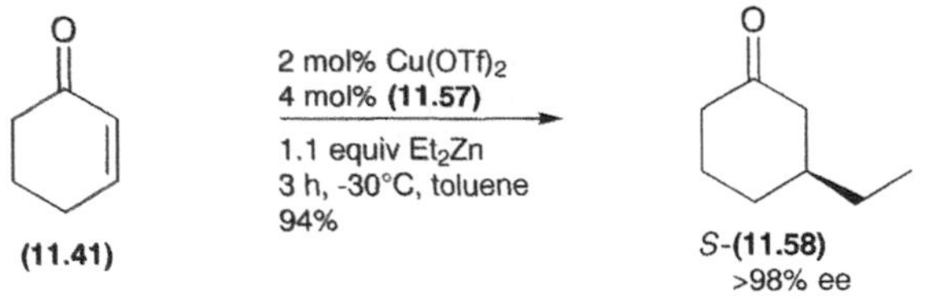

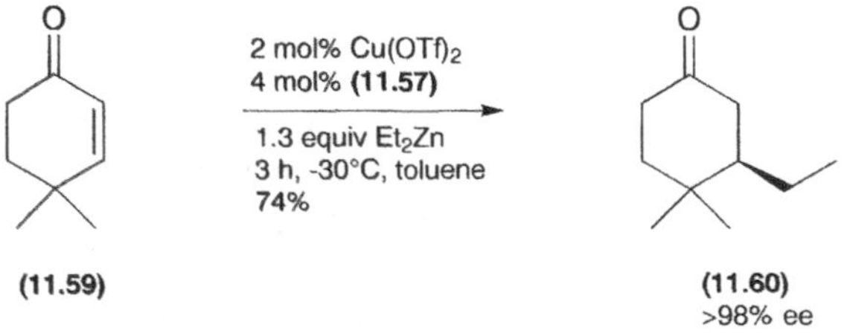

A recent paper disclosed a remarkable new approach to conjugate addition.[29] The reaction uses a rhodium/BINAP catalyst and boronic acids as the nucleophilic coupling partners. Thus, enones (**11.41**) and (**11.61**) react with phenylboronic acid (**11.64**) to yield the products of conjugate addition (**11.62**) and (**11.63**), with good enantioselectivity. It remains to be seen whether future developments will allow this reaction to rival the more conventional copper-catalysed conjugate addition of organometallic reagents.

Conjugate addition of hydride, using sodium borohydride and a cobalt semicorrin complex has been achieved, with high enantioselectivity.[30]

(**11.41**)

3 mol% Rh(acac)(C₂H₄)₂
3 mol% (S)-BINAP
5 equiv PhB(OH)₂ (**11.64**)

5 h, 100°C, 10:1 dioxane:H₂O
>99%

(**11.62**) 97% ee

(**11.61**)

3 mol% Rh(acac)(C₂H₄)₂
3 mol% (S)-BINAP
5 equiv PhB(OH)₂ (**11.64**)

5 h, 100°C, 10:1 dioxane:H₂O
>99%

(**11.63**) 97% ee

11.4 Conjugate addition with nitrogen-based nucleophiles and electrophiles

Nitrogen-based nucleophiles have not been widely investigated in enantio-selective conjugate addition reactions under catalytic conditions. In 1998, the magnesium/bisoxazoline (**11.64**) catalysed addition of O-benzyl-hydroxylamine (**11.65**) to the Michael acceptor (**11.66**) was reported, with good enantioselectivity.[31] Higher enantioselectivities (up to 96% ee) were observed when 100 mol% of the catalyst was employed.

The imide (**11.68**) is an electrophilic aminating agent, which can be thought of as a Michael acceptor. The magnesium catalyst (**11.69**) serves to activate the oxazolidinone (**11.70**) to enolisation and promotes the amination process.[32] The aminated product (**11.71**) was formed, with good enantioselectivity (80–90% ee).

(**11.64**)

(**11.69**)

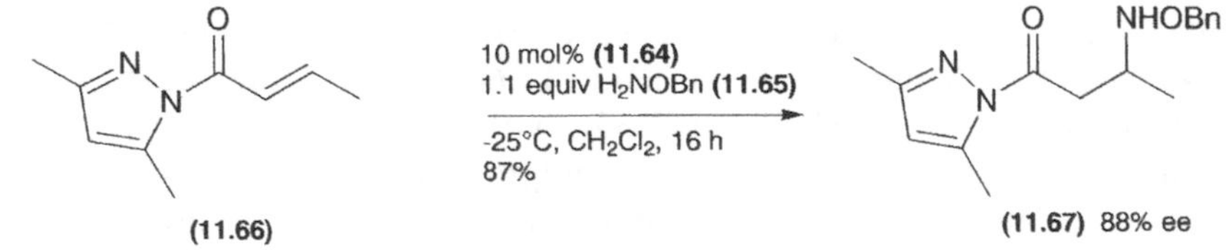

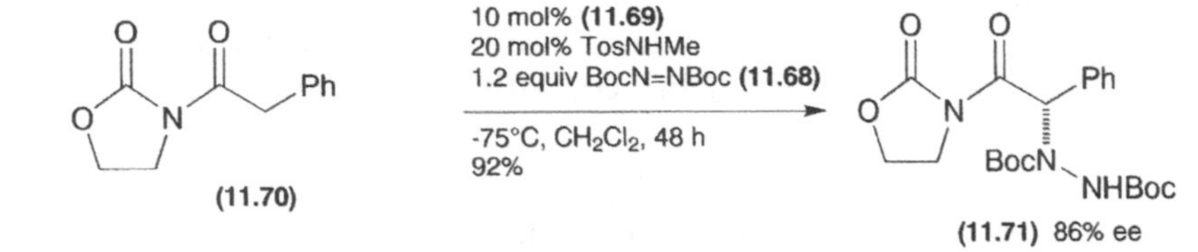

References

1. M. Sawamura, H. Hamashima and Y. Ito, *J. Am. Chem. Soc.*, **1992**, *114*, 8295.
2. M. Sawamura, H. Hamashima, H. Shinoto and Y. Ito, *Tetrahedron Lett.*, **1995**, *36*, 6479.
3. G. Desimoni, G. Dusi, G. Faita, P. Quadrelli and P.P. Righetti, *Tetrahedron*, **1995**, *51*, 4131.
4. D.J. Cram and G.D.Y. Sogah, *J. Chem. Soc., Chem. Commun.*, **1981**, 625.
5. S. Aoki, S. Sasaki and K. Koga, *Tetrahedron Lett.*, **1989**, *30*, 7229.
6. L. Tõke, P. Bakó, G.M. Keserü, M. Albert and L. Fenichel, *Tetrahedron*, **1998**, *54*, 213.
7. M. Yamaguchi, T. Shiraishi and M. Hirama, *Angew. Chem., Int. Ed. Engl.*, **1993**, *32*, 1176.
8. T. Arai, H. Sasai, K.K. Yamaguchi and M. Shibasaki, *J. Am. Chem. Soc.*, **1998**, *120*, 441.
9. E. Keller, N. Veldman, A.L. Spek and B.L. Feringa, *Tetrahedron: Asymmetry*, **1997**, *8*, 3403.
10. G. Manickam and G. Sundararajan, *Tetrahedron: Asymmetry*, **1997**, *8*, 2271.
11. S. Kobayashi, S. Suda, M. Yamada and T. Mukaiyama, *Chem. Lett.*, **1994**, 97.
12. (a) H. Nishikori, K. Ito and T. Katsuki, *Tetrahedron: Asymmetry*, **1998**, *9*, 1165. (b) H. Kitajima, K. Ito and T. Katsuki, *Tetrahedron*, **1997**, *53*, 17015.
13. For an earlier example of the use of Cu(II) bis-oxazoline complexes in conjugate addition reactions, see: A. Bernardi, G. Colombo and C. Scolastico, *Tetrahedron Lett.*, **1996**, *37*, 8921.
14. K.-I. Yamada, T. Arai, H. Sasai and M. Shibasaki, *J. Org. Chem.*, **1998**, *63*, 3666.
15. K. Tomioka, M. Okuda, K. Nishimura, S. Manabe, M. Kanai, Y. Nagaoka and K. Koga, *Tetrahedron Lett.*, **1998**, *39*, 2141.
16. K. Nishimura, M. Ono, Y. Nagaoka and K. Tomioka, *J. Am. Chem. Soc.*, **1997**, *119*, 12974.
17. E. Emori, T. Arai, H. Sasai and M. Shibasaki, *J. Am. Chem. Soc.*, **1998**, *120*, 4043.
18. B.E. Rossiter and N.M. Swingle, *Chem. Rev.*, **1992**, *92*, 771.
19. K. Tanaka, J. Matsui and H. Suzuki, *J. Chem. Soc., Perkin Trans. 1*, **1993**, 153.
20. M. van Klaveren, F. Lambert, D.J.F.M. Eijkelkamp, D.M. Grove and G. van Koten, *Tetrahedron Lett.*, **1994**, *35*, 6135.
21. Q.-L. Zhou and A. Pfaltz, *Tetrahedron Lett.*, **1993**, *34*, 7725.
22. M. Kanai and K. Tomioka, *Tetrahedron Lett.*, **1995**, *36*, 4275.
23. A. Alexakis, J. Frutos and P. Mangeney, *Tetrahedron: Asymmetry*, **1993**, *4*, 2427.
24. M. Asami, K. Usui, S. Higuchi and S. Inoue, *Chem. Lett.*, **1994**, 297.
25. A.K.H. Knöbel, I.H. Escher and A. Pfaltz, *Synlett*, **1997**, 1429.

26. B.L. Feringa, M. Pineschi, L.A. Arnold, R. Imbos and A.H.M. de Vries, *Angew. Chem. Int., Ed. Engl.*, **1997**, *36*, 2620.
27. N. Krause, *Angew. Chem., Int. Ed. Engl.*, **1998**, *37*, 283.
28. N. Sewald and V. Wendisch, *Tetrahedron: Asymmetry*, **1998**, *9*, 1341.
29. Y. Takaya, M. Ogasawara, T. Hayashi, M. Sakai and N. Miyaura, *J. Am. Chem. Soc.*, **1998**, *120*, 5579.
30. (a) U. Leutenegger, A. Madin and A. Pfaltz, *Angew. Chem., Int. Ed. Engl.*, **1989**, *28*, 60.
    (b) P. von Matt and A. Pfaltz, *Tetrahedron: Asymmetry*, **1991**, *2*, 691.
31. M.P. Sibi, J.J. Shay, M. Liu and C. P. Jasperse, *J. Am. Chem. Soc.*, **1998**, *120*, 6615.
32. D.A. Evans and S.G. Nelson, *J. Am. Chem. Soc.*, **1997**, *119*, 6452.

12 Further catalytic reactions

The final chapter contains a collection of interesting reactions, which have not found a comfortable home elsewhere in the book. Some of the following sections describe reactions which are currently topical and progressing well, especially deprotonations, ester formation and epoxide opening. It will be interesting to see, in a few years' time, how far these reactions will have developed.

12.1 Isomerisation

The isomerisation of allylic amines into the corresponding enamines is an excellent example of asymmetric catalysis, which has been exploited on a commercial basis. The isomerisation of the allylamine (**12.01**) with a rhodium/BINAP complex occurs, with excellent yield and enantioselectivity, to give the enamine (**12.02**), as the initial product.[1] The story of the development of the laboratory scale reaction into a 1 000 ton per year process has been told in several reviews,[2] and detailed information on the mechanism of the reaction has been published previously.[3] The enamine (**12.02**) formed initially is converted into citronellal (**12.03**) by hydrolysis, and undergoes subsequent cyclisation to isopulegol (**12.04**) and reduction to menthol (**12.05**). The whole process is performed by Takasago International Corporation, and represents the biggest application (so far) of an enantioselective reaction catalysed by a transition metal complex.

Interestingly, the alternative geometry of starting material (the (Z)-isomer) affords the opposite enantiomer of product.

1 mol% Rh[(*R*)-BINAP]ClO$_4$

40°C, THF, 23 h
100%

(12.01)

(12.02) >96% ee

H$_3$O$^+$

ZnBr$_2$

CHO

(12.03)

H$_2$/Ni

(12.04)

menthol

(12.05)

Similar isomerisation reactions have been applied to other substrates, with very high enantioselectivities for many trisubstituted allylic amines. In general, the rearrangement of allylic alcohols[4] and ethers[5,6] provides lower enantioselectivity. Unfunctionalised alkenes have been isomerised enantioselectively using a titanocene catalyst.[7]

The catalysed rearrangement of allylic imidates, such as substrate (12.06), into the corresponding allylic amide (12.07) was reported, initially, by Overman and co-workers, who used the palladium catalyst (12.08).[8,9] A subsequent report by Uozumi and co-workers in 1998, records higher selectivities in some cases (up to 81% ee) using a phosphino-oxazoline ligand.[10]

(12.08)

(12.06) → 5 mol% (12.08), 40°C, CH$_2$Cl$_2$, 48 h, 69% → (12.07) 55% ee

12.2 Deprotonation reactions

The use of enantiomerically pure bases to catalyse asymmetric deprotonations is an exciting idea, which has recently been shown to be technically feasible. The major difficulty is that the catalytic base must be continuously deprotonated under the reaction conditions. In order to be effective, whatever achiral base provides the continuous deprotonation must not directly deprotonate the substrate. This is conceptually similar to catalytic protonation reactions, which are described in more detail in the next section. Using the catalytic base (12.09), Duhamel and co-workers demonstrated that elimination of HBr from substrate (12.10) can be achieved, with remarkable enantioselectivity, in the formation of the alkene (12.11).[11]

Sparteine has been widely used in conjunction with butyllithium as an enantiomerically pure base. Early results using catalytic amounts of sparteine, especially isosparteine (12.12), have given good results, including the deprotonation of the bicyclic epoxide **(12.13)** and formation of the product **(12.14)** by a cyclisation/rearrangement.[12]

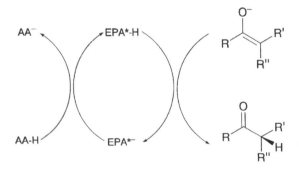

12.3 Protonation reactions

The protonation of prochiral enolates can be achieved using a stoichiometric enantiomerically pure proton source to provide a ketone or ester product, often with good control of enantioselectivity.[13]

The catalytic version of this reaction is more complex, since the kinetics in the catalytic cycle must be such that the achiral acid does not directly protonate the enolate at a significant rate. A catalytic cycle is represented in Figure 12.1, showing how the enantiomerically pure acid (EPA-H) protonates the enolate. The achiral acid (AA-H) recharges the enantiomerically pure acid, but, ideally, does not protonate the enolate directly.

Figure 12.1 Catalytic cycle for protonation reactions. Abbreviations: AA = achiral acid; EPA = enantiomerically pure acid.

The achiral acid needs to be a kinetically slow acid and is, typically, a hindered alcohol, an imide or a carbon acid (e.g. malonate). Asymmetric protonation of lithium enolates from ketones has been achieved using catalytic enantiomerically pure proton sources.[14] Enantiomerically pure imides (12.15),[15] diamine (12.16),[16] as well as a tetradentate amine have been used to protonate lithium enolates, including enolates (12.17) and (12.18). In each of these cases, the stoichiometric achiral acid is kinetically slow, and is also added slowly to the reaction mixture.[17] Samarium enolates have also been protonated enantioselectively (up to 93% ee) using catalytic amounts of a C_2-symmetric diol.[18]

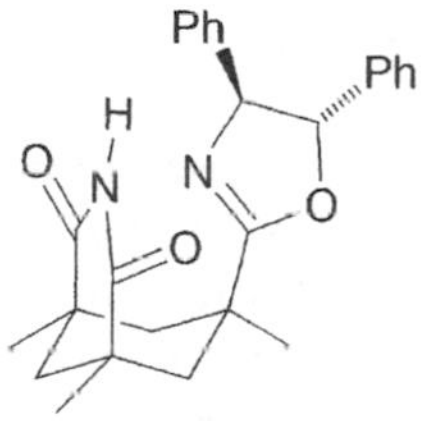

(12.15)

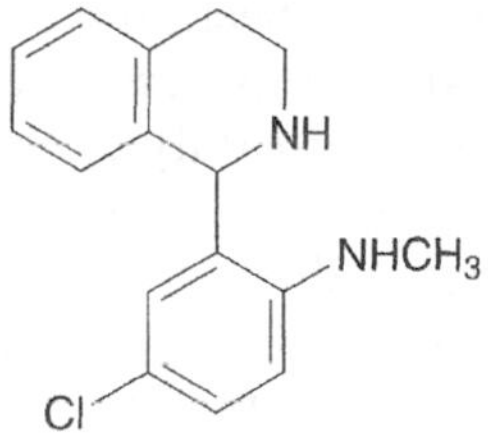

(12.16)

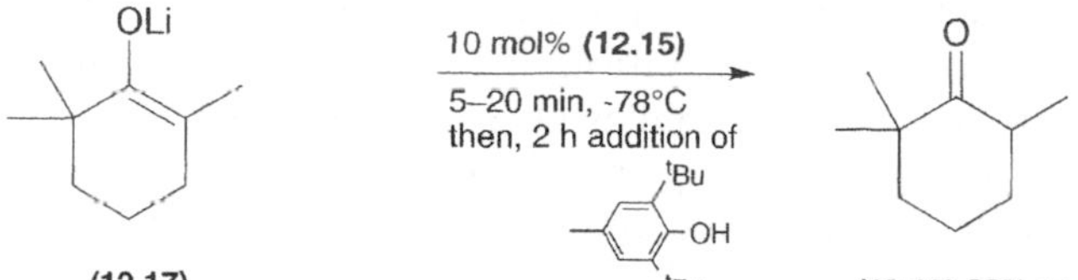

(12.17)

10 mol% (12.15)

5–20 min, -78°C
then, 2 h addition of

72 85% yield

(12.19) 90% ee

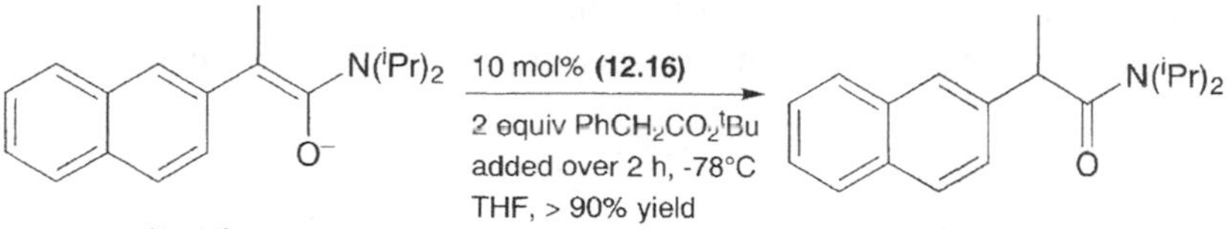

(12.18)

10 mol% (12.16)

2 equiv PhCH₂CO₂ᵗBu
added over 2 h, -78°C
THF, > 90% yield

(12.20) 94% ee

The enantioselective protonation of silyl enol ethers, such as (12.21), by a catalyst has been achieved using 2 mol% of the proton source (12.22).[19] The acidity of (12.22) is enhanced by coordination to the Lewis acid. An alternative approach to the protonation of silyl enol ethers involves the use of palladium catalysts, which proceed *via* intermediate palladium enolates. The asymmetry can be provided either by ligands on the palladium[20] or from an enantiomerically pure acid.[21]

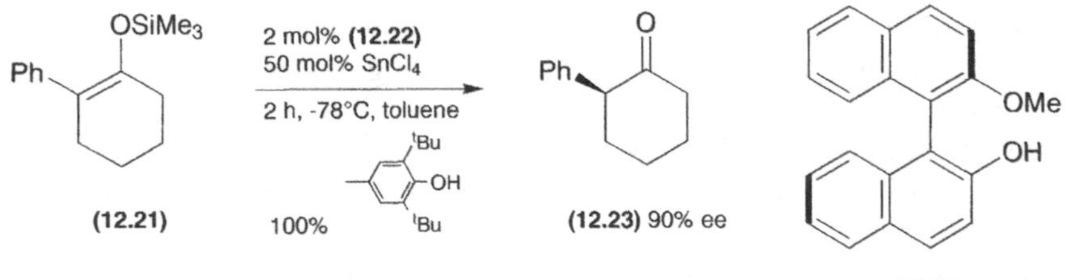

(12.21) 100% (12.23) 90% ee

(12.22)

12.4 Phase-transfer-catalysed alkylation

Phase-transfer reactions have featured in several sections of this book, including epoxidation (Section 4.5), Darzens condensation (Section 7.4) and Wadsworth-Emmons reactions (Section 12.5). Another important area of phase-transfer-catalysed reactions are those involving alkylation.[22] The use of O'Donnell's imine ester (12.24) is particularly noteworthy, since the products are readily converted into amino acids.[23]

Corey and co-workers developed the original work, and have shown that the quaternary ammonium salt (12.25) is remarkably capable as an asymmetric phase-transfer catalyst.[24] Alkyl iodides and Michael acceptors have been used as the electrophile in these reactions.[25] Thus, alkyl iodide (12.26) affords the alkylation product (12.27), with excellent enantioselectivity. Other electrophiles also work well (92–99.5% ee). The anthracenyl unit has also been incorporated into phase-transfer catalysts by Lygo and Wainwright for the enantioselective catalysis of alkylation reactions of imine ester (12.24).[26]

Phase-transfer reactions are not limited to quaternary ammonium salts for a successful outcome. Even the diol TADDOL (see Section 8.1) has been shown to be effective in the alkylation of related imines.[27]

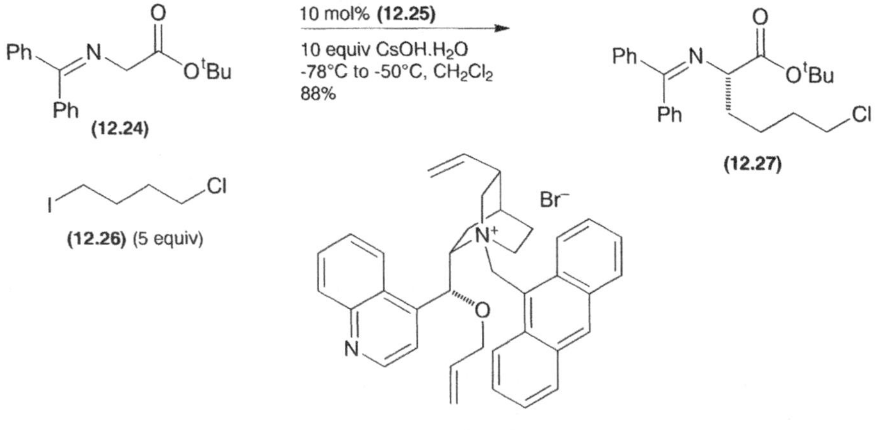

(12.24)

(12.26) (5 equiv)

(12.27)

(12.25)

12.5   Formation of alkenes

The formation of alkenes is not an obvious reaction to explore for enantioselective catalysis. Nevertheless, some examples of such reactions have already been described in previous sections (e.g. Section 12.2).

The synthetically useful ring-closing metathesis of dienes[28] has been investigated using enantiomerically pure molybdenum complexes (12.28) and (12.29). The kinetic resolution of acyclic dienes has been achieved with some success using such catalysts.[29, 30] Thus, the racemic diene (12.30) undergoes a kinetic resolution, such that the product (12.31) and recovered starting material can both be obtained with good enantioselectivity.

Fujimura and Grubbs have also shown that the ring-closing metathesis of an achiral starting material can be achieved with limited asymmetric induction (15% ee), using catalyst (12.28), where the reaction selectively involves one of two enantiotopic alkene groups.[31]

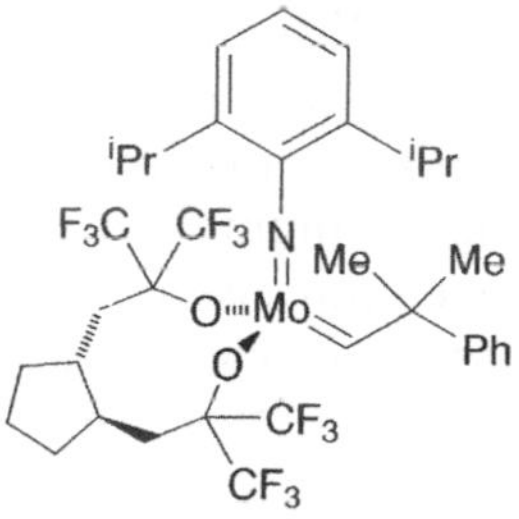

(12.28)

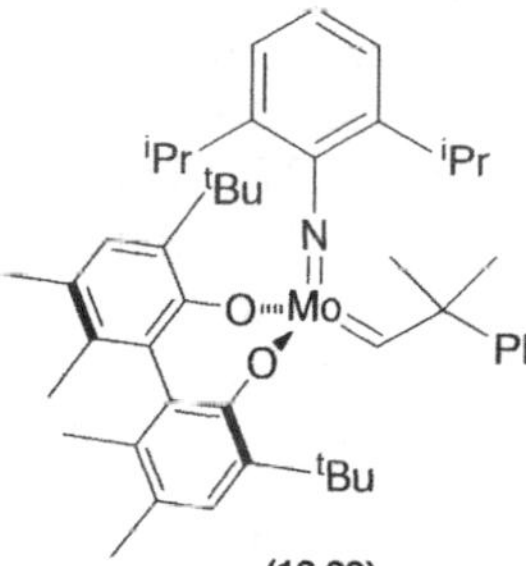

(12.29)

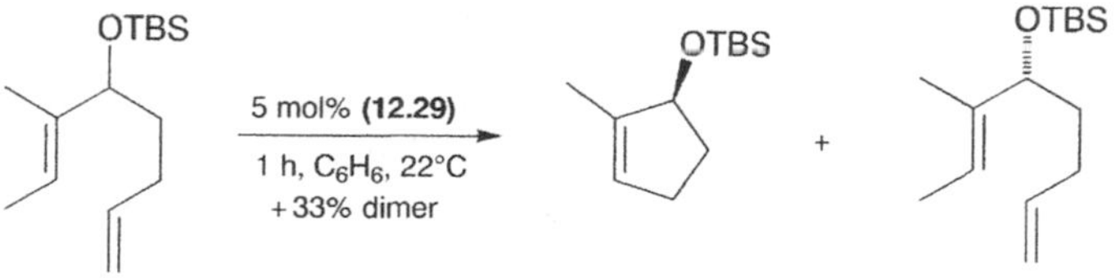

(±)-(12.30)

5 mol% (12.29)
1 h, C₆H₆, 22°C
+ 33% dimer

(12.31) 93% ee
42%

(12.30) >99% ee
25%

As well as ring-closing metathesis reactions, the catalytic asymmetric Horner-Wadsworth-Emmons reaction has been achieved using phase-transfer catalysts, including ammonium salt (12.32), with rubidium hydroxide as base. The achiral ketone (12.33) is converted into the chiral alkene (12.34), with reasonable enantioselectivity. Currently, the long reaction time is a drawback.[32]

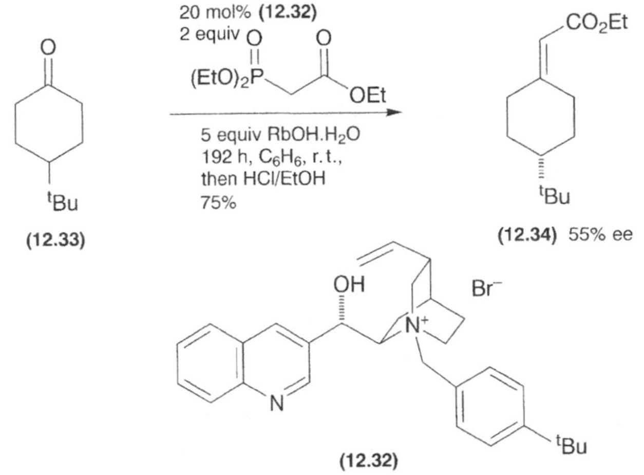

(12.33)   (12.34) 55% ee

(12.32)

12.6   Oxyselenylation-elimination reactions

A catalytic oxyselenylation-elimination sequence has been devised, which provides enantiomerically-enriched allyl ethers.[33, 34] In one example,[35] the diselenide (12.35) is used to effect the transformation of alkene (12.36) into the allyl ether (12.37). The catalytic pathway proceeds *via* initial oxyselenylation, followed by oxidation of the selenium, and elimination to provide the product, together with a selenium reagent capable of repeating the catalytic cycle. This unusual reaction is one of the few which employs a *p*-block element as a catalyst (see also Section 9.6).

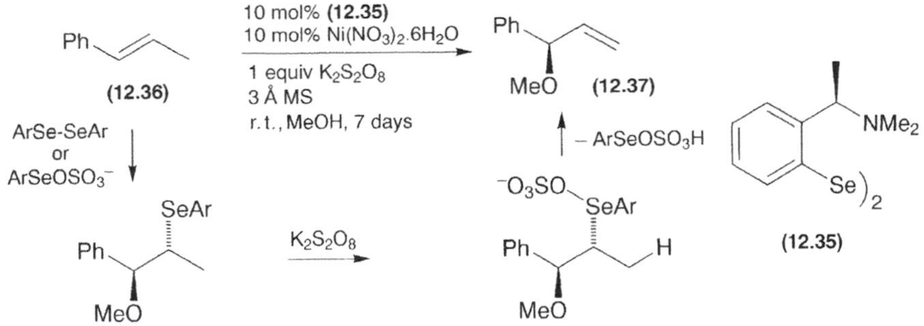

12.7   The Benzoin condensation

Thiazolium salts (12.38) are able to catalyse the conversion of benzaldehyde (12.39) into benzoin (12.40). The mechanism involves deprotonation of the thiazolium salt to give the true catalytic species (12.41), which acts as a nucleophile towards benzaldehyde, and subsequently promotes the formation of benzoin.

Several enantiomericaly pure thiazolium catalysts have been reported, generally with an enantiomerically pure moiety attached to the nitrogen of the thiazolium ring, for example salt (12.42).[36] Recently, reports of conformationally-restricted, bicyclic thiazolium salts (12.43)[37] and (12.44)[38] have increased optimism that a truly efficient enantioselective catalyst may be prepared, but this has not been achieved so far. The triazolium salt (12.45) has proved to be more selective than any of the thiazolium catalysts.[39]

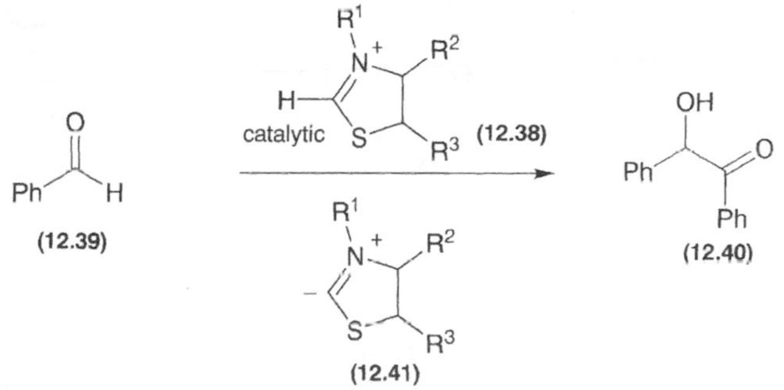

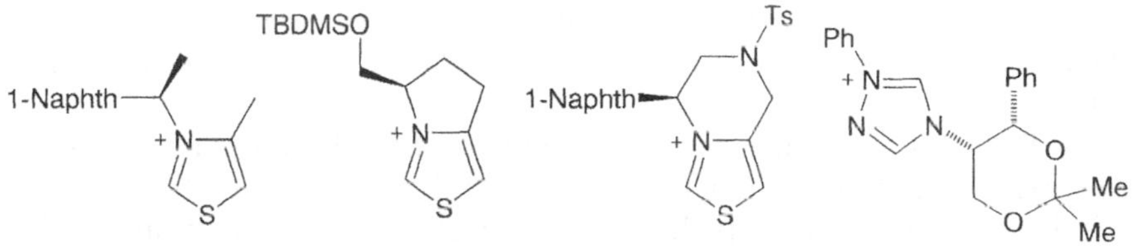

| (12.42) | (12.43) | (12.44) | (12.45) |
| provides 48% ee | provides 20.5% ee | provides 27% ee | provides 95% ee |

12.8 Ester formation and hydrolysis

The enantioselective formation and hydrolysis of esters has been very thoroughly investigated using enzymatic methods, and has been well reviewed previously.[40] The use of non-enzymatic methods for the kinetic resolution of ester formation and hydrolysis is becoming increasingly important, as new catalysts are beginning to show good levels of enantioselectivity.[41]

Fu and co-workers have developed planar chiral nucleophilic catalysts, including complex (12.46), which is a versatile and efficient acylation catalyst.[42, 43, 44] This catalyst functions like the achiral catalyst, 4-dimethylaminopyridine (DMAP). The selectivity factor, S, is a good measure of the discrimination between enantiomers. It is defined as the rate of the faster-reacting enantiomer divided by the rate of the slower-reacting enantiomer (see Section 4.1).[45]

Racemic alcohols (12.47) undergo acylation with acetic anhydride using the Fu catalyst (12.46), with very good selectivity factors. Other non-enzymatic catalysts have also been examined in kinetic resolution of alcohols, but it is unusual to see selectivity factors greater than 10.[46, 47] The desymmetrisation of *meso*-diols with non-enzymatic catalysts has

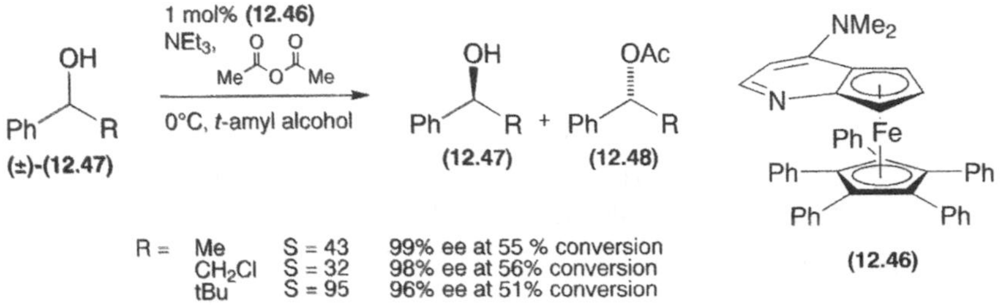

R =	Me	S = 43	99% ee at 55 % conversion
	CH$_2$Cl	S = 32	98% ee at 56% conversion
	tBu	S = 95	96% ee at 51% conversion

received less attention. Fu reported one example,[42] using *meso*-diol (12.49), which is monoacylated with remarkable selectivity. Oriyama and co-workers reported the use of enantiomerically pure diamines, including structure (12.51), as asymmetric acylation catalysts for 1,2-diols.[48, 49] The desymmetrisation of diol (12.52) occurs, with good yield and enantio-selectivity.

Ti-TADDOLates, such as complex (12.54), have been used catalytically to open cyclic meso-anhydrides.[50] The catalytic variant of this reaction still has to be fully developed, but early results are very encouraging. The *meso*-anhydride (12.55) is converted into the ring-opened product (12.56) with very good enantiocontrol, but the reaction time needs to be shortened in order for this process to become synthetically useful.

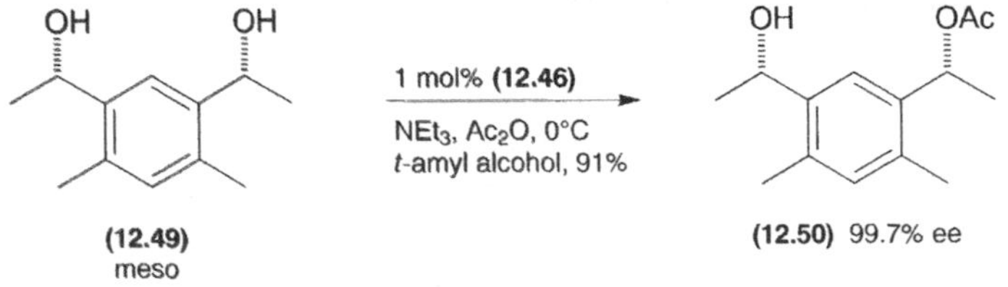

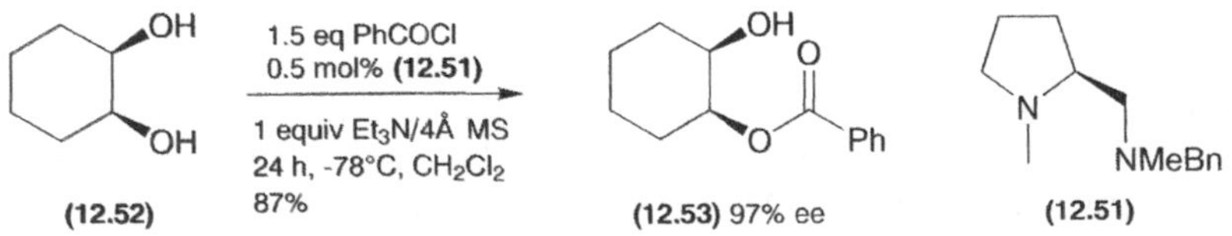

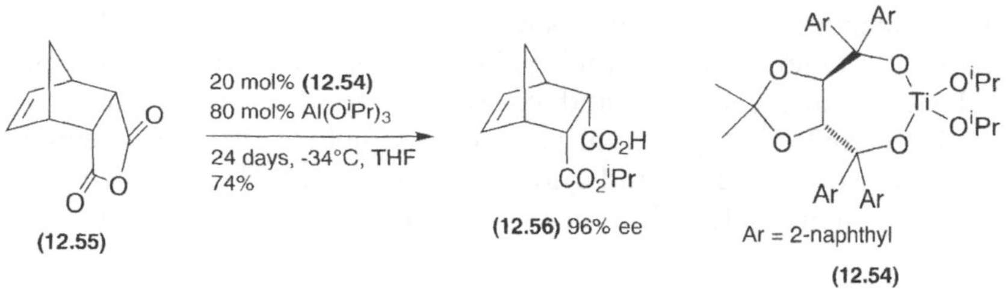

20 mol% (12.54)
80 mol% Al(O^{i}Pr)$_3$

24 days, -34°C, THF
74%

(12.55)

CO$_2$H
CO$_2^i$Pr

(12.56) 96% ee

Ar = 2-naphthyl

(12.54)

Fu and co-workers demonstrated that a dynamic kinetic resolution can be achieved using the non-enzymatic catalyst (12.46).[51] The azalactone (12.57) is very prone to racemisation, whilst the ring-opened product (12.58) is stable under the reaction conditions. Thus, the product is formed by methanolysis under dynamic resolution conditions (see Section 3.1), albeit with moderate enantioselectivity so far.

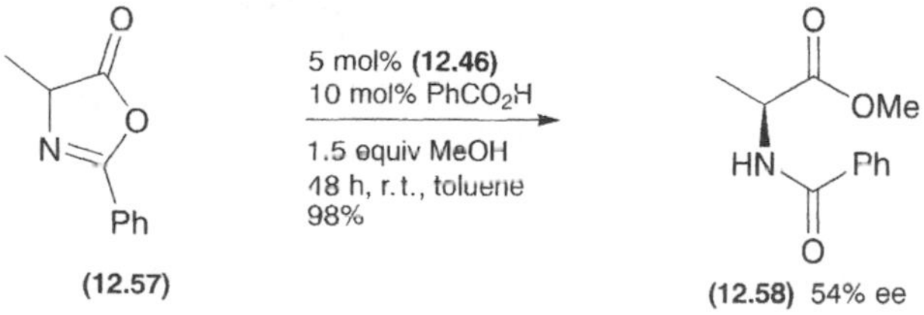

5 mol% (12.46)
10 mol% PhCO$_2$H

1.5 equiv MeOH
48 h, r.t., toluene
98%

(12.57)

(12.58) 54% ee

12.9 Ring-opening of epoxides

The nucleophilic ring opening of *meso*-epoxides leads to chiral products. Control over the position of attack will provide an enantiomerically-enriched product, as shown in Figure 12.2, where cyclohexene oxide (12.59) undergoes ring-opening to produce either compound (12.60) or its enantiomer (ent-12.60).[52]

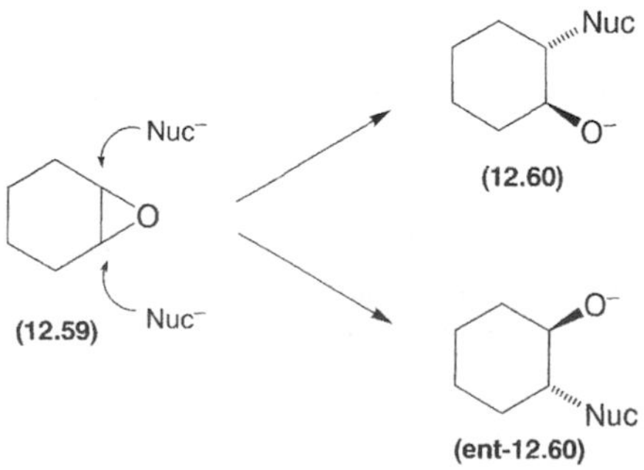

Nuc

(12.60)

(12.59) Nuc⁻

O⁻

Nuc

(ent-12.60)

Figure 12.2 Asymmetric ring-opening of epoxides.

Ring-opening of *meso*-epoxides with azide can be a synthetically useful process.[53] Nugent used ligand **(12.61)** associated with a zirconium catalyst,[54] whilst Jacobsen and co-workers used the chromium salen complex **(12.62)**.[55] Representative reactions include the ring-opening of cyclohexene oxide **(12.59)** and cylopentene oxide **(12.63)**, which give good yields and enantioselectivities in the azide ring-opening reactions.

(12.61)

(12.62-X)
X = Cl or N_3

(12.59) **(12.64)** 93% ee

(12.63) **(12.65)** 94% ee

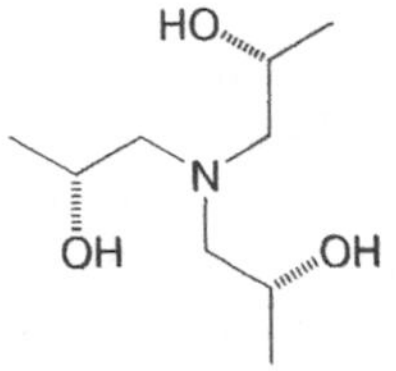

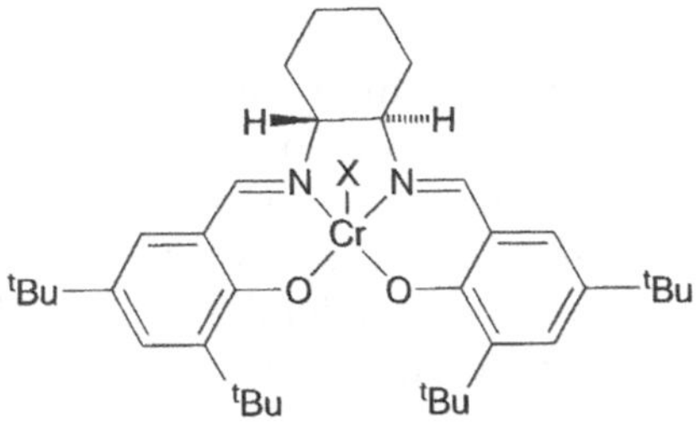

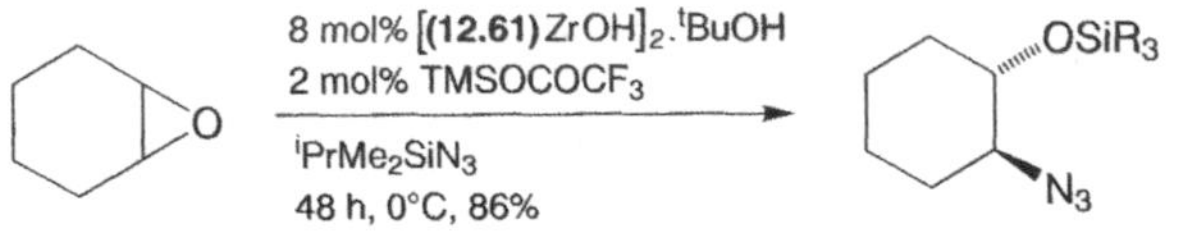

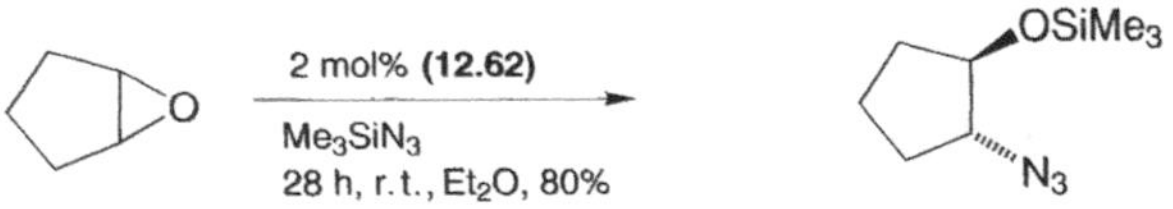

The ring-opening of epichlorohydrin **(12.66)** with trimethylsilyl azide using catalyst **(ent-12.62)** merits special consideration.[56] The starting material is racemic and, on first consideration, the product **(12.67)** should either be racemic or formed under kinetic resolution conditions. However, reversible ring-opening of the starting material by chloride provides a mechanism for its racemisation, thereby providing a dynamic kinetic resolution. The formation of symmetrical by-products lends support to the mechanism.

(±)-(12.66) **(12.67)** 97% ee

+ 12% bis-chloro compound
+ 12% bis-azido compound

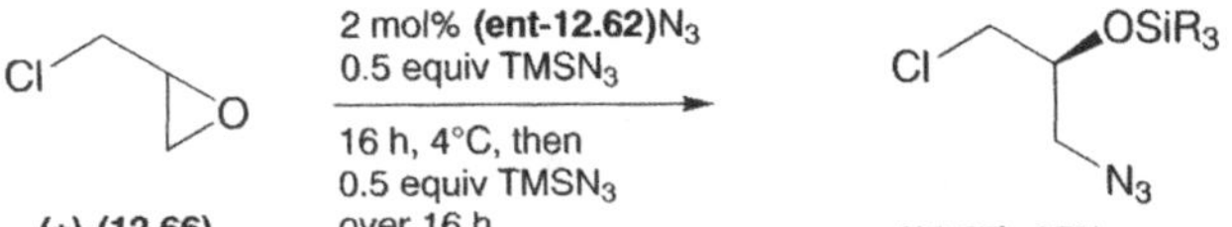

Snapper and co-workers used a ligand diversity approach to find a catalyst for the addition of trimethylsilyl cyanide to epoxides.[57] The basic ligand structure (**12.68**) was varied in an iterative fashion using a resin bound analogue, which allowed the parallel synthesis of over 20 ligands a day. Variation of the first amino acid residue (AA1) identified tert-leucine as the best candidate for the conversion of cyclohexene oxide (**12.59**) into the ring-opened adduct (**12.69**) with a titanium catalyst. With AA1 fixed, the AA2 residue was varied, which identified L-threonine (tert-butyl ether) as the best candidate. Finally, the Schiff base moiety was varied by the use of different aldehydes, and 3-fluorosalicaldehyde was found to be the best choice. In this way, ligand (**12.70**) was identified through rapid screening, and the formation of ring-opened adduct (**12.69**) was found to occur with 86% ee and 65% yield, using this ligand.

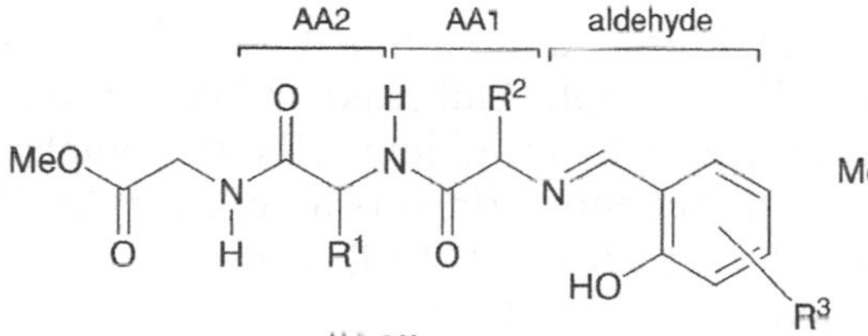

(12.68)

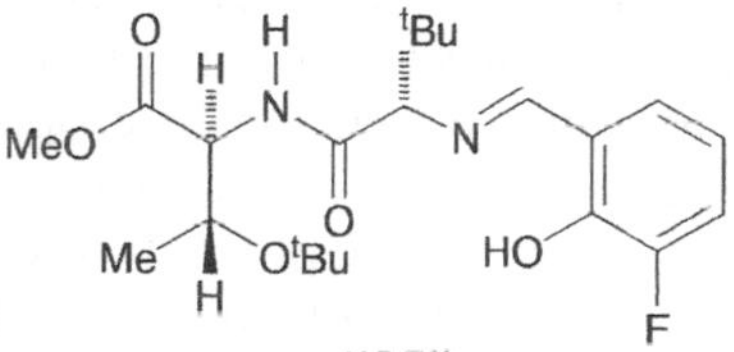

(12.70)

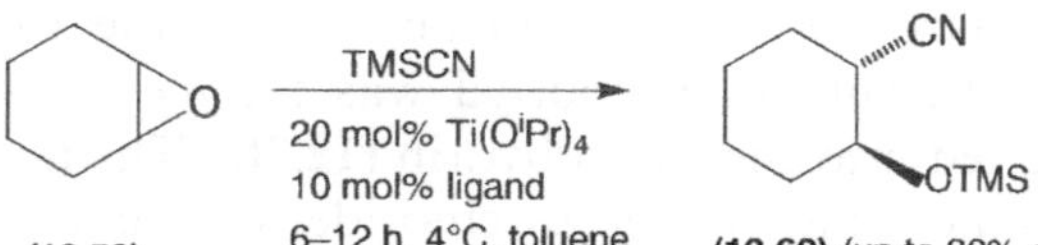

(12.59)

TMSCN

20 mol% Ti(O^{i}Pr)$_4$
10 mol% ligand
6–12 h, 4°C, toluene

(12.69) (up to 86% ee)

The use of oxygen nucleophiles for the asymmetric ring-opening of epoxides has been catalysed by the cobalt(salen) complex (**12.71**).[58, 59] Cyclohexene oxide (**12.59**) undergoes ring-opening with benzoic acid to give the mono-ester (**12.72**), with reasonable enantioselectivity. The kinetic resolution of propene oxide (**12.73**) by ring-opening with water is particularly noteworthy as an efficient route to either the epoxide or the diol (**12.74**), which are readily separated, and the catalyst can be re-cycled. The ring-opening of epoxides with water is also well known with epoxide hydrolase enzymes.[60]

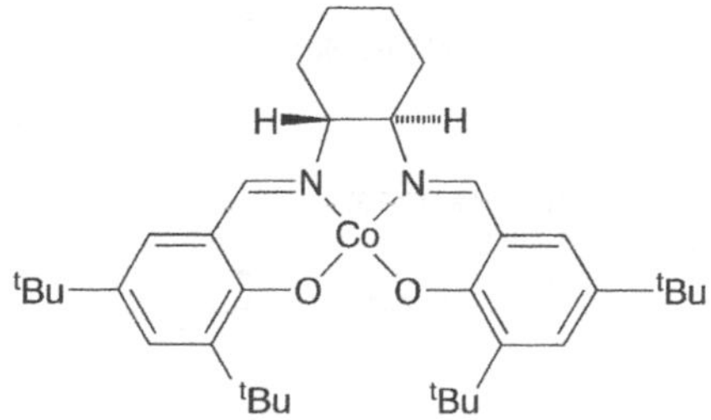

(12.71)

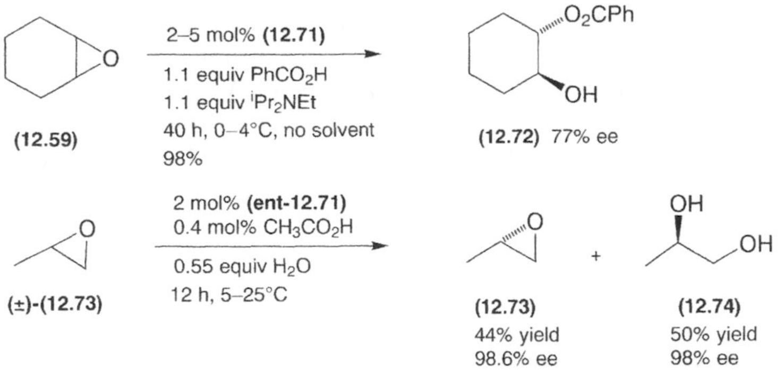

(12.59)

2–5 mol% (12.71)

1.1 equiv PhCO$_2$H
1.1 equiv ^{i}Pr$_2$NEt
40 h, 0–4°C, no solvent
98%

(12.72) 77% ee

(±)-(12.73)

2 mol% (ent-12.71)
0.4 mol% CH$_3$CO$_2$H

0.55 equiv H$_2$O
12 h, 5–25°C

(12.73)
44% yield
98.6% ee

+

(12.74)
50% yield
98% ee

The formation of β-bromohydrins[61] and β-thioalcohols[62] from epoxides has also been reported. For example, Shibasaki and co-workers have shown that the heterobimetallic gallium lithium BINOL complex (12.75) is a good catalyst for nucleophilic ring-opening epoxides,[63] including the use of thiols. Cyclohexene epoxide (12.59) is converted into the sulfide (12.76) with very good enantioselectivity.

In a recent example, Denmark and co-workers have shown that an enantiomerically pure Lewis base can be used in the ring-opening of epoxides.[64] The best substrate reported was the *meso*-epoxide (12.77), which was converted into the chlorohydrin (12.78). The authors suggested that the phosphoramide displaces chloride to give a reactive silicon cation, which activates the epoxide to nucleophilic attack by chloride.

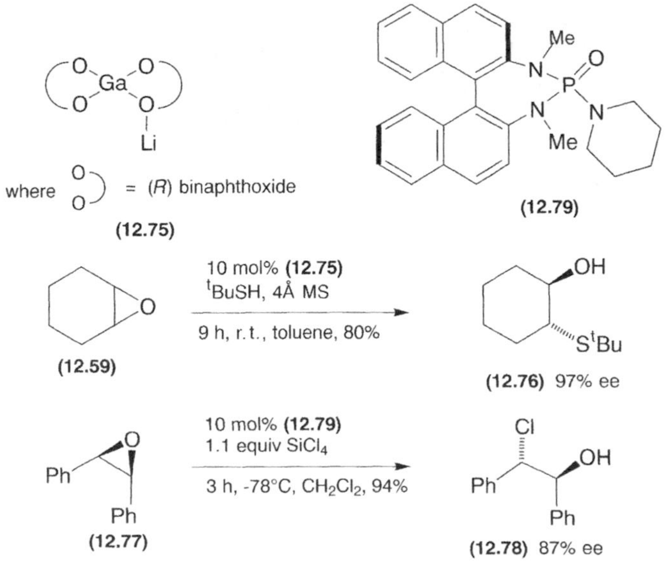

where $\overset{O}{\underset{O}{\big)}}$ = (R) binaphthoxide

(12.75)

(12.79)

(12.59)

10 mol% (12.75)
^{t}BuSH, 4Å MS

9 h, r.t., toluene, 80%

(12.76) 97% ee

(12.77)

10 mol% (12.79)
1.1 equiv SiCl$_4$

3 h, -78°C, CH$_2$Cl$_2$, 94%

(12.78) 87% ee

References

1. K. Tani, T. Yamagata, S. Otsuka, S. Akutagawa, H. Kumobayashi, T. Taketomi, H. Takaya, A. Miyashita, and R. Noyori, *J. Chem. Soc., Chem. Commun.*, **1982**, 600.
2. (a) R. Noyori, in *Asymmetric Catalysis in Organic Synthesis*, John Wiley and Sons, New York, **1994**, 95. (b) S. Akutagawa and K. Tani, in *Catalytic Asymmetric Synthesis*, (I. Ojima, ed.) VCH, New York, **1993**, Chapter 2.
3. (a) K. Tani, T. Yamagata, S. Akutagawa, H. Kumobayashi, T. Taketomi, H. Takaya, A. Miyashita, R. Noyori, and S. Otsuka, *J. Am. Chem. Soc.*, **1984**, *106*, 5208. (b) K. Tani, *Pure and Appl. Chem.*, **1985**, *57*, 1845.
4. M. Kitamura, K. Manabe, R. Noyori and H. Takaya, *Tetrahedron Lett.*, **1987**, *28*, 4719.
5. H. Frauenrath and M. Kaulard, *Synlett*, **1994**, 517.
6. H. Frauenrath, S. Reim and A. Wiesner, *Tetrahedron: Asymmetry*, **1998**, *9*, 1103.
7. Z. Chen and R.L. Halterman, *J. Am. Chem. Soc.*, **1992**, *114*, 2276.
8. T.K. Hollis and L.E. Overman, *Tetrahedron Lett.*, **1997**, *38*, 8837.
9. M. Calter, T.K. Hollis, L.E. Overman, J. Ziller and G.G. Zipp, *J. Org. Chem.*, **1997**, *62*, 1449.
10. Y. Uozumi, K. Kato and T. Hayashi, *Tetrahedron: Asymmetry*, **1998**, *9*, 1065.
11. M. Amadji, J. Vadecard, J.-C. Plaquevent, L. Duhamel and P. Duhamel, *J. Am. Chem. Soc.*, **1996**, *118*, 12483.
12. D.M. Hodgson, G.P. Lee, R.E. Marriott, A.J. Thompson, R Wisedale and J. Witherington, *J. Chem. Soc., Perkin Trans.*, *1*, **1998**, 2151.
13. For a review, see: C. Fehr, *Angew. Chem., Int. Ed. Engl.*, **1996**, *35*, 2566.
14. C. Fehr and J. Galindo, *Angew. Chem., Int. Ed. Engl.*, **1994**, *33*, 1888.
15. A. Yanagisawa, T. Kikuchi, T. Watanabe, T. Kuribayashi and H. Yamamoto, *Synlett*, **1995**, 372.
16. E. Vedejs and A.W. Kruger, *J. Org. Chem.*, **1998**, *63*, 2792.
17. P. Riviere and K. Koga, *Tetrahedron Lett.*, **1997**, *38*, 7589.
18. Y. Nakamura, S. Takeuchi, A. Ohira and Y. Ohgo, *Tetrahedron Lett.*, **1996**, *37*, 2805.
19. K. Ishihara, S. Nakamura, M. Kaneeda and H. Yamamoto, *J. Am. Chem. Soc.*, **1996**, *118*, 12854.
20. M. Sugiura and T. Nakai, *Angew. Chem., Int. Ed. Engl.*, **1997**, *36*, 2366.
21. S.J. Aboulhoda, I. Reiners, J. Wilken, F. Hénin, J. Martens and J. Muzart, *Tetrahedron: Asymmetry*, **1998**, *9*, 1847.
22. M.J. O'Donnell in *Catalytic Asymmetric Synthesis*, (I. Ojima, ed.) VCH, New York, **1993**, 389.
23. M.J. O'Donnell, W.D. Bennett and S. Wu, *J. Am. Chem. Soc.*, **1989**, *111*, 2353.
24. E.J. Corey, F. Xu and M.C. Noe, *J. Am. Chem. Soc.*, **1997**, *119*, 12414.
25. E.J. Corey, M.C. Noe and F. Xu, *Tetrahedron Lett.*, **1998**, *39*, 5347.
26. B. Lygo and P.G. Wainwright, *Tetrahedron Lett.*, **1997**, *38*, 8595.
27. Y.N. Belokon, K.A. Kochetkov, T.D. Churkina, N.S. Ikonnikov, A.A. Chesnokov, O.V. Larionov, V.S. Parmár, R. Kumar and H.B. Kagan, *Tetrahedron: Asymmetry*, **1998**, *9*, 851.
28. For a review see: (a) S.K. Armstrong, *J. Chem. Soc., Perkin Trans. 1*, **1998**, 371. (b) M. Schuster and S. Blechert, *Angew. Chem., Int. Ed. Engl.*, **1997**, *36*, 2036.
29. O. Fujimura and R.H. Grubbs, *J. Am. Chem. Soc.*, **1996**, *118*, 2499.
30. (a) J.B. Alexander, D.S. La, D.R. Cefalo, A.H. Hoveyda and R.R. Schrock, *J. Am. Chem. Soc.*, **1998**, *120*, 4041. (b) D.S. La, J.B. Alexander, D.R. Cefalo, D.D. Graf, A. H. Hoveyda and R.R. Schrock, *J. Am. Chem. Soc.*, **1998**, *120*, 9270.
31. O. Fujimura and R.H. Grubbs, *J. Org. Chem.*, **1998**, *63*, 824.
32. S. Arai, S. Hamaguchi and T. Shioiri, *Tetrahedron Lett.*, **1998**, *39*, 2997.
33. K. Fujita, M. Iwaoka and S. Tomoda, *Chem. Lett.*, **1994**, 923.

34. S.-I. Fukuzawa, K. Takahashi, H. Kato and H. Yamazaki, *J. Org. Chem.*, **1997**, *62*, 7711.
35. T. Wirth, S. Häuptli and M. Leuenberger, *Tetrahedron: Asymmetry*, **1998**, *9*, 547.
36. J.C. Sheehan and T. Hara, *J. Org. Chem.*, **1974**, *39*, 1196.
37. R.L. Knight and F.J. Leeper, *Tetrahedron Lett.*, **1997**, *38*, 3611.
38. C.A. Dvorak and V.H. Rawal, *Tetrahedron Lett.*, **1998**, *39*, 2925.
39. D. Enders, K. Breuer and J.H. Teles, *Helv. Chim. Acta*, **1996**, *79*, 1217.
40. C.-H. Wong and G.M. Whitesides, *Enzymes in Synthetic Organic Chemistry*, Tetrahedron Organic Chemistry Series, Volume 12, Pergamon, Oxford, **1994**, Chapter 2.
41. P. Somfai, *Angew. Chem., Int. Ed. Engl.*, **1997**, *36*, 2731.
42. J.C. Ruble, J. Tweddell and G.C. Fu, *J. Org. Chem.*, **1998**, *63*, 2794.
43. J.C. Ruble, H.A. Lathan and G.C. Fu, *J. Am. Chem. Soc.*, **1997**, *119*, 1492.
44. J.C. Ruble and G.C. Fu, *J. Org. Chem.*, **1996**, *61*, 7230.
45. H.B. Kagan and J.C. Fiaud, *Top. Stereochem.*, **1988**, *18*, 249.
46. S.J. Miller, G.T. Coperland, N. Papaioannou, T.E. Horstmann and E.M. Ruel, *J. Am. Chem. Soc.*, **1998**, *120*, 1629.
47. E. Vedejs, O. Daugulis and S.T. Diver, *J. Org. Chem.*, **1996**, *61*, 430.
48. T. Oriyama, K. Imai, T. Sano and T. Hosoya, *Tetrahedron Lett.*, **1998**, *39*, 3529.
49. T. Oriyama, K. Imai, T. Hosoya and T. Sano, *Tetrahedron Lett.*, **1998**, *39*, 397.
50. G. Jaeschke and D. Seebach, *J. Org. Chem.*, **1998**, *63*, 1190.
51. J. Liang, J.C. Ruble and G.C. Fu, *J. Org. Chem.*, **1998**, *63*, 3154.
52. For a review dealing with catalytic and stoichiometric aspects of epoxide desymmetrisation, see: D.M. Hodgson, A.R. Gibbs and G.P. Lee, *Tetrahedron*, **1996**, *52*, 14361.
53. M.H. Wu and E.N. Jacobsen, *Tetrahedron Lett.*, **1997**, *48*, 1693.
54. W.A. Nugent, *J. Am. Chem. Soc.*, **1992**, *114*, 2768.
55. L.E. Martínez, J.L. Leighton, D.H. Carsten and E.N. Jacobsen, *J. Am. Chem. Soc.*, **1995**, *117*, 5897.
56. S.E. Schaus and E.N. Jacobsen, *Tetrahedron Lett.*, **1996**, *37*, 7937.
57. B.M. Cole, K.D. Shimizu, C.A. Krueger, J.P.A. Harrity, M.L. Snapper and A.M. Hoveyda, *Angew. Chem., Int. Ed. Engl.*, **1996**, *35*, 1668.
58. E.N. Jacobsen, F. Kakiuchi, R.G. Konsler, J.F. Larrow and M. Tokunaga, *Tetrahedron Lett.*, **1997**, *38*, 773.
59. M. Tokunaga, J.F. Larrow, F. Kakiuchi and E.N. Jacobsen, *Science*, **1997**, *277*, 936.
60. I.V.J. Archer, *Tetrahedron Lett.*, **1997**, *53*, 15617.
61. W.A. Nugent, *J. Am. Chem. Soc.*, **1998**, *120*, 7139.
62. T. Iida, N. Yamamoto, H. Sasai and M. Shibasaki, *J. Am. Chem. Soc.*, **1997**, *119*, 4783.
63. T. Iida, N. Yamamoto, S. Matsunaga, H.-G. Woo and M. Shibasaki, *Angew. Chem., Int. Ed. Engl.*, **1998**, *37*, 2223.
64. S.E. Denmark, P.A. Barsanti, K.-T. Wong and R.A. Stavenger, *J. Org. Chem.*, **1998**, *63*, 2428.

Index

Printed and bound by CPI Group (UK) Ltd, Croydon, CR0 4YY

27/10/2024

14580190-0004